**Advances in Experimental Medicine and Biology**

Volume 957

More information about this series at http://www.springer.com/series/5584

Jozsef Laczko · Mark L. Latash
Editors

# Progress in Motor Control

## Theories and Translations

Springer

*Editors*
Jozsef Laczko
Department of Information Technology and Biorobotics, Faculty of Science
University of Pécs
Pecs
Hungary

Mark L. Latash
Department of Kinesiology
Pennsylvania State University
University Park, PA
USA

ISSN 0065-2598 ISSN 2214-8019 (electronic)
Advances in Experimental Medicine and Biology
ISBN 978-3-319-47312-3 ISBN 978-3-319-47313-0 (eBook)
DOI 10.1007/978-3-319-47313-0

Library of Congress Control Number: 2016954602



Printed on acid-free paper

This Springer imprint is published by Springer Nature
The registered company is Springer International Publishing AG
The registered company address is: Gewerbestrasse 11, 6330 Cham, Switzerland

# Preface

This volume continues the tradition of the series *Progress in Motor Control* started in the previous millennium with the publication of the first volume subtitled "Bernstein's Traditions in Movement Studies" (Latash 1998). That volume contained chapters written by speakers at the first conference with the same name held in 1996 in State College, Pennsylvania. Over the past 20 years, the field of motor control has grown substantially, which is reflected in the creation of the journal "Motor Control" and the International Society for Motor Control (ISMC). The conferences *Progress in Motor Control* have become biennial meetings of ISMC, and so far seven volumes have been published forming a kind of encyclopedic compendium of motor control that is updated every other year.

Most of the chapters in this volume were written by speakers who attended the conference *Progress in Motor Control X* in Budapest, held during July 22–25, 2015. In addition, we invited several additional groups of authors who have performed new, exciting studies over the past two years. The subtitle of this volume, *Theories and Translations*, reflects the two actively developed directions of research. One of them tries to develop theoretical approaches to biological movement that would make motor control a subfield of natural science, physics of living systems. The other applies recent advances in motor control to areas such as robotics, movement disorders, brain–computer interface, and rehabilitation.

The volume consists of six parts that are focused on specific aspects of motor control. Part I, "Theoretical Motor Control", opens with a chapter by Andrea d'Avella entitled "Modularity for Motor Control and Motor Learning". The author addresses a central issue of how the central nervous system (CNS) overcomes the complexity of multi-joint and multi-muscle control and suggests that modular architecture may simplify control by embedding features of both the dynamic behavior of the musculoskeletal system and of the task into a small number of modules. Recent studies, in which human subjects used myoelectric control to move a mass in a virtual environment, have suggested that recombination of modules may be more efficient than learning or adapting a skill by acquiring new modules. According to the view of Jeroen Smeets and Eli Brenner in their chapter "Synergies in Grasping", both transport of the hand and formation of the grip emerge from a combination of independent movements of individual digits towards the surface of the object. The chapter presents three experiments designed to test to what extent movements of individual digits can be considered the building blocks of the reach-to-grasp movement. The third chapter by Katja Kornysheva, "Encoding Temporal Features of Skilled Movements—What, Whether and How?" addresses the issue of storing memories of complex temporal dynamics of movement in the brain. It outlines the constraints that determine whether and how the timing of skilled movements is represented in the central nervous system and offers a schematic model of how these different representations complement and interact with each other in fast feedback loops to achieve precise motor timing. The chapter by Dagmar Sternad and Christopher Hasson "Predictability and Robustness in the Manipulation of Dynamically Complex Objects" explores the hypothesis that humans learn strategies that make the interactions in neural representations of object dynamics predictable and robust to inaccuracies. The chapter describes a virtual reality task that simulates a cart-and-pendulum system and shows, that with practice, subjects develop movement strategies that are predictable and robust. This approach may be a promising platform to gain insights into a variety of neurological diseases and healthy aging.

Part II of the book is dedicated to the equilibrium-point (EP) hypothesis. This hypothesis was suggested in the mid-1960s by Anatol Feldman. Fifty years later, the hypothesis remains hotly debated. It has not been rejected despite multiple claims of disproving the EP hypothesis. Neither has it been accepted by most researchers. Over the past few years, more and more experimental evidence has accumulated in favor of the EP hypothesis, addressing such diverse aspects as the neurophysiological mechanisms underlying the control of movements, relations between the EP hypothesis and motor synergies, and applications of this hypothesis to movement disorders. This part opens with a chapter by Mark Latash, which presents a brief review of the EP hypothesis and its relation to the synergic control of multi-effector systems. It suggests that the EP hypothesis represents an example of a physical approach to human movements making it a subfield of the physics of living systems. The chapter describes how the EP hypothesis can be naturally combined with the idea of hierarchical control of movements and of synergic organization of the abundant systems involved in all actions. The following chapter by Anatol Feldman "The Relationship Between Postural and Movement Stability" starts with summarizing approaches to the control of posture and movement from the middle of the nineteenth century to our times. Further, the chapter describes the basics of the control with referent coordinates and their neurophysiological mechanisms. Then, the chapter addresses issues of stability of posture and movement and the role of coactivation command with an emphasis on standing and stepping. This part ends with a chapter by Mindy Levin "Principles of Motor Recovery After Neurological Injury Based on a Motor Control Theory". The chapter describes how physiologically well-established principles in the control of actions, such as those outlined in the EP hypothesis, can help advance the understanding of deficits that may limit recovery at two levels: Body structure and function level and Activity level. In particular, the chapter addresses spasticity as a reflection of disordered control of the threshold of the tonic stretch reflex and offers practical lessons for motor rehabilitation.

Part III of the book addresses neurophysiological mechanisms of motor control. The chapter by Richard Carson and colleagues, "What Do TMS-Evoked Motor Potentials Tell Us About Motor Learning?" reviews the role of transcranial magnetic stimulation (TMS) in studies regarding the effects of motor learning. The authors emphasize the restricted explanatory scope of the TMS technique and consider a specific example of cross education: the interlimb transfer of functional capacity. Winfried Mayr and colleagues cover in their chapter, "Motor Control of Human Spinal Cord Disconnected from the Brain and Under External Movement", the role of the spinal cord in motor control and coordination. They describe a model of the human spinal cord with reduced and altered motor control and discuss how knowledge about human motor control as well as neurophysiology teach us to perform external modification of upper motor neurons by electrical stimulation and external control of afferents to spinal cord. Anticipatory adjustments during object manipulation are described in the chapter by Thomas Schneider and Joachim Hermsdörfer entitled "Anticipation in Object Manipulation: Behavioral and Neural Correlates". The authors review studies using brain functional imaging and examining the deficits of patients with localized brain damage to provide an insight into the basic principles of anticipatory motor control and their underlying neural substrates.

Part IV dedicated to problems of learning skilled behaviors opens with a chapter "Brain Plasticity and the Concept of Metaplasticity in Skilled Musicians" by Eckart Altenmüller and Shinichi Furuya, which explores the importance of brain plastic adaptations for enhanced sensory, motor, and cognitive functions. In particular, the authors focus on plastic changes in neuroplastic functions, so called metaplasticity, in musicians. The potential role of this mechanism for prevention of developing maladaptive changes in the nervous system, possibly leading to focal dystonia in musicians, is discussed. The next chapter, "The Coordination Dynamics of Observational Learning: Relative Motion Direction and Relative Phase as Informational Content Linking Action-Perception to Action-Production" by John Buchanan, emphasizes identifiable movement features that constrain and inform action-perception and action-production processes. The author puts forth relative phase as an informational variable that links perception to action. Across a series of tasks, it is shown that the relative motion and relative phase between limbs and joints are picked up through visual processes and support the observational learning of motor skills. Elizabeth Torres reviews new technological advances and new analytical methods in the study of movements and their changes in the clinical setting in a chapter titled "Rethinking the Study of Volition for Clinical Use". She emphasizes the importance of variability in the emergence of movement patterns and presents examples of solutions amenable to the habilitation and rehabilitation of volition in patient populations.

Part V of the book covers the field of impaired motor control and rehabilitation. Sainburg and colleagues in their chapter titled "Motor Lateralization Provides a Foundation for Predicting and Treating Non-paretic Arm Motor Deficits in Stroke" address clinical implications of the dynamic dominance hypothesis. This bilateral hemispheric model of motor control has successfully predicted hemisphere-specific motor control and motor learning deficits in the ipsilesional, or non-paretic, arm of patients with unilateral stroke. The chapter reviews a series of studies about the effects of intense practice of virtual reality and real-life tasks that lead to improved control of the ipsilesional arm in functional tasks. The chapter by Jozsef Laczko, Mariann Mravcsik, and Peter Katona "Control of Cycling Limb Movements: Aspects for Rehabilitation" addresses two aspects in the research on kinematics and muscle activation during cycling lower and upper limb movements. One of them deals with the effects of external load and resistance on the variance of movement patterns at different levels, from muscles to joint configurations and to limb end-points. The comparison of the variance indices in the dominant and nondominant arms drives attention to a special feature of arm cycling that is common for both arms. The second aspect is related to functional electrical stimulation as a means to drive cycling movements in individuals with a spinal cord injury. The advantages of applying and controlling these types of movements in rehabilitation of people with paralyzed limbs are discussed. The chapter by Andrew Gordon entitled "Impaired Voluntary Movement Control and Its Rehabilitation in Cerebral Palsy" reviews the pathophysiology and mechanisms underlying impaired upper extremity control in cerebral palsy. Further, the author shows that the developing corticospinal tract can reorganize its connectivity depending on the timing and location of the CNS injury, with implications for the severity of hand impairment and rehabilitation. The chapter ends by describing evidence for motor learning-based therapies and out-lining the future directions for rehabilitation. The chapter by John Rothwell addresses the effects of transcranial magnetic stimulation (TMS) on motor behavior, motor learning and on outcomes of presently applied rehabilitation therapies. It analyzes the question of whether non-invasive brain stimulation can enhance motor recovery after stroke. At the end of the chapter, new approaches are discussed that may lead to reliable and effective therapeutic treatments in medical rehabilitation.

Part VI of the book summarizes the recent progress in the field of the human–machine interface. Rajiv Ranganathan and Robert Scheidt address the learning of skilled behaviors in their chapter "Organizing and Reorganizing Coordination Patterns". They discuss how a new coordination pattern is acquired and refined when one learns a novel motor task. To examine this issue, the authors describe a body–machine interface paradigm. Then, the lessons of this paradigm for motor learning are outlined, especially for learning of motor patterns in high-dimensional spaces.

Davide Piovesan addresses the use of robot-assisted rehabilitation in the chapter titled "A Computational Index to Describe Slacking During Robot-Therapy". Robot-assisted arm movements were examined in stroke survivors; with training, the patient became able to execute voluntary movements with lower force levels and followed a minimum effort trajectory. This study offers new important insights into the rehabilitation of stroke survivors.

Tucker Tomlinsom and Lee Miller address the remarkable ability of paralyzed patients to control movement of a prosthetic limb or even their own hand with cortical signals in their chapter "Toward a Proprioceptive Neural Interface That Mimics Natural Cortical Activity". They emphasize the importance of somatosensation, including proprioception, for the natural control of movement and review studies focused on refining these sensations by stimulating the somatosensory cortex (S1) directly. Further, they describe the recent efforts to develop afferent neural interfaces through spatiotemporally precise intracortical microstimulation.

This volume is written for well-versed readers of the field. It presents a wealth of up-to-date material on various issues in the field of motor control and is designed as a reference book. It can also be used as an additional reading for graduate-level courses in such fields as physiology, psychology, kinesiology, engineering, physical therapy, and movement disorders.

We would like to thank the organizers and participants of the "Progress in Motor Control X" conference as well as the authors of the chapters in this book. We are also grateful to the Faculty of Science of the University of Pecs, the Hungarian Society of Sport Science and the Wigner Research Centre for Physics in Hungary for their help in organizing the meeting, which formed the foundation of this book.

Pécs, Hungary Jozsef Laczko
University Park, PA, USA Mark L. Latash

*The original version of the book was revised:*
*For detailed information please see Erratum.*
*The erratum to the book is available at*
*https://doi.org/10.1007/978-3-319-47313-0_21*

For detailed information please see Erratum.

# Contents

# Editors and Contributors

## About the Editors

**Jozsef Laczko** is Head of Department of Information Technology and Biorobotics at University of Pecs in Hungary. He also holds positions in the Wigner Research Centre for Physics in Budapest and an adjunct assignment at the Northwestern University in Chicago. Dr. Laczko earned his Ph.D. in Mathematics at the Eotvos Lorand University in Budapest, Hungary. He has been a trainee of the European Space Agency and has participated in a NASA funded research project about the effect of gravity on motor development. He later received a Fulbright Fellowship at the New York University and was a recipient of a Senior Leaders and Scholars Fellowship of the Hungarian-American Enterprise Fund. He is experienced in computational approaches for motor control. He has published over 40 papers in refereed journals and proceedings, four book chapters and has presented over 50 professional talks. He was the chair of the "Progress in Motor Control X" conference of the Society for Motor Control. His chief research interest is in the area of biomechanics, and in the last decade, he became engaged in the control of Functional Electrical Stimulation driven limb movements for spinal cord injured individuals.

**Mark L. Latash** is a Distinguished Professor of Kinesiology and Director of the Motor Control Laboratory at the Pennsylvania State University. He studied Physics, Physics of Living Systems, and Physiology in Moscow and Chicago. His research is focused on the control and coordination of human voluntary movements and movement disorders. He is the author of several books including, **Control of Human Movement** (1993), **The Neurophysiological Basis of Movement** (1998, 2008), **Synergy** (2008), **Fundamentals of Motor Control** (2012), and **Biomechanics and Motor Control: Defining Central Concepts** (with Vladimir Zatsiorsky, 2016). In addition, he edited nine books and published over 350 papers in refereed journals. Mark Latash served as the Founding Editor of the journal

"Motor Control" and as President of the International Society of Motor Control. He continues to serve as Director of the annual Motor Control Summer School series. He is a recipient of the Bernstein Prize in motor control.

## Contributors

**Eckart Altenmüller** Institute for Music Physiology and Musicians' Medicine (IMMM), Hannover University of Music, Drama and Media, Hannover, Germany

**Eli Brenner** Department of Human Movement Sciences, VU University Amsterdam, Amsterdam, BT, The Netherlands

**John J. Buchanan** Human Performance Labs: Perception-Action Dynamics, Department of Health and Kinesiology, Texas A&M University, College Station, TX, USA

**Richard G. Carson** Trinity College Institute of Neuroscience and School of Psychology, Trinity College Dublin, Dublin 2, Ireland; School of Psychology, Queen's University Belfast, Northern Ireland, UK

**Andrea d'Avella** Department of Biomedical and Dental Sciences and Morphofunctional Imaging, University of Messina, Messina, Italy; Laboratory of Neuromotor Physiology, Santa Lucia Foundation, Rome, Italy

**Milan R. Dimitrijevic** Baylor College of Medicine Houston, Houston, TX, USA

**Anatol G. Feldman** Department of Neuroscience and Institute of Biomedical Engineering, University of Montreal, Montreal, QC, Canada; Center for Interdisciplinary Research in Rehabilitation of Greater Montreal (CRIR), Institut de réadaptation Gingras-Lindsay-de-Montréal, CIUSSS CSMTL, Montreal, QC, Canada; Jewish Rehabilitation Hospital, CISSS de Laval, Laval, Canada

**Shinichi Furuya** Institute for Music Physiology and Musicians' Medicine (IMMM), Hannover University of Music, Drama and Media, Hannover, Germany; Department of Information and Communication Sciences, Musical Skill and Injury Center (MuSIC), Sophia University, Tokyo, Japan

**David Good** Department of Neurology, Penn State Milton S. Hershey College of Medicine, Hershey, USA

**Andrew M. Gordon** Department of Biobehavioral Sciences, Teachers College, Columbia University, New York, NY, USA

**Christopher J. Hasson** Department of Physical Therapy, Movement and Rehabilitation Sciences, Northeastern University, Boston, MA, USA

**Joachim Hermsdörfer** Institute of Human Movement Science, Department of Sport and Health Sciences, Technical University of Munich, Munich, Germany

**Peter Katona** University of Physical Education, Budapest, Hungary

**Katja Kornysheva** Institute of Cognitive Neuroscience, University College London, London, UK; Department of Neuroscience, Erasmus MC, Rotterdam, The Netherlands; School of Psychology, Bangor University, Bangor, Wales, UK

**Matthias Krenn** Center for Medical Physics and Biomedical Engineering, General Hospital Vienna, 4L, Medical University of Vienna, Vienna, Austria

**Jozsef Laczko** Department of Information Technology and Biorobotics, Faculty of Science, University of Pécs, Pecs, Hungary; Laboratory of Rehabilitation-Technology, Wigner Research Centre for Physics, Hungarian Academy of Sciences, Budapest, Hungary

**Mark L. Latash** Department of Kinesiology, The Pennsylvania State University, University Park, PA, USA

**Mindy F. Levin** School of Physical and Occupational Therapy, McGill University, Montreal, QC, Canada

**Candice Maenza** Department of Kinesiology, The Pennsylvania State University, University Park, PA, USA; Department of Neurology, Penn State Milton S. Hershey College of Medicine, Hershey, USA

**Winfried Mayr** Center for Medical Physics and Biomedical Engineering, General Hospital Vienna, 4L, Medical University of Vienna, Vienna, Austria

**Emmet McNickle** Trinity College Institute of Neuroscience and School of Psychology, Trinity College Dublin, Dublin 2, Ireland

**Lee E. Miller** Department of Physiology, Feinberg School of Medicine, Northwestern University, Chicago, IL, USA; Department of Physical Medicine and Rehabilitation, Northwestern University, Chicago, IL, USA; Department of Biomedical Engineering, Northwestern University, Evanston, IL, USA

**Mariann Mravcsik** Faculty of Science, University of Pécs, Pecs, Hungary; Laboratory of Rehabilitation-Technology, Wigner Research Centre for Physics, Hungarian Academy of Sciences, Budapest, Hungary

**Davide Piovesan** Department of Mechanical Engineering, Gannon University, Erie, PA 16501, USA

**Rajiv Ranganathan** Department of Kinesiology, Michigan State, 203 IM Circle, East Lansing, MI 48824, USA

**John C. Rothwell** Institute of Neurology, University College London, London, UK

**Kathy L. Ruddy** Neural Control of Movement Lab, ETH, Zurich, Switzerland

**Robert L. Sainburg** Department of Kinesiology, The Pennsylvania State University, University Park, PA, USA; Department of Neurology, Penn State Milton S. Hershey College of Medicine, Hershey, USA

**Robert A. Scheidt** Department of Biomedical Engineering, Marquette University, Milwaukee, WI, USA

**Thomas Schneider** Institute of Human Movement Science, Department of Sport and Health Sciences, Technical University of Munich, Munich, Germany

**Jeroen B.J. Smeets** Department of Human Movement Sciences, VU University Amsterdam, Amsterdam, BT, The Netherlands

**Dagmar Sternad** Department of Biology, Electrical and Computer Engineering, and Physics, Northeastern University, Boston, MA, USA

**Tucker Tomlinson** Department of Physiology, Feinberg School of Medicine, Northwestern University, Chicago, IL, USA

**Elizabeth B. Torres** Department of Psychology, Cognitive Science, Neuroscience, Biomedical Engineering, Computational Biomedical Imaging and Modeling, Rutgers University, Piscataway, NJ, USA

**Carolee Winstein** Department of Biokinesiology and Physical Therapy, University of Southern California, Los Angeles, USA

# Part I
# Theoretical Motor Control

# Modularity for Motor Control and Motor Learning

**Andrea d'Avella**

**Abstract** How the central nervous system (CNS) overcomes the complexity of multi-joint and multi-muscle control and how it acquires or adapts motor skills are fundamental and open questions in neuroscience. A modular architecture may simplify control by embedding features of both the dynamic behavior of the musculoskeletal system and of the task into a small number of modules and by directly mapping task goals into module combination parameters. Several studies of the electromyographic (EMG) activity recorded from many muscles during the performance of different tasks have shown that motor commands are generated by the combination of a small number of muscle synergies, coordinated recruitment of groups of muscles with specific amplitude balances or activation waveforms, thus supporting a modular organization of motor control. Modularity may also help understanding motor learning. In a modular architecture, acquisition of a new motor skill or adaptation of an existing skill after a perturbation may occur at the level of modules or at the level of module combinations. As learning or adapting an existing skill through recombination of modules is likely faster than learning or adapting a skill by acquiring new modules, compatibility with the modules predicts learning difficulty. A recent study in which human subjects used myoelectric control to move a mass in a virtual environment has tested this prediction. By altering the mapping between recorded muscle activity and simulated force applied on the mass, as in a complex surgical rearrangement of the tendons, it has been possible to show that it is easier to adapt to a perturbation that is compatible with the muscle synergies used to generate hand force than to a similar but incompatible perturbation. This result provides direct support for a modular organization of motor control and motor learning.

A. d'Avella (✉)
Department of Biomedical and Dental Sciences and Morphofunctional Imaging, University of Messina, Messina, Italy
e-mail: andrea.davella@unime.it

A. d'Avella
Laboratory of Neuromotor Physiology, Santa Lucia Foundation, Rome, Italy



J. Laczko and M.L. Latash (eds.), *Progress in Motor Control*,
Advances in Experimental Medicine and Biology 957,
DOI 10.1007/978-3-319-47313-0_1

**Keywords** Electromyography (EMG) · Muscle synergies · Coordinated recruitment · Joint angles trajectory · Inverse kinematics · Inverse dynamcis · Degrees-of-freedom (DOF) · Iterative algorithm

## Introduction

Any goal-directed limb movement, even one that appears simple and effortless such as reaching for a cup on a table, requires complex control processes. Mechanically, limbs are open kinematic chains with multiple segments and their dynamic behavior is complex due to the interactions between the different segments (Hollerbach and Flash 1982). A single segment rotates around a joint with an angular acceleration proportional to the applied torque. However, the angular accelerations of two or more segments in a chain depend on torques, angular velocities, and angular displacements at all joints. Moreover, muscles are redundant, i.e., there are more muscles in a limb than joints, and they generate force with complex dependences on neural activation, muscle length, and shortening velocity. How the CNS transforms sensory information about a goal into coordinated activations of many muscles necessary to achieve the goal is a fundamental and open question in neuroscience.

In robotics, the torques necessary to generate a desired limb movement can be computed from the equations of motion for the limb. However, the complexity of the equations of motion increases dramatically with the number of joints. As it is unlikely that the CNS explicitly represents the equation of motion and performs analytical computations, the CNS may instead rely on an implicit and possibly approximate knowledge of the dynamic behavior of the limb in response to muscle activation. A long-standing hypothesis in motor control is that motor commands are generated by a modular control architecture (Sherrington 1948; Bernstein 1967). Thus, motor modules may store approximate yet sufficient knowledge to construct adequate motor commands.

In the following sections, I will first address why and how modularity may simplify motor control. I will then review experiments showing that, in many tasks and conditions, motor commands are generated by the combination of a small number of muscle synergies, coordinated recruitment of group of muscles with specific amplitude balances or activation waveforms, suggesting that the CNS organizes muscle synergies as motor control modules. Finally, I will discuss how modularity may also help understanding motor learning. In a modular control architecture, new motor skills may be acquired by learning new combinations of existing modules or by learning new modules. As learning new modules is likely a slower adaptive process than learning new combinations of existing modules, modularity predicts learning difficulty. A recent study in which human subjects used myoelectric control to move a mass in a virtual environment has shown that the difficulty in adapting to a perturbation depends on the compatibility of the perturbation with the muscle synergies used to generate hand force. This study provides direct evidence for muscle synergies as motor control modules and suggests that motor learning relies on two adaptive processes: a fast learning of muscle synergy combinations and a slow learning of muscle synergy structure.

## Modularity to Simplify Motor Control

The processes underlying the control of a goal-directed limb movement can be described as sensorimotor transformations, i.e., transformations of sensory information about the goal and the state of the limb into motor commands. Reaching for a cup on a table, for example, requires mapping visual information about the position of the cup with respect to the body as well as visual and proprioceptive information about the initial posture of the arm into muscle activations. The activation profiles of all the muscles acting on the arm have to be accurately orchestrated in order to ensure that muscular tensions result in appropriate joint torques. Such torques must accelerate and decelerate the joint angles along a trajectory in joint space corresponding to a hand trajectory that brings the hand from its initial position to the target. In robot control, similar sensorimotor transformations must be implemented to generate the commands to the actuators of a robotic arm. Such transformations may be performed sequentially, starting from planning a trajectory that will bring the end effector to the target, then transforming the end-effector trajectory into a joint angles trajectory (inverse kinematics), then computing the torques necessary to generate the joint angles trajectory (inverse dynamics), finally determining the commands to the actuators needed to generate the desired torques. Inverse dynamics can be computed directly if the equations of motion are known. However, for a limb with many degrees-of-freedom (DOF), i.e., variables required for describing the rotations around each joint, the equations of motion are complex because they have a large number of terms that depend on the interactions between different DOF. Even when the equations of motions can be derived analytically, accurate estimation of all the geometrical and inertial parameters involved may be challenging and inverse dynamics may provide torques only approximating those required to generate a desired trajectory. Robots may, however, compensate for inaccuracies in the planned torques, as well as for noise in the actuators, using feedback control. Because of substantial delays in transmission and processing of sensory information, the CNS cannot rely on feedback control as much as robots, especially when performing fast movements. In many cases, the CNS has to perform accurate sensorimotor transformations and, in particular, it has to predict the torques needed for a given goal-directed movement accurately. How does the CNS implement accurate sensorimotor transformations?

As it is unlikely that the CNS explicitly represents the equation of motion and performs analytical computations, implicit knowledge of the dynamic behavior of the limb, sufficient to implement task-specific sensorimotor transformations, may be stored in a set of motor modules used to construct motor commands. Explicit knowledge such as that captured by the equation of motion can be exploited to compute the torques necessary to achieve any goal in any task, i.e., it represents the most general knowledge of the dynamic behavior of the limb. However, such high degree of generality has a high cost in terms of acquisition. How would the CNS acquire and represent complex functions of many variables? As a large recurrent neural network may represent complex functions, the CNS might be able to learn the most general sensorimotor transformations through practice. However, given the large number of variables and the functional complexity involved, learning sensorimotor transformations by trial-and-error might require an enormous number of trials, which seems at odds with the existence of both sophisticated innate motor behaviors and fast motor skill learning. At the opposite extreme in terms of generalization capabilities, a one-to-one mapping of a goal and an initial limb state into a specific motor command, i.e., a specific set of muscle activation profiles, would be easy to acquire and to represent. A specific motor program can be learned and used to reach a target at a given position from a given initial arm configuration, However, a different motor program would have to be stored and retrieved for each goal. As goals in most tasks vary continuously, e.g., the position in space of the target to reach, a mapping for each one of possibly infinity many goals would potentially require infinite storage. The solution for implementing adequate sensorimotor transformations that evolution may have endowed the CNS with or that the CNS may discover through practice and learning (Giszter et al. 2010) is likely a compromise between a hard to learn but very general mapping and a very extensive set of easy to learn one-to-one mappings. Motor modules may represent such a compromise. A long-standing hypothesis in motor control is that motor commands are generated by a modular control architecture. While different types of modules have been proposed, ranging from spinal reflexes (Sherrington 1948) to unit burst generators (Grillner 1981), from spinal force fields (Bizzi et al. 1991) to muscle synergies (Tresch et al. 1999; d'Avella et al. 2003; Ivanenko et al. 2004) and kinematic synergies (Santello et al. 1998), motor modules may be characterized by their ability to implement sensorimotor transformations with a small number of parameters. They may allow storing limited yet sufficient knowledge of the dynamic behavior of the limb to construct adequate motor commands by mapping goals into a small number of module combination parameters, thus simplifying control.

Considering again a reaching task, as the target position changes the muscle patterns required to reach the target will also change. However, the muscle patterns for reaching two targets close to each other will only have small differences and most of their structure will be shared. Differences will increase with distance but there will still be structure shared among muscle patterns for all targets. Thus, rather than computing each muscle pattern on the fly or storing it as a separate mapping, shared structure may be exploited to generate the muscle patterns for each possible goal of a task as the combination of a small number of task-specific motor modules. Once an appropriate set of motor modules are stored, the sensorimotor transformations will simply map goals into a small number of parameters that determine the contribution of each module for the generation of the appropriate muscle pattern. Thus, motor modules reduce the dimensionality of the output of the sensorimotor transformations.

How a modular control architecture may reduce the dimensionality of the control problem has been investigated with a simulation in which a planar kinematic chain actuated by torque profiles was used as a model of an arm performing a reaching task (Alessandro et al. 2013). Despite the nonlinearity of the equations of motion, the torque profiles bringing the arm endpoint from a given starting position to any target can be generated by a linear combination of a small set of basic actuation profiles, i.e., motor modules. Instead of planning a desired trajectory and using inverse dynamics to compute torque profiles, motor modules allow mapping directly a target into motor commands by selecting a few combination coefficients. In the simulation, interpolation of the kinematic trajectories generated by the dynamic responses to each basic activation profile was used to determine the combination coefficients but a neural network can easily learn a low-dimensional mapping between target positions and combination coefficients by trial-and-error. Importantly, the number and choice of motor modules affects performance. Good task performance can be achieved with a small number of modules if the modules are selected as torque profiles that achieve the task in a rich enough set of prototypical conditions, i.e., if they embed features of the system dynamics and of the task.

## Evidence for Modularity from EMG Decomposition into Muscle Synergies

What is the evidence that the CNS actually employs motor modules to simplify control? In the last two decades, an increasing number of studies have investigated whether muscle synergies, coordinated recruitment of groups of muscles with specific amplitude balances or activation waveforms, are motor modules organized by the CNS. Muscle synergies may store structure in the muscle activity patterns shared across tasks and task conditions. Thus, if muscle synergies are motor modules, one expects a small number of synergies to be able to reconstruct the muscle patterns observed across a variety of conditions. Indeed, studies based on the decomposition of the EMG activity recorded from multiple muscles in different species during the performance of a variety of motor tasks have shown that muscle patterns can be reconstructed by the combination of a small number of muscle synergies, thus supporting a modular organization of motor control. In this section, I will first present the different quantitative definition of muscle synergies that have been introduced in the last two decades and I will then briefly review some of the studies that have used EMG decomposition approaches to identify muscle synergies, focusing in particular on reaching movements and locomotion. A number of review articles can be consulted for more details on recent studies on muscle synergies (Bizzi et al. 2008; Tresch and Jarc 2009; Lacquaniti et al. 2012; d'Avella and Lacquaniti 2013; Giszter 2015; Ting et al. 2015).

The idea that muscles are activated together, either as a desirable strategy to simplify control or because of pathological loss of control due to neural damage, has a long history and the term "muscle synergy" has been used in different contexts with different meanings. In the context of motor control, muscle synergies have been defined as coordinated activations of groups of muscles allowing the CNS to control movements through a smaller number of variables than those required to control individual muscles (Bernstein 1967) or as a neural organization that provides stability of an important performance variable by covaried adjustments of elemental variables such as coordinated muscle activations (Latash 2012). The existence of a muscle synergy as the activation of a group of muscles with a fixed activation balance was first investigated by analyzing the activations of pairs of muscles (Lee 1984; Macpherson 1991). Muscles recruited synergistically were expected to have correlated activations. These investigations failed to fully support the existence of muscle synergies, as pairwise correlations were often observed but not as frequently as expected (Buchanan et al. 1986; Soechting and Lacquaniti 1989; Maier and Hepp-Reymond 1995). However, strong pairwise correlations provide a clear signature of a muscle synergy defined as a fixed balance of muscle activation, i.e., capturing spatial (across muscles) structure, only when a synergy is recruited by itself. If the same muscles belong to more than one synergy, each with a different fixed activation balance, and different synergies are combined flexibly across task conditions, pairwise correlations may be weak (Bizzi et al. 2002). Thus, the prediction of a model that allows for flexible combinations of fixed spatial muscle synergies is the existence of a low-dimensional covariance structure among all muscles rather than high pairwise correlations. To test quantitatively such prediction, a model based on the linear combination of spatial muscle synergies was introduced and a method to identify synergies from the EMG signals recorded from many muscles was developed (Tresch et al. 1999). Thus, spatial muscle synergies (also known as time-invariant or synchronous synergies) have been defined as basis vectors in muscle activation space:

$$\mathbf{m}(t) = \sum_{i=1}^{N} c_i(t)\,\mathbf{w}_i, \qquad (1)$$

where $\mathbf{m}(t)$ is a vector of activations in a set of muscles sampled at time $t$, $\mathbf{w}_i$ is a (time-invariant) vector representing a fixed balance of activation within the muscles in the $i$-th spatial synergy, and $c_i(t)$ is a time-varying combination coefficient scaling in amplitude the $i$-th synergy and allowing to flexibly combine $N$ fixed synergies. As muscle activation is a nonnegative variable, the elements of the synergy vectors and the combination coefficients are constrained to be nonnegative. The extraction of such synergy vectors from muscle patterns recorded in many different conditions essentially entails identifying the subspace of the muscle activation space, i.e., the abstract vector space in which each axis represent the level of activation of one muscle that contains most of the variation of the muscle patterns. Such subspace could be readily identified by principal component analysis, which consists in the diagonalization of the data covariance matrix, but the basis vectors extracted with this procedure do not satisfy the nonnegativity constraint and they are instead constrained to be mutually orthogonal. An iterative optimization algorithm based on gradient descent of the reconstruction error was then developed for muscle synergy identification (Tresch et al. 1999). At the same time, a more efficient iterative algorithm for nonnegative matrix factorization (NMF) was developed for image and language decomposition (Lee and Seung 1999) and later became the standard approach for EMG decomposition.

In addition to being defined in the spatial domain, i.e., as groups of muscles with specific activation balances, muscle synergies have also been defined in the temporal domain, as specific activation waveforms shared across groups of muscles. Temporal muscle synergies (also known as temporal components or basic activation patterns) are defined by the same equation (Eq. 1) used to define spatial muscle synergies but the time-varying coefficients $[c_i(t)]$ rather than by the muscle synergy vectors or weights $(\mathbf{w}_i)$ are taken as the modular elements. Thus, the two models can only be distinguished when considering muscle patterns for multiple task conditions. In the case of spatial synergies, one set of synergy vectors is combined by different time-varying coefficients in different conditions while for temporal synergies one set of temporal components is multiplied by condition-dependent weights (Russo et al. 2014). The same matrix factorization algorithms, such as NMF, can be used to identify both types of modules, once the data matrix is constructed by stacking data for individual conditions either along the spatial or along the temporal dimension [see Fig. 2 in Russo et al. (2014)].

Muscle synergies may also capture more complex coordination patterns, such as the recruitment of different muscles in a sequence, which may be described by specific collections of muscle activation waveforms. Such spatiotemporal muscle synergies (also known as time-varying muscle synergies) may be defined as sequences of vectors in muscle activation space (d'Avella and Tresch 2002; d'Avella et al. 2003):

$$\mathbf{m}(t) = \sum_{i=1}^{N} c_i \, \mathbf{w}_i(t - t_i), \tag{2}$$

where $\mathbf{m}(t)$, as in Eq. 1, is a vector of activations in a set of muscles sampled at time $t$, $\mathbf{w}_i(t)$ is now a vector representing a set of muscle activation waveforms sampled at time $t$, i.e., $w_{ji}(t)$ is the waveform for the $j$-th muscle in the $i$-th synergy, and $c_{\mathrm{i}}$ is a scalar combination coefficient scaling in amplitude the $i$-th synergy and allowing to flexibly combine $N$ synergies. Additionally, each sequence of vectors may be shifted in time according to an onset parameter $t_i$. If no time-shifts are included in the model, $\mathbf{m}(t)$ can be considered as a single spatiotemporal vector, i.e., different time samples of the same muscles are treated as different dimensions. Then spatiotemporal muscle synergies may be identified from a data matrix obtained by stacking spatiotemporal vectors for different conditions using the same matrix factorization algorithms as for spatial and temporal synergies (Klein Breteler et al.

2007; Russo et al. 2014). If time-shifts are included, the decomposition of the data into spatiotemporal synergies requires identifying onset parameters in addition to scaling coefficients and synergies. An iterative decomposition algorithm was developed specifically for this purpose (d'Avella and Tresch 2002; d'Avella et al. 2003). Finally, in a spatiotemporal synergy the activation waveform is the same for all muscles it can be expressed as the product of a spatial times a temporal synergy, i.e., as a space-by-time synergy (Delis et al. 2014).

Spatial, temporal, and spatiotemporal synergy models have been used to decompose the EMG activity recorded from multiple muscles in different species during the performance of a variety of motor tasks. The initial studies were performed in frogs (Tresch et al. 1999; Saltiel et al. 2001; d'Avella et al. 2003; Hart and Giszter 2004), cats (Ting and Macpherson 2005), and humans (Ivanenko et al. 2003). Those and a growing number of additional studies investigating spinal reflexes, postural control, walking, running, pedaling, reaching, hand force generation, grasping, finger movements, and many other tasks have provided support to the modular hypothesis showing that in most conditions the muscle patterns can be reconstructed by the combination of a small number of muscle synergies. I will now briefly present the results of two of these studies: one investigating reaching using spatiotemporal synergies (d'Avella et al. 2006) and a second investigating the development of locomotion using temporal synergies (Dominici et al. 2011).

A number of kinematic and kinetic features are preserved across reaching movement conditions, such as the straightness of the hand path and the shape of the tangential velocity profiles for different planar movements (Morasso 1981), the specific hand path for movements at different speeds or with different loads (Soechting and Lacquaniti 1981; Lacquaniti et al. 1982; Atkeson and Hollerbach 1985), the linear relation between torques at different joints (Soechting and Lacquaniti 1981; Lacquaniti et al. 1986; Gottlieb et al. 1997). Such invariances suggest that the CNS uses simple rules for controlling reaching. In contrast, the EMG patterns for reaching show complex changes in the shape and timing of the muscle activation waveforms as a function of movement direction and speed (Flanders et al. 1994, 1996). The apparent discrepancy between regularities in the kinematic and kinetic features and variability of the muscle patterns for reaching may be resolved at the level of muscle synergy organization. The decomposition of the EMG patterns for reaching in different directions and at different speeds shows that the complex dependences of the muscle activation waveforms on movement parameters result from the combination of a small number of spatiotemporal muscle synergies according to simple rules (d'Avella et al. 2006, 2008). Combinations of four or five spatiotemporal synergies could accurately reconstruct the phasic EMG waveforms recorded from up to 19 shoulder and arm muscles during fast point-to-point movements between a central location and eight peripheral targets in both a frontal and a sagittal plane (d'Avella et al. 2006). Phasic waveforms, responsible for accelerating and decelerating the arm, were computed by subtracting the tonic components used for balancing gravitational forces and maintaining postural stability (Flanders 1991). The extracted synergies involved specific subsets of muscles, acting at multiple joints and with synchronous and asynchronous waveforms. In many cases, individual muscles were recruited by more than one synergy. The good reconstruction of the muscle patterns for movements in different directions with five synergies [see Fig. 8 in d'Avella et al. (2006)] demonstrates the accuracy of the spatiotemporal synergy model and the parsimony of the synergy representation of the muscle patterns. The entire set of waveforms constituting a muscle pattern for a specific movement could be generated by selecting five amplitude coefficients and five onset times. Moreover, movement direction and speed modulated the amplitude coefficients according to simple rules. The synergy amplitude coefficients depended on the movement direction in most cases according to a cosine function. When movement speed varied in addition to movement direction, phasic spatiotemporal synergies maintained the same directional tuning and were modulated in amplitude by speed (d'Avella et al. 2008). These simple modulation rules support the notion of muscle synergies as a mechanism for implementing a direct mapping of movement goals into motor commands.

Modular decomposition of EMG patterns may also provide insights on motor development. Whether the muscle activation patterns for locomotion are innate, or whether they are acquired during development is an important question. A study of temporal synergies underlying locomotion (locomotor primitives) at different developmental stages demonstrated that there are both innate and learned modules (Dominici et al. 2011). Human newborns, supported and with their feet in contact with a surface, generate stepping-like movements that generally disappears at ~2 months after birth. Decomposition of EMG activity during newborn stepping by NMF revealed two primitives that were roughly similar to two of the four primitives observed in the adult, but more prolonged in duration. As in adults, one primitive was related to body support during stance, while the other to limb flexion during swing. In newborns, however, there were no specific primitive related to either touch-down or lift-off. In toddlers (~1-year-old) at their first unsupported steps, in addition to the same two primitives seen in the newborn, two new primitives timed at touch-down and lift-off appeared and they were similar to the two other primitives seen in the adult. In preschoolers (2–4 years), all four primitives changed shape with increasing age, becoming narrower and closer to the adult waveforms. Moreover, comparing the development of locomotor patterns in humans with that in other vertebrates it appears that locomotion of several animal species is built starting from common modular elements, perhaps related to ancestral neural networks. However, with development, the motor patterns may be adjusted to satisfy the specific biomechanical requirements of a given animal species, such as those for erect bipedal locomotion in humans.

The studies briefly reviewed above and many other studies based on the decomposition of multi-muscle EMG recordings indicate that muscle synergies capture regularities in the spatial, temporal, and spatiotemporal organization of the muscle patterns, supporting the notion that the CNS organizes muscle synergies to simplify control. However, the observed regularities might also derive from biomechanical constraints. The low dimensionality of the muscle patterns captured by muscle synergies might simply reflect the limited number of ways in which a task can be performed by the musculoskeletal system. For example, the dimensionality of the set of muscle activation vectors capable of generating all feasible forces at a limb endpoint is smaller than the number of muscles involved (Kutch and Valero-Cuevas 2012), suggesting that some of the observed reduction in the dimensionality of muscle patterns might not come from neurally organized synergies. Direct evidence for a neural organization of muscle synergies thus requires going beyond descriptive approaches and testing muscle synergies as a causal model.

## Modularity Predicts Learning Difficulty

In a modular control architecture, acquisition of a new motor skill or adaptation of an existing skill after a perturbation may occur at the level of modules or at the level of module combinations. Considering a modular architecture based on spatial muscle synergies, the sensorimotor transformations mapping a goal **g** and an initial limb state **x** into a muscle activation pattern **m**(t), can be written (see Eq. 1) as

$$\mathbf{m}(t;\mathbf{g},\mathbf{x}) = \sum_{i=1}^{N} c_i(t;\mathbf{g},\mathbf{x})\ \mathbf{w}_i \tag{3}$$

assume that the synergies $\mathbf{w}_i$ do not depend on the goal and the initial state. Then, for a set of tasks for which those synergies provide an adequate basis for the generation of the motor commands, different task goals are achieved by selecting a different combination of the existing synergies. Figure 1a represents schematically such mapping in baseline condition, i.e., before attempting to learn a new task or to adapt to a perturbation. A new motor skill is compatible with the existing synergies if it can be acquired by learning a new mapping of goals and initial state onto the existing synergies (Fig. 1b). Similarly, a perturbation to a task is compatible with the existing synergies if it can be compensated by adapting the mapping between goals and initial states onto the existing synergies (Fig. 1c). Throwing while wearing prism glasses (Martin et al. 1996) provides an example of a perturbation and a new motor skill likely compatible with existing synergies. Immediately after wearing prism glasses, which deviate the light path laterally so that the direction of gaze deviates from the target direction, subjects throw in the direction of gaze and show large errors. With practice, subjects gradually adapt to the perturbation of the mapping from visual targets to muscle patterns induced by the prisms and throw accurately again. As there is no need to change the throwing movement and, thus, to adapt the muscle synergies underlying the throwing skill, adaptation likely occurs at the level of the mapping of visual targets onto synergy combinations. Indeed, invariance of synergy structure has been observed during adaptation to a visuomotor rotation in a force reaching task (Gentner et al. 2013). After removal of the glasses, errors are in the opposite direction with respect to the initial error (aftereffect), indicating that the original mapping has been altered, and additional practice is required to gradually return to baseline performance. However, if subjects continue practicing throwing with and without prism glasses after a few weeks they are able to throw accurately immediately after wearing and removing the glasses. A new throwing skill, throwing with prism glasses, is therefore acquired with extensive practice and it can be recalled immediately without interfering with the unperturbed throwing skill. In terms of modular architecture, both skills rely on the original muscle synergies for throwing but a new mapping of goals and initial states onto synergy combinations is acquired. Finally, new skills and perturbations may also be incompatible with the existing synergies and may require new synergies. New synergies may be added to the set of existing synergies (Fig. 1d) or they may be obtained by adapting one of multiple existing synergies (Fig. 1e). For example, learning to walk, as discussed above, requires reusing existing synergies as well as organizing new synergies.

One advantage of a modular control architecture for motor learning and motor adaptation is that it may achieve flexibility without affecting stability of motor performance. In the CNS, acquisition of new motor skills and adaptation of existing skills involve changes in synaptic connections. The more plastic the CNS is, the easier is to learn and adapt the required sensorimotor transformations. However, if the CNS is too plastic and the pattern of synaptic connectivity is constantly changing,

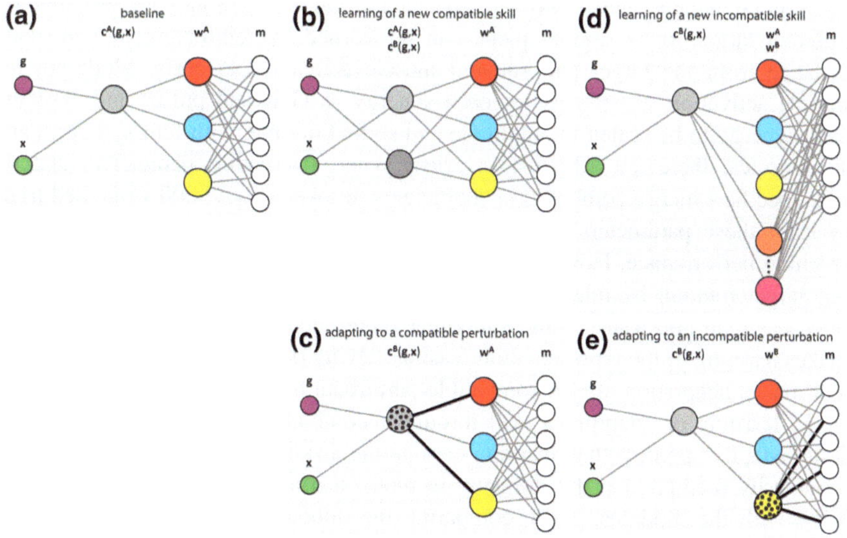


**Fig. 1** Learning in a modular control architecture. **a** Schematic representation of the sensorimotor transformations of goal (**g**) and initial limb state (**x**) onto muscle patterns through combinations $[c^A(t; \mathbf{g}, \mathbf{x})]$ of muscle synergies ($\mathbf{w}^A$). **b** Learning a new compatible skill by acquisition of a new mapping of goals and initial states onto synergy combinations ($c^A \rightarrow \{c^A, c^B\}$). **c** Adapting to a compatible perturbation by adapting the original mapping ($c^A \rightarrow c^B$). **d** Learning a new incompatible skill by adding new synergies ($\mathbf{w}^A \rightarrow \{\mathbf{w}^A, \mathbf{w}^B\}$). **e** Adapting to an incompatible perturbation by changing the existing synergies ($\mathbf{w}^A \rightarrow \mathbf{w}^B$)

the storage of sensorimotor transformations may become unstable. New motor skills may erase existing ones. In a noisy and highly redundant neural network such as the CNS, a way to address such stability–plasticity dilemma may be to segregate synaptic weights corresponding to different skills in different regions of the huge synaptic weight space that, even when weights are hyperplastic, represent and maintain stable sensorimotor transformations (Ajemian et al. 2013). By operating distinct adaptive processes at the level of modules and module combinations, the CNS may achieve a trade-off between stability and plasticity and facilitate the acquisition of segregated synaptic representations of different skills. Acquisition of skills and adaptation to perturbations compatible with the existing modules may require adjusting a smaller number of synaptic weights than acquiring skills or adapting to perturbations incompatible with the modules, which must represent in their synaptic weights all necessary knowledge of the dynamic behavior of the musculoskeletal system to implement adequate sensorimotor transformations. However, a larger synaptic weight space has more capacity for storing different modules in segregated and noninterfering (i.e., orthogonal) regions of the space. Thus, learning or adapting an existing skill through recombination of modules is likely a faster adaptive process than learning or adapting a skill by acquiring new modules.

Faster learning of module combinations than of module structure has been demonstrated in a simulation of a reaching task with a musculoskeletal model (Ruckert and d'Avella 2013). Spatiotemporal muscle synergies and synergy combination coefficients generating shoulder and elbow joint torques for reaching six different targets on a sagittal plane can be learned by reinforcement with hand distance from the target and squared muscle activation as costs. Modeling the muscle activation waveforms for each synergy as Gaussian pulses and allowing each synergy to be scaled in amplitude and shifted in time, with four synergies and 11 muscles, there are 132 task-invariant synergy structure parameters and 48 task-specific synergy combination coefficients. It takes about 3000 simulated trials to learn these parameters with a stochastic search method and achieve a good reaching performance. However, once synergy structure is learned, adaptation to a 30° target rotation by relearning only the synergy combination coefficients takes only less than 200 trials. Thus, a modular representation of the motor commands makes learning of the representation feasible relying only on a reinforcement signal and makes adaptation after a compatible perturbation very fast.

As learning or adapting a skill through recombination of modules is likely a faster adaptive process than learning or adapting a skill by acquiring new modules, modularity predicts that the difficulty in motor learning depends on the compatibility with the modules. Thus, comparing the difficulty for learning or adapting skills which are either compatible or incompatible with the modules allows testing a prediction of the modular hypothesis. Consider a perturbation of the way in which the activation of the muscles acting on the arm generates force at the hand, such as a surgical rearrangement of the tendons. Imagine a tendon transfer that alters the amplitude and direction of the forces generated by the muscles involved in a specific spatial muscle synergy. If the forces are altered such that they cancel each other when the muscles are activated according the balance prescribed by the synergy, recruiting that synergy would generate zero force [see Fig. 1 in Berger et al. (2013)]. Thus, to compensate such incompatible perturbation it would be necessary to generate muscle activation patterns that are not in the direction of the synergy made ineffective by the tendon transfer. If synergies are only a description of the regularities in the muscle patterns generated by a control architecture that has direct access to individual muscles, it should be equally difficult to generate new muscle patterns when they can be expressed as synergy combinations as when they cannot. In contrast, if the CNS relies on a set of muscle synergies to generate muscle patterns, if the new muscle patterns required to overcome the perturbation cannot be generated by the synergies, the synergies must be adapted or new synergies must be learned. Thus, if muscle synergies are organized by the CNS, adapting to an incompatible perturbation is expected to be more difficult than adapting to a compatible perturbation. In contrast, if synergies are just a description of regularities generated by a controller that has access to individual muscles, adapting to an incompatible perturbation is expected to be equally difficult than adapting to a compatible perturbation, provided that the required changes in the activation of individual muscles are comparable in the two cases.

A recent study in which human subjects used myoelectric control to move a mass in a virtual environment has tested the prediction that in a modular architecture adaptation to incompatible perturbations must be harder than adaptation to compatible perturbation (Berger et al. 2013). Subjects were instructed to perform a reaching task by displacing a cursor according to either the isometric force applied on a transducer attached to a forearm, wrist, and hand splint (force control) or the force estimated from the EMG activity recorded from many shoulder and arm muscles (myoelectric or EMG control). The cursor and the reach targets were spheres displayed by a flat monitor occluding the subject's hand and appearing on a virtual desktop matching the appearance and position the real desktop. Initially the reaching task was performed using force control and, for each individual participant, the force and EMG data collected were used to estimate, using multiple linear regression, a linear mapping between the recorded EMG activity of each muscle and the force generated at the hand [EMG-to-force matrix, see Fig. 3 in Berger et al. (2013)]. EMG data were also used to identify spatial synergies using NMF and to estimate, through the EMG-to-force matrix, the force generated by each synergy. Subjects then performed the rest of the experiment using EMG control. In baseline condition, i.e., when the movement of the cursor depended on the force computed with the unperturbed EMG-to-force matrix, subjects were able to perform the task immediately after switching from force control, showing that the linear mapping was sufficiently accurate. Because the forces were computed in real-time by "virtual" muscles, it was possible to arbitrarily modify the EMG-to-force mapping and to perform the type of tendon transfers discussed above, i.e., a "virtual surgery" on the musculoskeletal system. This allowed to compare the adaption rate after a surgery compatible with the muscle synergies identified in force control with the adaptation rate after an incompatible surgery. Both compatible and incompatible surgeries were generated by transforming the instantaneous muscle activity vector through a multidimensional rotation in muscle space, which affected the amplitude and direction of the force generated by each muscle in a complex way but by an equal amount, on average across muscles, in the two cases. However, after a compatible surgery, the forces generated by the synergies, also transformed in a complex way, could still span the force space, i.e., could be combined to generate any force. In contrast, after an incompatible surgery, the forces generated by the synergies were aligned along a single direction and did not span the force space. Forces in any other direction required generating muscle patterns that could not be obtained by synergy combinations.

Task performance, quantified by the angular error of the cursor's initial movement direction with respect to the target direction and by the fraction of trials in which the cursor did not reach and hold the target position in the available time [see Fig. 6 in Berger et al. (2013)], dropped significantly in the first block after both types of surgeries but it then improved faster after a compatible surgery than after an incompatible surgery. The performance in the last block after the compatible surgery was significantly better than after the incompatible surgery. Performance improvements after incompatible surgeries, even if they occurred significantly more slowly than after compatible surgeries, were associated with changes in the muscle patterns that could not be captured by the original muscle synergies. There was a significant reduction of the muscle pattern reconstruction quality [see Fig. 8 in Berger et al. (2013)] at the end of the exposure to incompatible virtual surgery with respect to compatible virtual surgery. In sum, differences in adaptation rates after compatible and incompatible virtual surgeries support a neural organization of muscle synergies and suggest that a slower synergy adaptation process is active when the set of synergies usually employed for a task becomes ineffective.

## Take Home Message

In this chapter, I have suggested that the CNS relies on modularity to reduce the complexity of motor control by embedding features of both the dynamic behavior of the musculoskeletal system and of the task into a small number of modules and by directly mapping task goals into module combination parameters. Evidence for muscle synergies as modules has come from the decomposition of EMG patterns recorded in many species and tasks. For example, the combinations of small number of spatiotemporal synergies explain the organization of the muscle patterns across reaching movement directions and speeds and synergies are modulated in amplitude according to simple rules. Finally, in a modular architecture, acquisition of a new motor skill or adaptation of an existing skill after a perturbation may occur at the level of modules or at the level of module combinations. As learning or adapting an existing skill through recombination of modules is likely faster than learning or adapting a skill by acquiring new modules, compatibility with the modules predicts learning difficulty. The observation of slower adaptation after incompatible than after compatible virtual surgeries provides direct evidence for modularity in motor control and motor learning.

## References

Ajemian R, D'Ausilio A, Moorman H, Bizzi E (2013) A theory for how sensorimotor skills are learned and retained in noisy and nonstationary neural circuits. Proc Natl Acad Sci USA 110: E5078–E5087. doi:10.1073/pnas.1320116110

Alessandro C, Carbajal JP, d'Avella A (2013) A computational analysis of motor synergies by dynamic response decomposition. Front Comput Neurosci 7:191. doi:10.3389/fncom.2013.00191

Atkeson CG, Hollerbach JM (1985) Kinematic features of unrestrained vertical arm movements. J Neurosci 5:2318–2330

Berger DJ, Gentner R, Edmunds T, Pai DK, d'Avella A (2013) Differences in adaptation rates after virtual surgeries provide direct evidence for modularity. J Neurosci 33:12384–12394. doi:10.1523/JNEUROSCI.0122-13.2013

Bernstein N (1967) The co-ordination and regulation of movement. Pergamon, Oxford

Bizzi E, Cheung VC, d'Avella A, Saltiel P, Tresch M (2008) Combining modules for movement. Brain Res Rev 57:125–133. doi:10.1016/j.brainresrev.2007.08.004

Bizzi E, D'Avella A, Saltiel P, Tresch M (2002) Modular organization of spinal motor systems. Neuroscientist 8:437–442

Bizzi E, Mussa-Ivaldi FA, Giszter SF (1991) Computations underlying the execution of movement: a biological perspective. Science 253:287–291

Buchanan TS, Almdale DPJ, Lewis JL, Rymer WZ (1986) Characteristics of synergic relations during isometric contractions of human elbow muscles. J Neurophysiol 56:1225–1241

d'Avella A, Tresch MC (2002) Modularity in the motor system: decomposition of muscle patterns as combinations of time-varying synergies. In: Dietterich TG, Becker S, Ghahramani Z (eds) Advances in neural information processing systems, vol 14. MIT Press, Cambridge, MA, pp 141–148

d'Avella A, Lacquaniti F (2013) Control of reaching movements by muscle synergy combinations. Front Comput Neurosci 7:42. doi:10.3389/fncom.2013.00042

d'Avella A, Saltiel P, Bizzi E (2003) Combinations of muscle synergies in the construction of a natural motor behavior. Nat Neurosci 6:300–308

d'Avella A, Portone A, Fernandez L, Lacquaniti F (2006) Control of fast-reaching movements by muscle synergy combinations. J Neurosci 26:7791–7810. doi:10.1523/JNEUROSCI.0830-06.2006

d'Avella A, Fernandez L, Portone A, Lacquaniti F (2008) Modulation of phasic and tonic muscle synergies with reaching direction and speed. J Neurophysiol 100:1433–1454. doi:10.1152/jn.01377.2007

Delis I, Panzeri S, Pozzo T, Berret B (2014) A unifying model of concurrent spatial and temporal modularity in muscle activity. J Neurophysiol 111:675–693. doi:10.1152/jn.00245.2013

Dominici N, Ivanenko YP, Cappellini G et al (2011) Locomotor primitives in newborn babies and their development. Science 334:997–999. doi:10.1126/science.1210617

Flanders M (1991) Temporal patterns of muscle activation for arm movements in three-dimensional space. J Neurosci 11:2680–2693

Flanders M, Pellegrini JJ, Soechting JF (1994) Spatial/temporal characteristics of a motor pattern for reaching. J Neurophysiol 71:811–813

Flanders M, Pellegrini JJ, Geisler SD (1996) Basic features of phasic activation for reaching in vertical planes. Exp Brain Res 110:67–79

Gentner R, Edmunds T, Pai DK, d'Avella A (2013) Robustness of muscle synergies during visuomotor adaptation. Front Comput Neurosci 7:120. doi:10.3389/fncom.2013.00120

Giszter SF (2015) Motor primitives—new data and future questions. Curr Opin Neurobiol 33: 156–165. doi:10.1016/j.conb.2015.04.004

Giszter SF, Hart CB, Silfies SP (2010) Spinal cord modularity: evolution, development, and optimization and the possible relevance to low back pain in man. Exp Brain Res 200:283–306. doi:10.1007/s00221-009-2016-x

Gottlieb GL, Song Q, Almeida GL, Hong DA, Corcos D (1997) Directional control of planar human arm movement. J Neurophysiol 78:2985–2998
Grillner S (1981) Control of locomotion in bipeds, tetrapods, and fish. In: Brooks VB (ed) Handbook of physiology: section I. The nervous system. American Physiological Society, Bethesda, MD, pp 1179–1236
Hart CB, Giszter SF (2004) Modular premotor drives and unit bursts as primitives for frog motor behaviors. J Neurosci 24:5269–5282. doi:10.1523/JNEUROSCI.5626-03.2004
Hollerbach MJ, Flash T (1982) Dynamic interactions between limb segments during planar arm movement. Biol Cybern 44:67–77
Ivanenko YP, Grasso R, Zago M et al (2003) Temporal components of the motor patterns expressed by the human spinal cord reflect foot kinematics. J Neurophysiol 90:3555–3565
Ivanenko YP, Poppele RE, Lacquaniti F (2004) Five basic muscle activation patterns account for muscle activity during human locomotion. J Physiol 556:267–282
Klein Breteler MD, Simura KJ, Flanders M (2007) Timing of muscle activation in a hand movement sequence. Cereb Cortex 17:803–815. doi:10.1093/cercor/bhk033
Kutch JJ, Valero-Cuevas FJ (2012) Challenges and new approaches to proving the existence of muscle synergies of neural origin. PLoS Comput Biol 8:e1002434. doi:10.1371/journal.pcbi.1002434. PCOMPBIOL-D-11-01006 [pii]
Lacquaniti F, Ivanenko YP, Zago M (2012) Patterned control of human locomotion. J Physiol 590:2189–2199. doi:10.1113/jphysiol.2011.215137
Lacquaniti F, Soechting JF, Terzuolo CA (1982) Some factors pertinent to the organization and control of arm movements. Brain Res 252:394–397. doi:10.1016/0006-8993(82)90410-3
Lacquaniti F, Soechting JF, Terzuolo SA (1986) Path constraints on point-to-point arm movements in three-dimensional space. Neuroscience 17:313–324
Latash ML (2012) Fundamentals of motor control. Academic Press
Lee WA (1984) Neuromotor synergies as a basis for coordinated intentional action. J Mot Behav 16:135–170
Lee DD, Seung HS (1999) Learning the parts of objects by non-negative matrix factorization. Nature 401:788–791
Macpherson JM (1991) How flexible are muscle synergies? In: Humphrey DR, Freund H-J (eds) Motor control: concepts and issues. Wiley, Chichester, pp 33–47
Maier MA, Hepp-Reymond MC (1995) EMG activation patterns during force production in precision grip. II. Muscular synergies in the spatial and temporal domain. Exp Brain Res 103:123–136
Martin TA, Keating JG, Goodkin HP, Bastian AJ, Thach WT (1996) Throwing while looking through prisms. II. Specificity and storage of multiple gaze-throw calibrations. Brain 119(Pt 4):1199–1211
Morasso P (1981) Spatial control of arm movements. Exp Brain Res 42:223–227
Ruckert E, d'Avella A (2013) Learned parametrized dynamic movement primitives with shared synergies for controlling robotic and musculoskeletal systems. Front Comput Neurosci 7:138. doi:10.3389/fncom.2013.00138
Russo M, D'Andola M, Portone A, Lacquaniti F, d'Avella A (2014) Dimensionality of joint torques and muscle patterns for reaching. Front Comput Neurosci 8:24. doi:10.3389/fncom.2014.00024
Saltiel P, Wyler-Duda K, D'Avella A, Tresch MC, Bizzi E (2001) Muscle synergies encoded within the spinal cord: evidence from focal intraspinal NMDA iontophoresis in the frog. J Neurophysiol 85:605–619
Santello M, Flanders M, Soechting JF (1998) Postural hand synergies for tool use. J Neurosci 18:10105–10115
Sherrington CS (1948) The integrative action of the nervous system. Univeristy Press, Cambridge
Soechting JF, Lacquaniti F (1981) Invariant characteristics of a pointing movement in man. J Neurosci 1:710–720
Soechting JF, Lacquaniti F (1989) An assessment of the existence of muscle synergies during load perturbations and intentional movements of the human arm. Exp Brain Res 74:535–548

Ting LH, Chiel HJ, Trumbower RD, Allen JL, McKay JL, Hackney ME, Kesar TM (2015) Neuromechanical principles underlying movement modularity and their implications for rehabilitation. Neuron 86:38–54. doi:10.1016/j.neuron.2015.02.042

Ting LH, Macpherson JM (2005) A limited set of muscle synergies for force control during a postural task. J Neurophysiol 93:609–613. doi:10.1152/jn.00681.2004

Tresch MC, Jarc A (2009) The case for and against muscle synergies. Curr Opin Neurobiol 19:601–607. doi:10.1016/j.conb.2009.09.002

Tresch MC, Saltiel P, Bizzi E (1999) The construction of movement by the spinal cord. Nat Neurosci 2:162–167

# Synergies in Grasping

**Jeroen B.J. Smeets and Eli Brenner**

**Abstract** The reach-to-grasp movement is a prototype of human movement coordination. Since the pioneering work of Jeannerod (Attention and performance, ix. Erlbaum, Hillsdale, NJ, pp 153–169, 1981), this movement is generally considered to be a coordinated combination of hand transport and grip formation. One of the main theoretical reasons for choosing transport and grip as building blocks is that they are anatomically independent: one can determine whether each muscle, joint, or brain area belongs to transport or grip. We have proposed a different view on grasping, in which the coordination problem is formulated as one related to the movements of the digits (Smeets and Brenner in Motor Control 3:237–271, 1999). According to this view, both the transport of the hand and the formation of the grip emerge from the combination of independent digits' movements toward the objects' surface. This independency of the digits resembles the independence of synergies (as discussed in the chapter of d'Avella). Different synergies are activated independently, but a single muscle can be part of several synergies. In this chapter, we will present three types of experiments that were designed to test to what extent the individual digits' movements can be considered as the building blocks of the reach-to-grasp movement.

**Keywords** Reach-to-grasp · Digits · Finger-thumb asymmetry · Visuomotor · Prehension · Hand transport · Grasp control

## Introduction

Marc Jeannerod started the study of the grasping movement by postulating two independent visuomotor channels through which visual input controls the movement (Jeannerod 1981). The argumentation for this postulate was based on the

J.B.J. Smeets (✉) · E. Brenner
Department of Human Movement Sciences, VU University Amsterdam,
van der Boechorststraat 9, 1081 Amsterdam, BT, The Netherlands
e-mail: J.B.J.Smeets@vu.nl



J. Laczko and M.L. Latash (eds.), *Progress in Motor Control*,
Advances in Experimental Medicine and Biology 957,
DOI 10.1007/978-3-319-47313-0_2

observation that grasping can be functionally split into a reach component (bringing the hand to the location of the target) and a grasp component (opening the hand in accordance with the size and shape of the target). A corresponding distinction is present in the nature of visual information: information about intrinsic properties like shape and color ("what") is essentially different from extrinsic properties such as distance and orientation ("where"). Both distinctions are also thought to be present in the neural processing: the neuromuscular system for shaping the hand is to a large extent independent of the system that is used to transport the hand to the object (Brinkman and Kuypers 1973) and it has been argued that there are distinct neural pathways for "what" and "where" (Trevarthen 1968; Ungerleider and Haxby 1994).

An underlying assumption in the above argumentation is that we can understand human sensorimotor control better if we regard it as two (neuro-) anatomically distinct pathways, each serving a distinct function. This assumption that distinct structures are the building blocks of behavior is not specific to the reasoning of the papers cited above, but also underlies some other influential schemes, such as the two visual systems hypothesis (Goodale and Milner 1992; Milner and Goodale 2006, 2008). However, one could also bring order into distributed control systems by the concept of synergies (Lee 1984; Soechting and Lacquaniti 1989; d'Avella et al. 2003; Ting and Macpherson 2005; Latash et al. 2007; Tresch and Jarc 2009; Overduin et al. 2015). The cited studies all define synergy in their own way. For the present purpose, the most important commonality is that a single biomechanical element (e.g., muscle or joint) can be part of several synergies. Importantly, the concept of synergies implies that there is no need to have a strict separation between anatomical aspects of a task. The components could therefore be selected such that they could be combined with other components to perform quite different tasks. This is potentially a more efficient strategy, because it allows any combination of components of movements to be combined in different ways for different tasks. A consequence, however, is that the components might not be optimized for a single task.

We will concentrate on kinematic synergies: movements of joints that are controlled as a single unit. In our earlier work (Smeets and Brenner 1999, 2001, 2008; Verheij et al. 2012), we proposed that the tips of the individual digits are controlled in grasping. In terms of synergies, we thereby assume the existence of two synergies: a thumb synergy and an index finger synergy. Given the fact that the index finger and thumb are part of the same hand, the two synergies' substrates overlap: they both contain the shoulder, elbow, and wrist; they differ in the more distal joints.

At the level of muscles, synergies have been defined as coherent time-varying patterns of muscle activations (d'Avella et al. 2003; Tresch and Jarc 2009). It has been shown for force control that the directional tuning of the synergies that are determined on the basis of the forces exerted at the end effector closely resemble the tuning of synergies that are based on an analysis of the EMG of more than ten muscles that are involved (Ting and Macpherson 2005). Therefore, we limit our analysis to the movements of the end effectors: the index finger and thumb. The essence of the concept of synergies is that the same synergies are used in many tasks to simplify control. In other words, specific characteristics of synergies should be visible in more than a single task. If the thumb and finger synergies are indeed the building blocks of grasping movements, we should be able to see the signature of the same two synergies in other tasks like pointing (we use this term for pointing to an object by moving to touch it with a single digit). In this chapter, we will describe three ways to study the signature of synergies in kinematics. The first way is to study idiosyncratic differences in the kinematics of the end effector: if synergies vary between individuals, the same differences should be visible in all tasks that involve these synergies. The second way is to adapt the synergies: if a synergy is adapted in pointing, this adaptation should transfer to grasping. The third way is to observe fast responses: if a target changes during a goal-directed movements, the fastest responses should be a direct manifestation of the synergies, neglecting higher coordination.

## Idiosyncratic Kinematics

If the movements of the digits are the synergies in grasping movements, the characteristics of the movements of the digits during grasping movements should be similar to those of movements of a single digit. The typical grasping movement starts with both digits at a staring position, then moving the hand toward the object while at the same time moving the digits apart to open the hand to be able to grasp the object (Jeannerod 1981). At about 75 % of the movement time (at 95 % of the distance; Cuijpers et al. 2004), the digits start closing to contact the object (Jeannerod 1984). How can we compare the movements of the digits in this task with a movement of a single digit with similar constrains?

We argued that we could do so by looking at the difference in shape of the trajectory between finger and thumb. If a separate finger synergy and thumb synergy exist in grasping, the resulting trajectory formation is likely to be slightly different for the two digits. Of course, finding a difference does not prove that they are separate synergies. It might be that the trajectory of the thumb is less curved than that of the index finger because the thumb is transported during grasping, and the finger moves relative to the thumb, as has been proposed on the basis of such findings (Haggard and Wing 1997; Mon-Williams and McIntosh 2000; Galea et al. 2001). If this were the case, one would not expect to find the same difference between finger and thumb if one were to compare the trajectories of pointing movements with these digits. On the other hand, if the difference between finger and thumb are completely due to differences in the synergies, one would expect the same differences to be present in other tasks involving the same synergies. To test these predictions one could use a task that is expected to recruit the same synergies as in the reach-to-grasp movement, but while independently performing different movements at different times with the two digits.

We decided to compare the reach-to-grasp movement with the reach-to push movement: move a single digit toward an object's surface in order to push that object away. This comparison is a reanalysis of the data of two of the three tasks in an earlier study (Smeets et al. 2010). We evaluated how much the trajectories of the two digits deviated from each other's mirror image, both when moving together during the reach-to-grasp movement, and when moving independently during a reach-to-push movement. We did this by dividing each path of an active digit into 100 segments of equal length, and subsequently averaging the paths of all trials for each of the 101 ends of the segments. We subsequently mirrored the average trajectory of one of the digits, and aligned the start and end position with that of the other digit, and determined the difference between the two (Fig. 1). We refer to this difference as the finger-thumb asymmetry.

We have two hypotheses for this finger-thumb asymmetry. The first hypothesis is based on the view that this asymmetry is based on the specificities of the control of grasping (Haggard and Wing 1997; Mon-Williams and McIntosh 2000; Galea et al. 2001). If this "grasp control" hypothesis were correct, one would expect the finger-thumb asymmetry in grasping to be consistent across subjects, and one would not expect the asymmetry in grasping to be very consistent with the asymmetry in pushing. In both cases, a slight asymmetry might be caused by the underlying anatomy (reducing the similarity between subjects with different physique or introducing some similarity between tasks performed with the same arm, respectively). On the other hand, if the asymmetry were caused by the difference between the thumb synergy and the finger synergy ("digit control" hypothesis), one would expect little consistent asymmetry across subjects, but one would expect each subject to show a consistent finger-thumb asymmetry across the two tasks. Finding such idiosyncratic asymmetries would not tell us anything about the underlying reason for the asymmetry. It might be that small anatomical differences between subjects are responsible. However, finding similar asymmetries, even if due to anatomical differences, in both pushing and grasping would support the idea that the two tasks are based on similar control mechanisms.

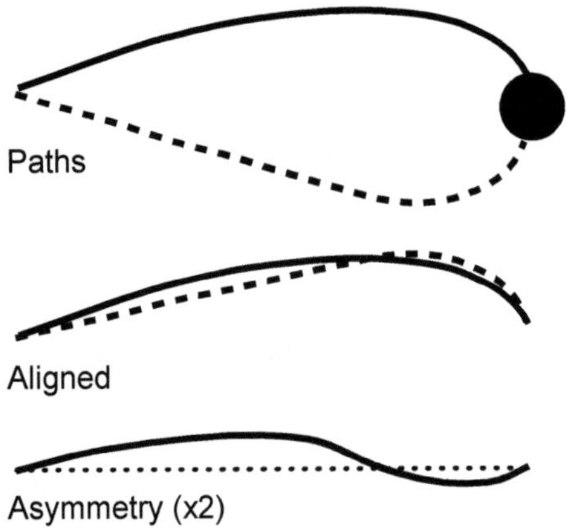


**Fig. 1** Determining the finger-thumb asymmetry from the (*average*) movement paths of the digits (*top*). The ends of the paths were aligned and one was mirrored (*middle*), after which the separation between the two was determined (shown magnified by two at the *bottom*)

We assessed the asymmetry by the Pearson's correlation coefficient between the asymmetry values across segments of the path (based on the average of 20 trials for a subject in a task). We did this both across tasks (within a subject), and across subjects (within a task). We chose this method because the two digits move independently (on different trials) in the pushing experiments, while they were physically connected when grasping. As this physical connection limits the possible curvature of the digits' paths, it might change the overall size of the asymmetry. We therefore chose to use the correlation along the path, rather than some measure of the distance itself (such as the sum of the squared differences) as our measure of the asymmetry. The predictions for the two hypotheses are plotted schematically, together with the experimentally obtained values in Fig. 2.

In line with both hypotheses, the asymmetries found when pushing are not highly correlated across subjects. This means that the differences between the shapes of the trajectories of finger and thumb are idiosyncratic, and thus not determined by the differences in anatomy between the two digits (which are common to all subjects). For the other two correlations, the hypotheses made opposite predictions. In line with the digit control hypothesis, we found that the asymmetries when grasping are not highly correlated across subjects, just as in pushing. So, also here, the differences in the trajectories are idiosyncratic, rather than being determined by a control mechanism or anatomic difference that is shared by all participants. The finding that the asymmetry in grasping is highly correlated with that in pushing is obviously also in line with the digit control hypothesis. This correlation means that if a certain difference between the trajectories of finger and thumb is found when moving together while grasping, a similar difference is found when comparing pushing with the thumb with pushing with the index finger.

The results are in line with the predictions made by assuming separate synergies for the control of finger and thumb that are used for both pushing and grasping. The overall correlation between subjects in the asymmetry is about 0.25 in both tasks. This means that the consistency in the difference between finger and thumb across subjects is not negligible, which is probably not surprising given the anatomical

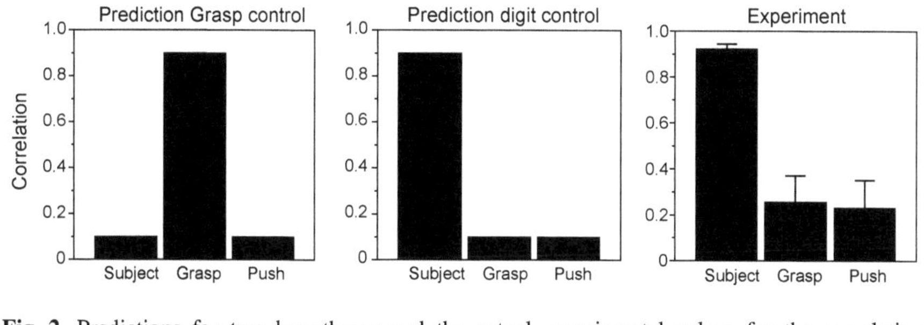


**Fig. 2** Predictions for two hypotheses and the actual experimental values for the correlation between asymmetries. The correlation is calculated between the two tasks within each *subject*, or between subjects within the grasp or push task. For the predictions, we plotted "strong correlation" as 0.9, and "weak correlation" as 0.1. For the experimental data (reanalysed from Smeets et al. 2010), the *error bars* show the standard error of the mean across subjects and pairs of subjects

differences between the digits. Most importantly for the present discussion is that the consistency in the asymmetry across subjects is the same for both tasks. In terms of the digit control hypothesis: apparently, despite the anatomical similarities between subjects, the finger synergy and the thumb synergy differ in a consistent way between subjects.

## Visuomotor Adaptation of Synergies

### *Pointing*

A second line of research that supports the existence of separate synergies for the thumb and index finger is that of visuomotor adaptation. We performed an experiment in which subjects were pointing alternatingly to the left and right side of a cube, using the thumb and the index finger, respectively (Schot et al. 2014). The cube could be placed at one of three positions at the far end of a wooden board. Subjects made their (curved) movements under the board, so the hand was invisible during most of the movement. Only the digit that made contact with the cube became visible when it was near the cube. We used the last position of the digit before it became visible to measure adaptation.

After a baseline phase with an unperturbed (binocular) view of the target, we gave the subjects a special pair of prism spectacles. These spectacles contained one leftward and one rightward deviating prism, combined with shutters. The shutters ensured that the subjects only looked through one of the prisms; which one depended on the movement they were asked to make. They experienced a leftward deviated view when pointing with their thumb and a rightward deviated view when pointing with their index finger. As subjects saw the target cube displaced 5 cm to the left of its actual position when moving the thumb, subjects initially moved their thumb to a position 5 cm more to the left than without a prism. In a next trial (with the index finger), subjects saw the cube 5 cm to the right, leading to an error in the opposite direction. After 45 trials with each digit, we removed the prisms, and let the subjects view binocularly again. Comparing the behavior in the post-adaptation phase with the baseline is a clean measure of the effect of adaptation of the digits' movements. In a second session, we reversed the pairing (Schot et al. 2014). What do we expect to happen to the arm movements?

Visuomotor adaptation can be divided in a visual and a motor (proprioceptive) component (Redding and Wallace 1988). As human vision combines the images of the two eyes from early vision on, the visual part of the adaptation will be common for the movements of both digits, and thus cancel each other. If one does not assume that separate synergies for the control of the digits' movements adapt independently, but assumes that each of the joints adapts independently, one would predict that adaptation will be distributed over the joints that are involved in the movement (depending on how much each joint contributes to the movement). As the shoulder, elbow, and wrist are responsible for most of the transport of the digits in space, this would mean that most of the adaptation would occur in these joints. The consequence would be that the effect of the two opposite perturbations would cancel each other. We would therefore expect very little adaptation. Only information about the orientation of the eyes and of some joints in the hand would adapt. If the synergies underlying the finger and thumb movements adapt, one would expect considerable adaptation of both synergies in opposite directions.

What we found is that both digits adapted to the visual displacement that was associated with their own movement (Fig. 3b, c). The adaptation that was obtained was about 60 % for both digits (asymptotes in Fig. 3). Each digit's adaptation resembled the conventional result for viewing through prisms (Martin et al. 1996) or moving through force fields (Shadmehr and BrashersKrug 1997). Such adaptation can be described by various models of (sensori-)motor learning (Smeets et al. 2006; Smith et al. 2006). The results are therefore consistent with the predictions based on independent synergies for the index finger and thumb.

The adaptation was incomplete. This could be interpreted as evidence that the adaptation of the two digits was not independent. However, it has been shown that in conditions with terminal feedback continuous forgetting can lead to incomplete adaptation (van der Kooij et al. 2015). If forgetting were the cause of the incomplete adaptation in Fig. 3, a similar incomplete adaptation would be obtained for blocks of trials with only movements of the thumb (or of the index finger), although the adaptation would probably be less incomplete because the trials with the same digit

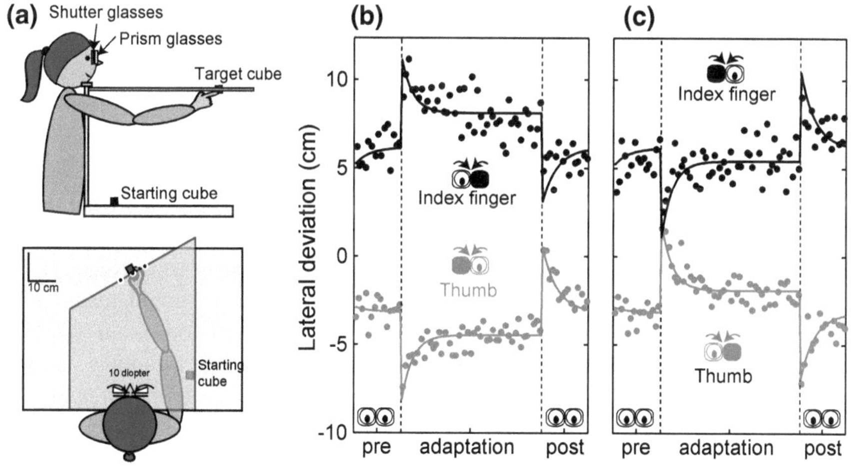


**Fig. 3** Simultaneous adaptation of index finger and thumb to opposite prism displacement; data are reanalysed from (Schot et al. 2014). **a** *Side* and *top* views of the subject performing a finger pointing movement. The hand remains invisible until just before the end of the movement. **b**, **c** behavior for the two sessions with opposite pairing between the digit that was moving and the viewing eye (and therefore the direction of the displacement). *Points* average response of the eight subjects. *Curves* exponential fit to the data points (constrained to have a change in deviation equal to the size of the prism-induced displacement at the time of addition/removal of the prism)

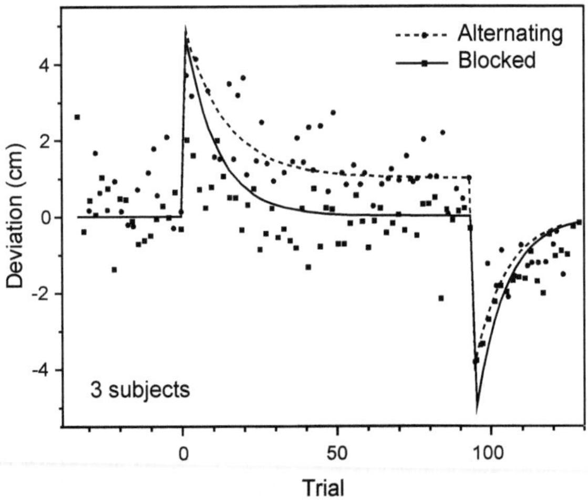


**Fig. 4** Adaptation with alternating prism-digit association compared with a blocked association. In the latter case there is complete adaptation

will be closer to each other in time (or number of movements). We tested this with a small number of subjects (3), and found *complete* adaptation (Fig. 4). Apparently, the present paradigm (involving real prisms and goal-directed movements that contact the targets) does not lead to trial-to-trial forgetting. We do not know why, but perhaps the presence of haptic feedback in the present experiments is important (Cuijpers et al. 2008; Schenk 2012). Thus, the adaptation that we found on its own does not provide conclusive evidence for adaptation within synergies related to the digits.

## *Transfer to Grasping*

In the previous section, we argued that we could adapt the synergies for the thumb and index finger separately. The main claim in this chapter is that the same synergies that are used in pointing are used in the reach-to-grasp movement. If this claim is correct, adaptation of the synergies during pointing should result in an aftereffect in the reach-to-grasp movement. As the adaptation is in opposite directions for the two digits, we predict that the aftereffect of adapting pointing movements will be an increase or decrease of grip aperture (depending on the direction of adaptation).

The best way to test a prediction for the transfer of an aftereffect from pointing to grasping would be to make it quantitative. However, based on Fig. 3b, we realized that such a prediction would be unrealistic. After removal of the prisms, the predicted deviation of each digit has changed by about 3 cm. For grasping, this would imply that the grip aperture would be either increased or reduced by 6 cm, depending on the pairing between prism and digit during the adaptation phase. As our objects' widths are less than 5 cm, the planned positions for the digits in the aftereffect of the thumb-left pairing would correspond to a crossed configuration of the thumb and the index finger. And the planned positions for the digits in the thumb-right pairing would correspond to the grip aperture for an extremely large object, bigger than any object used in research on the precision grip (Smeets and Brenner 1999). In both cases, the planned end configuration is outside the normal range (left part of Fig. 5), inevitably leading to limited transfer. Therefore, we cannot expect to have full transfer of the aftereffect of adaptation from pointing to grasping, so we limit our prediction to a qualitative one: there will be clear transfer from pointing to grasping.

To test this prediction, we repeated the adaptation experiment, and made one change relative to the experiment described above: we not only varied the position of the block, but also its size. In this way, we ensured that all parameters of grasping that might be controlled vary between trials: both the contact positions (our theory) and the position and size of the target (classical theory). Note that the visual perturbation that we applied in the adaptation phase only shifted positions, leaving the size of the objects unaltered.

The right panel of Fig. 5 shows that in all three conditions, grip aperture scales in a normal way with object size. Importantly, apertures are larger after adaptation of pointing with the thumb-right pairing than after adaptation with the opposite pairing. In this figure, we see a slightly smaller effect for the large object in the thumb-right pairing, presumably due to the ceiling effect that we predicted for this pairing. For the opposite pairing, we frequently observed that subjects' digits touched each other once or twice in the first few trials of grasping, in line with the prediction of the bottom-left of Fig. 1. The transfer from pointing to grip aperture

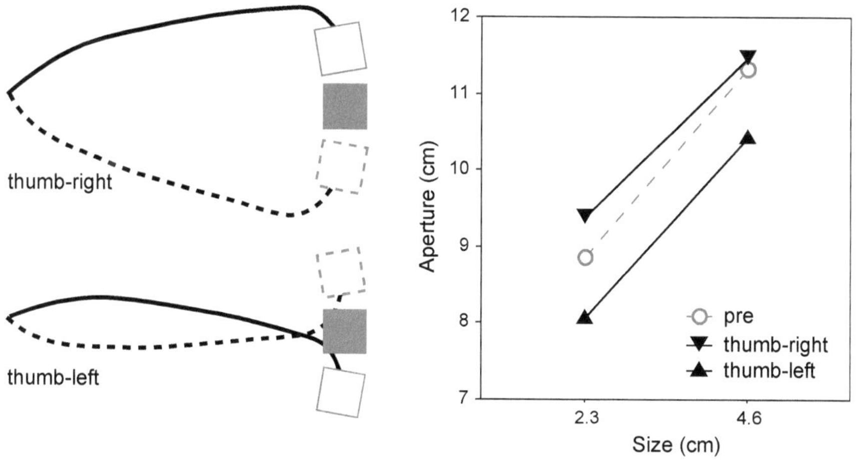


**Fig. 5** Transfer from pointing to grasping. *Left panel* the paths corresponding to 100 % transfer for both pairings. The *solid squares* represent the target, the *open squares* and *curves* represent the situation corresponding with a full aftereffect. These paths are impossible to produce. *Right panel* the actual grip apertures when the digits were approaching the border of the board (for the two target sizes). There is a clear transfer of the aftereffect

shows that grasping is not controlled fundamentally differently from moving the individual digits during pointing. Assuming that synergies regulate movements of the thumb and index finger separately, irrespective of the task, provides a parsimonious explanation for these findings. Assuming that grip aperture is regulated by a separate synergy involved in grasping cannot account for the transfer.

## Fast Responses to Perturbations

Humans are known to be able to adjust their movements to changes in the position of a target with a very short (~110 ms) latency (reviewed by Cluff et al. 2015; Smeets et al. 2016). Such fast responses are very interesting, as they must be based on the least amount of information processing possible: a fast link between visual information and the fundamental elements of control. As these fast responses do not take into account all information, the responses can sometimes be counterproductive. We showed this for a situation in which an obstacle is initially positioned to the left of the line connecting the hand to the target. In unperturbed trials, subject then veer slightly rightward. If this obstacle jumps to the right of the line, the optimal response would be to veer slightly leftward. This is however not what happens: subjects follow the target, and veer more to the right, and therefore hit the obstacle (Aivar et al. 2008). The reason is probably that the movement was planned to pass the obstacle on the right, and the first response to a change in position was a direct response to the change in the obstacle's position, without reconsidering one's options.

As fast responses link low-level elements of perception to those of motor control, they can be used to reveal the synergies that underlie grasping behavior. This has been done in paradigms that involved changes in the position and/or size of an object (Paulignan et al. 1991a, b; Smeets et al. 2002; van de Kamp and Zaal 2007; Hesse and Franz 2009; van de Kamp et al. 2009). For instance, the fact that a perturbation of the contact position for one digit sometimes has a small effect on the trajectory of the other digit (van de Kamp and Zaal 2007) has been used to argue that they cannot be controlled independently. However, the coupling between the digits can also be mechanical. We therefore present data on responses to object rotation.

There are relatively few studies on responses to object rotation when grasping (Desmurget et al. 1996; Voudouris et al. 2013). The oldest experiment showed that you can adjust the orientation of your hand during a whole-hand-grip grasping movement to a change in object orientation with a short latency (Desmurget et al. 1996). In our experiment (Voudouris et al. 2013), we let subjects grasp lightweight objects (a cube or a sphere) with a precision grip. The objects were magnetically connected to a motor that could rotate them very quickly over 12° (clockwise or counterclockwise) as soon as the subject initiated their grasping movement. For grasping the cube, it is clear that subjects should respond to the perturbation, otherwise their grip orientation will not coincide with the surfaces of the cube. For grasping the ball, rotating the hand in response to the ball's rotation does not provide any advantage. What would we expect for various direct couplings between visual information and controlled synergies?

According to the grasp control hypothesis, the presumed synergies are the transport and grip: the transport synergy, that includes the orientation (Jeannerod 1981; Desmurget et al. 1996), is coupled to the extrinsic object properties position and orientation, whereas the grip synergy is coupled to the intrinsic object property size. As none of these properties changes when a ball rotates, this hypothesis predicts no response (which is indeed optimal). According to the digit control hypothesis, the presumed synergies are the digit's movements. These movements are directed to suitable contact positions on the surface of the object. So, if the object rotates, the intended contact positions move, which will lead to fast adjustment of the digits' movements.

In line with the prediction of the digit control hypothesis, we always see a fast response to object rotations (Fig. 6). The sign of the response depends on the direction of the rotation: the digits follow the object's surface. The response is initially the same for the cube and the ball, but continues for more than 200 ms for the cube (left panel), whereas it disappears within 50 ms for the ball (central panel).

To test whether the short duration of the response for the ball is due to a reselection of grasping points, we performed an experiment in which the cube was placed in an ambiguous orientation that allowed for two grasping orientations (right panel of Fig. 6). We restrict our analysis to the subjects who changed their grip orientation in response to the perturbation. We see an initial following response that stops after about 50 ms and reverses its direction. This response is very similar to the response to the rotation of the ball. Thus the fastest response consists of digits following the local position on the surface, whereas only the later parts of the

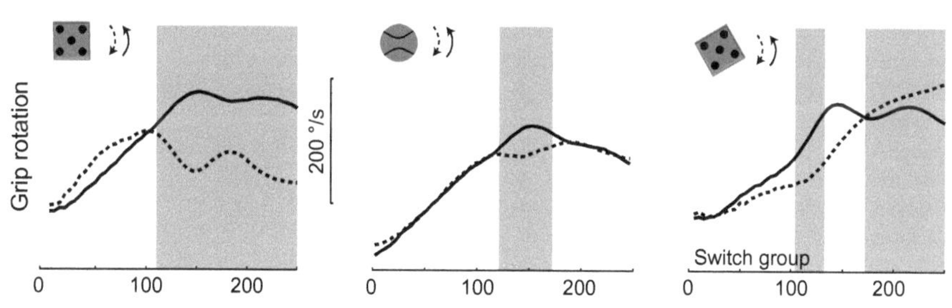


**Fig. 6** The velocity of rotation of the grip in response to target rotations in opposite directions. Data replotted from Voudouris et al. (2013). The *gray* area indicates times at which the rotation speed differed for the two directions. Three conditions are displayed: grasping a cube for which it is clear by which surfaces it can best be grasped (*left*), grasping a ball (*center*), and grasping a cube that is oriented so that it is not evident how best to grasp it (*right*). In the latter case, subjects sometimes switch their choice of surfaces in response to the perturbation. In all conditions we see an initial fast response in the direction of the rotation of the target (*solid curves* above *dashed* ones). For the ball and the cube with the ambiguous orientation, subjects chose new contact points within 50 ms of their initial response

response consider the object's shape (for the ball) and orientation (for the cube). The results of the fast responses thus suggest that the synergies are therefore related to the positions on the surface of the object (i.e., digit control), rather than to the object as a whole (grasp control).

## Take Home Message

We used three totally different paradigms to study whether the classical grip control (with synergies *transport* and *grip*) or digit control (with synergies for the individual digits) yield the most comprehensive description of the reach-to-grasp movement. We found that the peculiarities of the digits' movements varied considerably between subjects, but were (within a subject) remarkably consistent across grasping with two digits and pushing with a single digit. Second, we showed that by using prisms, we can adapt the pointing movements of finger and thumb in opposite directions, and that the aftereffect of these adaptations transfers to grasping. Third, we show that when grasping a ball with a precision grip, the digits show a fast following response to a (task-irrelevant) rotation of the ball. All these findings suggest that the elements that are controlled in grasping are synergies for the individual digits, rather than synergies for transport and grip.

## References

Aivar MP, Brenner E, Smeets JBJ (2008) Avoiding moving obstacles. Exp Brain Res 190: 251–264

Brinkman J, Kuypers HGJM (1973) Cerebral control of contralateral and ipsilateral arm, hand and finger movements in the split-brain rhesus monkey. Brain 96:653–674

Cluff T, Crevecoeur F, Scott SH (2015) A perspective on multisensory integration and rapid perturbation responses. Vision Res 110:215–222

Cuijpers RH, Brenner E, Smeets JBJ (2008) Consistent haptic feedback is required but it is not enough for natural reaching to virtual cylinders. Hum Mov Sci 27:857–872

Cuijpers RH, Smeets JBJ, Brenner E (2004) On the relation between object shape and grasping kinematics. J Neurophys 91:2598–2606

d'Avella A, Saltiel P, Bizzi E (2003) Combinations of muscle synergies in the construction of a natural motor behavior. Nat Neurosci 6:300–308

Desmurget M, Prablanc C, Arzi M, Rossetti Y, Paulignan Y, Urquizar C (1996) Integrated control of hand transport and orientation during prehension movements. Exp Brain Res 110:265–278

Galea MP, Castiello U, Dalwood N (2001) Thumb invariance during prehension movement: effects of object orientation. NeuroReport 12:2185–2187

Goodale MA, Milner AD (1992) Separate visual pathways for perception and action. Trends Neurosci 15:20–25

Haggard P, Wing A (1997) On the hand transport component of prehensile movements. J Mot Behav 29:282–287

Hesse C, Franz VH (2009) Corrective processes in grasping after perturbations of object size. J Mot Behav 41:253–273

Jeannerod M (1981) Intersegmental coordination during reaching at natural visual objects. In: Long J, Baddeley A (eds) Attention and performance ix. Erlbaum, Hillsdale, NJ, pp 153–169
Jeannerod M (1984) The timing of natural prehension movements. J Mot Behav 16:235–254
Latash ML, Scholz JP, Schoner G (2007) Toward a new theory of motor synergies. Mot Control 11:276–308
Lee WA (1984) Neuromotor synergies as a basis for coordinated intentional action. J Mot Behav 16:135–170
Martin TA, Keating JG, Goodkin HP, Bastian AJ, Thach WT (1996) Throwing while looking through prisms. 1. Focal olivocerebellar lesions impair adaptation. Brain 119:1183–1198
Milner AD, Goodale MA (2006) The visual brain in action, 2nd edn. Oxford University Press, Oxford
Milner AD, Goodale MA (2008) Two visual systems re-viewed. Neuropsychologia 46:774–785
Mon-Williams M, McIntosh RD (2000) A test between two hypotheses and a possible third way for the control of prehension. Exp Brain Res 134:268–273
Overduin SA, d'Avella A, Roh J, Carmena JM, Bizzi E (2015) Representation of muscle synergies in the primate brain. J Neurosci 35:12615–12624
Paulignan Y, Jeannerod M, MacKenzie C, Marteniuk R (1991a) Selective perturbation of visual input during prehension movements. 2. The effects of changing object size. Exp Brain Res 87:407–420
Paulignan Y, MacKenzie C, Marteniuk R, Jeannerod M (1991b) Selective perturbation of visual input during prehension movements. 1. The effects of changing object position. Exp Brain Res 83:502–512
Redding GM, Wallace B (1988) Components of prism adaptation in terminal and concurrent exposure: Organization of the eye-hand coordination loop. Percept Psychophys 44:59–68
Schenk T (2012) No dissociation between perception and action in patient df when haptic feedback is withdrawn. J Neurosci 32:2013–2017
Schot WD, Brenner E, Smeets JBJ (2014) Simultaneous adaptation of the thumb and index finger of the same hand to opposite prism displacements. J Neurophys 111:2554–2559
Shadmehr R, BrashersKrug T (1997) Functional stages in the formation of human long-term motor memory. J Neurosci 17:409–419
Smeets JBJ, Brenner E (1999) A new view on grasping. Mot Control 3:237–271
Smeets JBJ, Brenner E (2001) Independent movements of the digits in grasping. Exp Brain Res 139:92–100
Smeets JBJ, Brenner E (2008) Grasping weber's law. Curr Biol 18:R1089–R1090
Smeets JBJ, Brenner E, Biegstraaten M (2002) Independent control of the digits predicts an apparent hierarchy of visuomotor channels in grasping. Behav Brain Res 136:427–432
Smeets JBJ, Martin J, Brenner E (2010) Similarities between digits' movements in grasping, touching and pushing. Exp Brain Res 203:339–346
Smeets JBJ, Oostwoud Wijdenes L, Brenner E (2016) Movement adjustments have short latencies because there is no need to detect anything. Motor Control (in press)
Smeets JBJ, van den Dobbelsteen JJ, de Grave DDJ, van Beers RJ, Brenner E (2006) Sensory integration does not lead to sensory calibration. Proc Natl Acad Sci USA 103:18781–18786
Smith MA, Ghazizadeh A, Shadmehr R (2006) Interacting adaptive processes with different timescales underlie short-term motor learning. PLoS Biol 4:1035–1043
Soechting JF, Lacquaniti F (1989) An assessment of the existence of muscle synergies during load perturbations and intentional movements of the human arm. Exp Brain Res 74:535–548
Ting LH, Macpherson JM (2005) A limited set of muscle synergies for force control during a postural task. J Neurophys 93:609–613
Tresch MC, Jarc A (2009) The case for and against muscle synergies. Curr Opin Neurobiol 19:601–607
Trevarthen CB (1968) Two mechanisms of vision in primates. Psychologische Forschung 31: 299–348
Ungerleider LG, Haxby JV (1994) 'What' and 'where' in the human brain. Curr Opin Neurobiol 4:157–165

van de Kamp C, Bongers RM, Zaal FTJM (2009) Effects of changing object size during prehension. J Mot Behav 41:427–435

van de Kamp C, Zaal FTJM (2007) Prehension is really reaching and grasping. Exp Brain Res 182:27–34

van der Kooij K, Brenner E, van Beers RJ, Smeets JBJ (2015) Visuomotor adaptation: how forgetting keeps us conservative. PLoS ONE 10:e0117901

Verheij R, Brenner E, Smeets JBJ (2012) Grasping kinematics from the perspective of the individual digits: a modelling study. PLoS ONE 7:e33150

Voudouris D, Smeets JBJ, Brenner E (2013) Ultra-fast selection of grasping points. J Neurophys 110:1484–1489

# Encoding Temporal Features of Skilled Movements—What, Whether and How?

**Katja Kornysheva**

**Abstract** In order to reliably produce intelligible speech or fluently play a melody on a piano, learning the precise timing of muscle activations is essential. Surprisingly, the fundamental question of how memories of complex temporal dynamics of movement are stored across the brain is still unresolved. This review outlines the constraints that determine whether and how the timing of skilled movements is represented in the central nervous system and introduces different computational and neural mechanisms that can be harnessed for temporal encoding. It concludes by proposing a schematic model of how these different mechanisms may complement and interact with each other in fast feedback loops to achieve skilled motor timing.

**Keywords** Motor timing · Spatiotemporal control · Sequence learning · Modular representation · Cortico-subcortical loops

## Introduction ("What")

In the middle of the past century, the engineer and photographer Gjon Mili developed a technique to capture trajectories of movements in space such as those produced by musicians, athletes and painters using stroboscopic cameras. He was able to record skilled movement sequences by attaching a light to the subjects' effector of interest, such as the hand holding the violin bow, and letting the movement unfold in darkness

---

The original version of this chapter was revised: An erratum to this chapter can be found at https://doi.org/10.1007/978-3-319-47313-0_21.

---

K. Kornysheva (✉)
Institute of Cognitive Neuroscience, University College London, London, UK
e-mail: k.kornysheva@ucl.ac.uk

K. Kornysheva
Department of Neuroscience, Erasmus MC, Rotterdam, The Netherlands

K. Kornysheva
School of Psychology, Bangor University, Bangor, Wales, UK



J. Laczko and M.L. Latash (eds.), *Progress in Motor Control*,
Advances in Experimental Medicine and Biology 957,
DOI 10.1007/978-3-319-47313-0_3

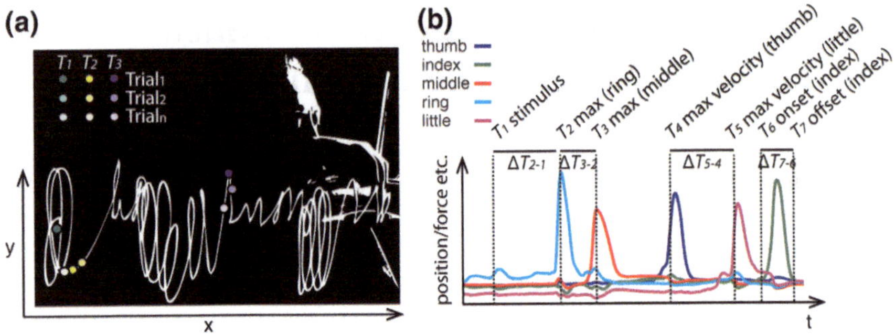


**Fig. 1** **a** Example of a skilled motor sequence depicted in two-dimensional Cartesian space ($x$ and $y$) (adapted from http://www.telegraph.co.uk/culture/culturepicturegalleries/7073785/On-the-Move-Visualising-Action-at-the-Estorick-Collection-of-Modern-Italian-Art.html?image=4). Repeating the skilled sequence can lead to the clustering of time points $T_2$ to $T_n$ following the onset of movement ($T_1$) respectively. Note that while here for illustrative purposes the variability of the spatial trajectory across trials is ignored, in reality the clustering across trials would take into account both space (position) and time (*colour*), cf. Laje and Buonomano (2013). **b** An example of a variable of interest during motor production such as dynamics (force) on a finger keyboard during a timed finger sequence task (adapted from Kornysheva and Diedrichsen 2014). Other variables of interest could be different kinematic measurements such as position and velocity depending on the motor task requirements. Accordingly motor timing can be quantified as time differences between task-relevant extrinsic stimuli and intrinsic states—such as maximum finger force after a go cue ($\Delta T_{2-1}$), eyelid position or velocity after a conditioning stimulus in eyeblink conditioning, the interval between two finger presses defined as the points of maximum velocity for each finger ($\Delta T_{5-4}$), or the movement duration, i.e. the difference between the offset and the onset of a movement ($\Delta T_{7-6}$)

with a long film exposure. The artist himself was only captured towards the end of the sequence when illuminating the room (Fig. 1a). Recording these trajectories revealed the skillful movement sequences humans are able to retrieve from memory and produce with their body in space. What remained invisible to Mili's lens is how the captured trajectory unfolded in time. It is left to the observer's imagination what velocity, acceleration and deceleration patterns the trajectory follows, how these spatial patterns emerged in time—its temporal features.

While traditionally the focus in motor neuroscience has been on the spatial dimension of movement sequences, such as the ordering or evolution of movements in space (Tanji and Shima 1994; Graybiel 1998; Hikosaka et al. 2002; Shenoy et al. 2012), the temporal dimension is equally crucial for the production of many skilled actions. Producing muscle activations in a correct order, but with inaccurate timing can have detrimental effects on performance in domains such as speech, complex tool use and music—a verbal utterance would become incomprehensible to the receiver, the tennis racket would miss the tennis ball and the violinist would desynchronize from the orchestra's pace.

At a purely descriptive level, skilled timing of a movement sequence in space entails that the movement has a reproducible temporal structure relative to an external stimulus or an internal motor state such as the occurrence of a movement onset. Here reproducibility entails that there is a certain level of temporal accuracy—typically within tens of milliseconds for most skilled motor sequences—relative to such a point of reference, when reaching a particular extrinsically (e.g. in Cartesian space) or intrinsically (e.g. in joint or muscle space) defined state of the body. Thus, when repeating a skilled spatial sequence of movements such as the position of the hand controlling the bow, the particular points in time ($T_2$, $T_3$) after movement onset time ($T_1$), cluster at the same extrinsic positions of the bow in two-dimensional space ($x$ and $y$ coordinates), respectively. In other words, a certain spatial configuration is reached at about the same time, with the degree of clustering reflecting the temporal precision of the movement. The temporal pattern of a movement trajectory becomes particularly evident with increased jerk, which reflects the strength of changes between acceleration and deceleration and whether the movement sequence contains activation pauses such as in a finger pressing task (Fig. 1b). Defining the motor points of interest is more straightforward for the latter type of actions (Fig. 1b), as they involve discrete kinematic events. When measuring motor timing, the timing of several kinematic and dynamic variables may be of interest depending on the motor task requirements, such as the variability of the spatial trajectory in time, the interval between an external stimulus and the maximum force, position, velocity, of a movement, etc., as well as between movements produced using the same or different effectors. Thus, in principle these variables may capture different aspects of temporal dynamics of skilled motor sequences as diverse as typing out a Morse code involving one effector and uttering a word or phrase which engages hundreds of muscles, both of which have to be executed with precise timing.

How does the nervous system represent and integrate the temporal features of such spatio-temporal sequences?

## Representation of Timing for Spatio-temporal Skills ("Whether")

Regularity or precision of a behavioural feature such as the temporal or spatial structure of a movement does not entail that the central nervous system (CNS) forms a dedicated representation or control mechanism for this feature. While goal directed and skilled movements have been shown to be sub-served by dedicated representations of force, direction, temporal order of muscle activations or a trajectory of movement in space (Evarts 1968; Georgopoulos et al. 1982; Hikosaka et al. 2002; Averbeck et al. 2002; Churchland et al. 2006; Shima et al. 2007; Shenoy et al. 2012) the presence of a dedicated substrate for encoding the timing for spatio-temporal motor skills is under debate.

In a series of experiments, Mussa-Ivaldi and colleagues demonstrated that the motor system is inherently biased to learn velocity-dependent over time-dependent representations during force field adaptations (Conditt and Mussa-Ivaldi 1999). Subjects performed reaching movements and were perturbed by force fields dependent either on the time after movement onset (time-dependent) or on the velocity (velocity-dependent, proportional to velocity) of the movement. Crucially, aftereffects and adaptation were evaluated in the context of generalization, when subjects were tested on circular instead of the trained reaching movements. These experiments revealed that after training on a time-dependent force field, generalization to a new movement was indistinguishable from the aftereffects and adaptation to velocity-dependent training. The authors concluded that there is an automatic bias to learn state-dependent instead of time-dependent representations during motor adaptation. Notably, the force field profile employed in the time-dependent condition was designed to be similar to a velocity-dependent force field, involving a bell-shaped perturbation with a maximum force in the middle of the movement when subjects produced the highest velocity. The primacy of state-dependent representations occurred when a perturbation environment was similar to a viscous field (water like environment). It is thus feasible that time-dependent force field profiles that are less correlated with movement velocity may override this bias.

However, in a follow-up study, Mussa-Ivaldi and colleagues (Karniel and Mussa-Ivaldi 2003) demonstrate that a time-dependent force field that is uncorrelated to movement velocity still produces no motor adaptation. Here the time-dependent force followed a sinusoidal amplitude at 3 Hz and was presented continuously during the experiment. This important study suggests that the CNS is unable to form a representation of a regular, temporally predictable force profile that is uncoupled from state-dependent representation. However, the employed time-dependent perturbation was not coupled to the onset of the movement as in the previous experiment (Conditt and Mussa-Ivaldi 1999), or at least to an external cue relevant to movement initiation. It can thus be hypothesized that this link may be a constraint for the acquisition of a time-dependent movement adaptation.

Indeed, Medina and colleagues demonstrated that learning motor timing during adaptation in smooth pursuit eye movements could be independent of state-dependent encoding (Medina et al. 2005). In training trials, a target moved horizontally for a fixed duration (500 ms) and deflected vertically from a horizontal to vertical movement. Probe trials were used to assess adaptation by looking at eye movement velocity into the vertical direction. Learning to time movements correctly was independent of the position of the eyes on the horizontal plane and of the distance/velocity of the movements. Importantly the adaptation effects were dependent on the predictive power of each variable. If both the time from target motion onset and the distance travelled were equally predictive, the adapted eye movements were a mixture of the two representations, whereas if only one variable was predictive of the vertical perturbation, the adaptation reflected the learning of time or distance only, respectively. This highlights the flexibility of motor adaptation with regard to the representation of time and space depending on which variable leads to task success.

Diedrichsen and colleagues showed that time- and state-dependent representation of spatio-temporal movements that involves the coordination of two effectors—the arm and the thumb—depends on whether their activation overlaps in time (Diedrichsen et al. 2007). Following a training phase in which the movements had to be timed precisely, the subjects were asked to reduce the speed of the arm movement. The thumb press was also timed and scaled in length proportionally to the arm movements, suggesting that the thumb movement was made dependent on the state (velocity) of the arm movement and not on absolute time since arm movement onset. Interestingly, absolute timing was employed when the movements were separated in time during training, that is when the thumb preceded the arm movement by 100–500 ms. This suggests that training temporally overlapping movements produces a bias to encode the movements of multiple effectors relative to their state, efficiently binding the effectors together to achieve well-timed coordination. Indeed, it would be detrimental to actions such as throwing a ball to a target to time arm and wrist movements based on independent time estimates. Independent noise levels or drifts would quickly lead to a decoupled motor state where the timing of muscle activations is disrupted, as in cerebellar ataxia, and may lead to a state resembling movement decomposition (Bastian et al. 1996; Timmann et al. 1999).

The impact of overlap between different motor activity states on their temporal encoding echoes the findings on discrete (non-overlapping) versus continuous (overlapping) timing tasks. Ivry and colleagues suggested a dichotomy of dedicated versus emergent encoding of timing for discrete versus continuous movements, respectively (Spencer et al. 2003; Ivry and Spencer 2004; Ivry and Schlerf 2008). Temporal variability on continuous tasks characterized by smooth transitions between different motor states (e.g. circle drawing) have been reported to be uncorrelated with the temporal variability on discrete tasks characterized by movement pauses in between boosts of motor activity (e.g. tapping) (Zelaznik et al. 2005). Moreover adjustment to timing perturbations is faster and more precise for discrete as opposed to continuous movements (Elliott et al. 2009; Repp and Steinman 2010; Studenka and Zelaznik 2011) and patient studies suggest that these movements might rely on different neural substrates (Spencer et al. 2003; Spencer and Ivry 2005). Yet, it is unlikely that movement kinematics alone determine whether temporal encoding is dedicated versus emergent: As discussed above, even continuous movements like smooth pursuit can be controlled using dedicated timing mechanisms and independently of parameters such as movement velocity, whenever the absolute timing predicts task success (Medina et al. 2005), or when a periodic circle drawing tasks contains a salient auditory cue marking the completion of a cycle (Zelaznik and Rosenbaum 2010; Braun Janzen et al. 2014).

When it comes to dissociating the spatial and temporal organization of sequential motor skills, the focus has been on learning the organization of sequences of movements rather than on learning the production of the constituent movements per se. Thus, typically subjects are trained to sequence simple overlearned movements like finger presses (Sakai et al. 2003; Ullen and Bengtsson 2003; O'Reilly et al. 2008; Kornysheva et al. 2013; Kornysheva and Diedrichsen 2014). With training the production of sequences becomes more accurate and is retrieved faster as evidenced by shorter sequence duration or reaction times (RT) depending on the task employed. In addition, a temporal grouping idiosyncratic to the subject or facilitated externally by the sequence structure emerges, such that certain movements in the sequence become closer in time than others creating so-called chunks. There is compelling evidence that breaking up the sequence within chunks as opposed to between chunks when reordering the sequence leads to losses in performance [for reviews see (Sakai et al. 2004)]. This suggests that a dedicated representation has been formed for each chunk of movements in space which facilitates performance—similar to chunks in working memory and cognitive control (Baddeley 2010). It has been hypothesized that this temporal grouping is a sign of a skill becoming automatic and pairing the sequence with a different temporal structure would lead to losses in performance as this automatic representation has not been formed (Hikosaka et al. 2002; Sakai et al. 2004).

Interestingly, there is evidence that while changing a chunking structure (externally induced) can lead to performance losses, these are not as pronounced as when performing a novel sequence (O'Reilly et al. 2008). This suggests some form of independence for the spatial organization of sequences, on top of the integrated spatio-temporal chunking structure. In contrast, many studies have shown that retaining the timing while changing the spatial feature of movement sequences does not provide any benefit as compared to a new sequence, which advocates that the temporal structure of these sequences is invariably bound to their sequential movements in space (Shin and Ivry 2002, 2003; O'Reilly et al. 2008).

This, however, has been challenged recently in a series of experiments (Kornysheva et al. 2013; Kornysheva and Diedrichsen 2014). Here the experimental test involved producing sequences following training of a single spatio-temporal sequence of finger presses in a timed SRT task (Penhune and Steele 2012). These were either repeated in a block of several trials or new on every trial. The results suggested that RT savings for a trained temporal feature paired with a new sequence of finger presses (spatial feature) could only emerge once the new spatial feature became more predictable through repetition (Fig. 2a, b). Note that the advantage for the trained temporal features is relative to the control condition in which the sequence was also repeated and the finger sequence became equally more predictable with repetition. In contrast when the finger order was new on each trial comparable to the random spatial sequence controls in the studies discussed above, there was no advantage related to learning the timing of the sequence. It is unlikely that this is an effect of whether these sequences were learned implicitly or had an explicit component, as both the presence and the absence of temporal transfer were found depending on the familiarity with the spatial feature.

More formally, drift diffusion modelling demonstrated that these results can be best approximated using a multiplicative integration of independent spatial and temporal sequence feature representations as follows such as $Z_{n+1} = Z_n + V + S + (S * T)$, rather then an additive integration ($Z_{n+1} = Z_n + V + S + T$), or a combined spatio-temporal term without a separate temporal representation ($Z_{n+1} = Z_n + V + S + C$). Here $Z$ is the selection layer corresponding to the five fingers, $V$ is the visual stimulus in the serial reaction time task (SRTT), $S$ the spatial, $T$ the temporal and $C$ the combined representation in which the temporal sequence feature is linked to a specific spatial feature (weights and noise terms are omitted for abbreviation purposes). Essentially, this means that while effects of the spatial feature representation act independently (additive integration) the temporal representation can only be expressed when $S > 0$, in other words there is some knowledge of the spatial representation. The difference between an integrated spatio-temporal versus an independent temporal representation which is multiplicatively combined with the spatial one is critical, as only the latter allows for temporal transfer which we could reliably observe across experiments (Fig. 2b, c).

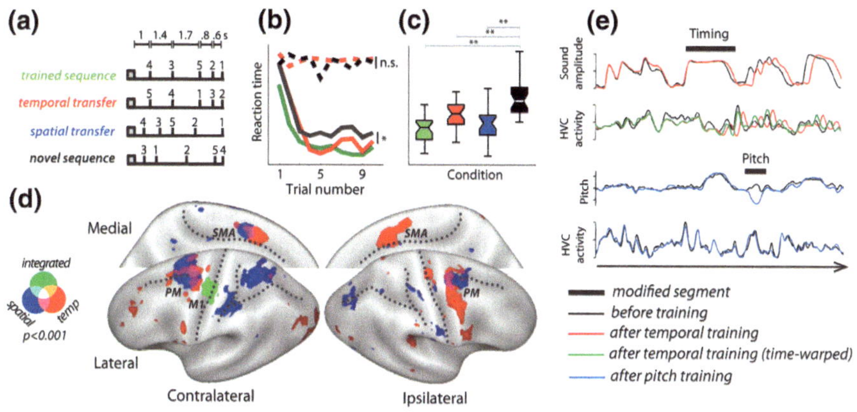


**Fig. 2** Evidence suggesting that spatial and temporal features of movement sequences are represented independently. **a** Participants were trained on a specific spatio-temporal finger sequence (*green*) and then tested on a novel sequence (*black*) or on sequences that retained either the temporal (*red*) or spatial (*blue*) structure (Kornysheva et al. 2013; Kornysheva and Diedrichsen 2014). The numbers 1–5 in exemplary sequences correspond to the thumb, index, middle, ring and little finger, respectively. **b** Reaction time advantages relative to a new sequence that are related to a learned trained temporal feature can only be expressed when the spatial feature becomes more predictable. Solid lines correspond to "trained", "temporal" and "novel" conditions in which the corresponding sequences are presented 10 times in a row, whereas the dashed lines correspond to conditions where the trained temporal feature is paired with a new spatial feature on every trial (*dashed red*) and compared to a sequence that changes both the temporal and the spatial feature on every trial. Stars indicate significant differences across trials (Kornysheva et al. 2013). **c** Reaction time results indicate independent transfer of spatial and temporal features to test conditions (Kornysheva and Diedrichsen 2014). **d** Separate, but partly overlapping spatial (*blue*) and temporal (*red*) representations of finger sequences can be revealed bilaterally in premotor cortex (PM and SMA) using multi-voxel pattern analysis. The two features are integrated in contralateral M1 only (*green*). In a series of behavioural and fMRI experiments employing (Kornysheva and Diedrichsen 2014) **e** The premotor nucleus HVC in zebra finches reflects changes in the temporal feature of a bird song (*red line*), such as a prolonged syllable, but not changes in its pitch feature (*blue line*). Both types of changes were acquired through aversive conditional auditory feedback (adapted from Ali et al. 2013)

A subsequent study investigated how independent and integrated spatial and temporal representations are represented across the neocortex and the cerebellum based on fine-grained local fMRI activity patterns (Kornysheva and Diedrichsen 2014). Despite the low resolution (fMRI voxels) these neural representations can be probed due to tiny, but systematic spatial activity biases which occur with trial repetition. Here instead of training one particular spatio-temporal sequence, subjects were trained to produce nine spatio-temporal finger sequences, which were unique combinations of three finger order (spatial feature) and temporal interval (temporal feature) sequences. This factorial design in combination with multivariate pattern analysis allowed to test for local voxel activity patterns related to the spatial feature across sequences with different temporal features, and orthogonally, patterns related to the temporal feature across different spatial features—feature transfer on the neural level. Moreover, subtracting out the main effects of independent spatial and temporal features from the overall activity patterns isolated residual patterns, which, if unique for each sequence, were taken as integrated neural representations.

The results revealed that fine-grained patterns in overlapping patches of the lateral (dorsal and ventral) and medial (SMA) premotor cortex carried information on the independent spatial as well as independent temporal patterns, whilst the only region informative of an integrated spatio-temporal representation was the contralateral primary motor cortex, the output stage of the neocortex (Fig. 2d). Thus, in M1 each sequence may recruit a subpopulation of neurons that controls a particular combination of spatiotemporal synergies (d'Avella et al. 2003). The latter cannot be synergies of individual finger movements as each finger movement occurred in each sequence, but particular spatio-temporal transitions within sequences. The same principle, but now for spatial and temporal parameters would apply for the premotor cortex—unique combinations of synergies capturing particular spatial (timing-invariant) or temporal transitions.

The alternative is that the encoding observed in M1 is not sequential encoding per se, but reflects the two spatial and temporal codes being combined nonlinearly. Also while the force level for each finger matched well across sequences, it cannot be completely excluded that tiny biases—thumb, index finger, etc., being more active in one sequence than in another—may have contributed to the encoding to some extent. Yet, this explanation is unlikely, since encoding in contralateral M1 correlated with sequence learning, but not with sequence classification accuracy based on the force at each finger.

The presence of independent spatial and temporal codes, as well as integrated representations suggests varied levels of abstraction from the actual motor response implementation. To be transferable across different temporal profiles, the spatial sequence in the premotor cortices has to lack specifics on the kinematics or dynamics of each effector involved during sequence production, and may carry more abstract information such as on sequential transitions between movements (Tanji and Shima 1994). Conversely, the temporal feature representation is bound to lack any information on the effectors and the dynamics such as force on each finger to be transferable across different finger movement sequences.

Interestingly, a similar dissociation in the control of spatial (pitch) and temporal sequences has been found in songbirds (Ali et al. 2013). Using aversive auditory conditioning, the authors taught the animals to selectively modify temporal and spectral features of their song, such as changing the length of a syllable, or its pitch which requires a different configuration of muscle activations controlling the syrinx (Fig. 2e). The basal ganglia analog was required for the modification of the spectral properties (pitch), but not for changes in the temporal structure. By contrast, the activity in HVC (an analog to the premotor cortex) reflected the temporal but not spectral features of the song. This dissociation and therefore modularity of spatial and temporal features in motor sequence control may thus be a universal property of the CNS.

These findings resonate with the hypothesis by d'Avella and colleagues suggesting that the control of movement may be modular during a variety of reaching movements (d'Avella 2017), as the variability of muscle activations recorded as EMG signals can be explained by three types of components, so called muscle synergies: (a) time-invariant spatial (S), (b) muscle-invariant temporal (T), (c) as well as muscle-specific spatio-temporal synergies (ST). S are the activation weights on each muscle required for the movement, which do not specify any change over time, T are the temporal activation profiles which are shared across different muscles and ST are activation waveforms for specific muscles which amount to an idiosyncratic dynamical trajectory of individual muscles. These results suggest that at the muscular level the underlying temporal features of movements are transferable across different muscle synergies, respectively. Although explaining variability of muscle activations by synergies does not provide direct evidence for the encoding of these synergies in the CNS, these results allow for the possibility of controllers somewhere in the corticoid-spinal pathway that impose this modular regularity on motor output. A recent analysis of premotor and primary motor units provided the first evidence that neural activity in the CNS can be explained by EMG synergies (Overduin et al. 2015).

A modular representation enables a radical simplification of motor control policies: Instead of controlling the spatio-temporal evolution of each individual muscle throughout the movement, the CNS triggers spatial and temporal synergies required for the skilled movement. Moreover, instead of encoding all combinations of movements, the brain utilizes temporal and spatial synergies or profiles which can be recombined flexibly into different combinations. If skilled movements did not in principle require a dedicated representation of their temporal dimension and were merely emergent from the encoding of the dynamics of the movement they are performed with, such learned movements would be rigid with regard to their temporal evolution beyond a simple speed up of slow down. It would entail that the temporal dimension could not be utilized across different effectors and motor states. Coming back to the musical example, the violinist would have to form an entirely new representation whenever the temporal structure of a sequence is modified or whenever a new sequence of movements is paired with a familiar temporal structure, which contradicts the findings above.

## Computational Models and Neural Mechanisms of Temporal Representation ("How")

It has been hypothesized that a variety of neural structures are capable of encoding the timing of movements, which corresponds to the widespread involvement of these areas in explicit or implicit motor timing tasks—in particular the cerebellum, the striatum and the lateral and medial premotor cortices (Lewis and Miall 2003; Buhusi and Meck 2005; Ivry and Schlerf 2008; Buonomano and Laje 2010; Teki et al. 2011; Laje and Buonomano 2013). This is surprising as these different parts of the nervous system have diverse neural architectures, as well as physiological and computational constraints. Conversely, such diversity suggests that these systems are unlikely to be redundant with respect to skilled motor timing, specializing on a particular neural computation which determines or co-varies with motor timing. Below I will present a hypothesis of how such parallel processes may operate and interact to enable precise motor timing based on results from computational modelling and current neuroscientific evidence.

The cerebellar cortex has been one of the first regions hypothesized in motor and more generally sub-seconds timing (Braitenberg 1967). In stark contrast to the neocortex, the architecture of the cerebellar circuitry is remarkably uniform across the different parts of the cerebellum (with the exception of the floccular cortex) with the main difference between regions being the origins of their inputs and the targets of their outputs. The circuitry is designed to integrate only two types of inputs from the rest of the nervous system, which converge in the cerebellum: The mossy fibre pathway that relays information from the cortex (via the pons), as well as the periphery (via the brainstem) and the climbing fibre pathway that carries signals from the inferior olive in the brainstem. The cerebellar output is sent to the neocortex via the thalamus or to the periphery via brainstem nuclei, and has been shown to form reciprocal multisynaptic cortico-cerebellar loops (Kelly and Strick 2003).

While the deep cerebellar nuclei (DCN) receive excitatory input *directly* via mossy and climbing fibre collaterals, the anatomical connections of the two fibre systems to the Purkinje cell (PC) layer is at the core of cerebellar architecture: Unlike to the DCN, the mossy fibre to PC projection is *indirect*, going through a layer of granule cells, which remarkably constitute the majority of neurons in the brain. Granule cells relay this information by parallel fibres that run transversally through flattened and orthogonally oriented dendritic trees of PCs with some of which they form direct excitatory connection on the way, and inhibit them indirectly via the inhibitory interneurons. Remarkably, Purkinje cells have a baseline firing rate of 50–100, sometimes up to 200 Hz (Zeeuw et al. 2011; Zhou et al. 2014), and inhibitory projections to the DCN as their only output (GABA). They act as a constant break on the DCN, which activity is released only when the PCs exhibit a firing pause that in turn disinhibits the DCN, the sole output of the cerebellum.

The granular layer has been hypothesized to act like a giant "filter" of the mossy fibre input (Dean et al. 2009, 2013) redistributing the mossy fibre inputs across granule cells (divergence), but at the same time mixing inputs from different channels—sensory and motor at the single cell level (Huang et al. 2013; Ishikawa et al. 2015). In classical eyeblink conditioning, which acts as a model for the learning of timed motor responses, time varying activity in a subset of granule cells activated by the conditioning stimulus (CS) has been hypothesized to produce a temporal code at the parallel fibre to PC synapses (Medina and Mauk 2000). This synaptic input to the PC can act as a clock, as each unique state of the synaptic input after a stimulus corresponds to the passage of time following the CS onset. In contrast, learning of the precisely time motor response (eyeblink) takes place based on an aversive stimulus, such as a short air-puff directed into the eye (unconditioned stimulus, US). The latter is transmitted by the climbing fibre system, and leads to the depression of those parallel fibre to PC synapses active just before the time of the aversive stimulus, partly mediated by plasticity in interneurons inhibiting the PC (Medina and Mauk 2000; Heiney et al. 2014). This eventually leads to decreased PC simple spike cell firing during the interval between the two stimuli with the most pronounced reduction timed just before the conditioned response (CR), the latter being initiated via the disinhibition of the DCN (Jirenhed et al. 2007; Ten Brinke et al. 2015). It has been repeatedly shown that the intact cerebellar cortex is necessary for a precisely timed response, as the intact DCN alone produces a short-latency response without any temporal features necessary for the task (Perrett et al. 1993; Koekkoek et al. 2003). Importantly, this notion advocates a distributed motor learning architecture across the cerebellum (Gao et al. 2012), and argues for a special role of the cerebellar cortex in motor timing.

More recently it has been proposed that the temporal profile of the response can be acquired locally in the PC (Johansson et al. 2014). Specifically, pairing a CS consisting of a direct stimulation of the parallel fibres (circumventing the granular cell layer) with a US consisting of direct climbing fibre stimulation led to a Purkinje cell CR that was adaptively timed. The cell reached maximum suppression of 75 ms before the onset of the US across different CS-US intervals. Importantly, even when blocking inhibition from inhibitory interneurons that are also innervated by parallel fibres and could have had an effect on the PC response, the learned timing was preserved. This led the authors to conclude that the encoding of the precisely timed response is located in the PC at the molecular level. Specifically, blocking mGluR7 receptor has been shown to disrupt timing in the direct stimulation paradigm above (Johansson et al. 2015). While the exact mechanism of molecular timing is still unknown, it has been hypothesized that the CS may initiate a predictable biochemical cascade while the US onset induces interval-specific changes to this cascade. This could take place in form of a selection of different molecular components with particular properties with regard to the duration of ion channel open states, so that the time course of the PC simple spike depression matches the CS-US interval.

Regardless of whether the timing mechanism is distributed or localized, the parts of the cerebellar cortex involved in classical conditioning project to a specific target effector in the periphery and cannot be expected to be transferable across different effectors, spatial configurations or motor states. For instance, the cerebellar cortical projection to the anterior interpositus of the DCN nucleus involved in eyeblink conditioning innervates periorbital muscles of the eye via the brain stem (Ten Brinke et al. 2015). However, a more abstract representation of timing for spatio-temporal movements is still conceivable in those regions of the cerebellum that project to the premotor and prefrontal cortices via the dentate nucleus (Kelly and Strick 2003), albeit only if they receive climbing fibre stimulation at the time of the US during learning which has not been investigated systematically so far.

Another timing mechanism has been attributed to the basal ganglia, the striatal beat frequency model (Matell et al. 2004; Buhusi and Meck 2005). Unlike the cerebellar timing mechanisms described in this chapter, the latter is relevant for interval timing that involves intervals of seconds-to-minutes. While even the lower range may appear too long to be relevant for motor timing many skilled movements like verbal utterances, musical and dance sequences, as well as the typing Morse code messages involve sequences of movements that unfold over the timescale of several seconds to tens of seconds. The basal ganglia is organized in cortico-basal ganglia-thalamo-cortical loops with the majority of the excitatory input coming from the cortex and then sent out to direct and indirect pathways of the basal ganglia which excite and inhibit the cortex, respectively, via the thalamus (Graybiel 1998). Here each medium spiny neuron in the striatum receives up to 30.000 separate axons from the cortex. Thus, it has been proposed that through learning the medium spiny neurons in the striatum act as coincidence detectors of neural oscillations that operate at different frequencies in the neocortex (Buhusi and Meck 2005). With trial onset the phase of the oscillations is reset ("start-gun"). During learning a reward signal at the end of the interval to be trained is conveyed by dopaminergic input from the substantia nigra pars compacta and the ventral tegmental area. Experience-dependent changes in cortico-striatal transmission (both LTP and LTD) lead to a ramp of striatal activity with a peak at the time of the expected reward, i.e. at the end of the interval. Accordingly, following training striatal neurons may be capable of detecting the unique coincidence of phases of the neural oscillators that project to these neurons, respectively. Interestingly such adaptively timed ramping activity has also been observed in the neocortex, such as in a motor synchronization-continuation task involving isochronous intervals performed at different speeds in the monkey supplementary motor area (SMA) (Merchant et al. 2013) and an interval reproduction task in the parietal cortex (Jazayeri and Shadlen 2015). Although there has been no direct experimental evidence from studies involving sub-second intervals, it is likely that such ramps reflect the striatal activity via the direct basal ganglia thalamic route to the neocortex. Indeed, imaging, lesion and pharmacological studies have confirmed the involvement of the striatum in interval timing (for a review cf. Buhusi and Meck 2005).

Finally, the neocortex could be regarded as most closely related to models involving random recurrent networks (Thomson and Bannister 2003; Buonomano and Laje 2010). Recent concurrent multiunit recordings from premotor and primary motor cortices suggest that the trajectory of a movement is not represented in terms of its features such as position, velocity, direction, force and timing as suggested before, but rather as a compound of variables correlated leading to the performed trajectory in space (Churchland et al. 2006; Shenoy et al. 2012; Kaufman et al. 2015). Here the timing is merely an emergent feature of the evolution of the multiunit activity which controls the spatial movement trajectory. Accordingly, a model of randomly connected networks can be trained to produce skilled sequential movements and have perfectly reproducible temporal dynamics without any dedicated encoding of the temporal dimension in the model (Laje and Buonomano 2013). Such a network of interconnected units can be trained to represent the spatio-temporal evolution of a trajectory as complex as handwriting (in two-dimensional space).

Central to the function of this model is a random recurrent network of interconnected firing-rate nodes with a multiunit firing rate that learn to follow a particular innate trajectory depending on the input trigger. Learning consists of the reduction of the variability in this innate trajectory in space by adjusting the network weights enabling the firing rate activity to be robust to noise and perturbations, so that the trajectory can return to a carved out path. This network activity can be read out continuously by an output module that maps its multiunit state into external variables like an x and y position for complex motor trajectories and could in principle also guide movements in muscle space. The timing of this movement is also reliable after training, such that a certain position in space clusters equally tightly in time. This is despite the temporal features of the movement not having a dedicated representation, but emerging from the dynamics of the trajectory dedicated to the spatial position of the movement.

While the dynamical systems view focuses on the representation of a movement in space with timing being an emergent property of the trajectory, Buonomano and colleagues proposed that the dynamical trajectories produced by random recurrent networks could also be utilized to encode discrete timing of movements (Buonomano and Laje 2010). These networks could be trained to control a simple timing task, producing a phasic pulse after a specific interval (activity in one-dimensional space y), analogous to a discrete button press in a finger tapping task or eyelid closure in eyeblink conditioning. Computationally the mechanisms of such dedicated temporal representations are equivalent to the encoding of the continuous spatio-temporal trajectory. What is crucial here is the mapping of the network output to a readout unit controlling a motor response. This mapping determines whether the timing is a by-product of the spatial trajectory or whether the network activity which is consolidated after training essentially acts as a population clock, triggering a discrete response once the network activity reaches a particular state. The latter can be extrapolated to sequential representations of finger movement sequences. Thus, from the perspective of the neocortex discrete event timing and continuous emergent timing which have been tied to distinct neural substrates as discussed earlier (cf. Spencer et al. 2003) could in principle be encoded in the same way.

This flexibility of temporal encoding in the networks resembling the neocortex resonates with the imaging results showing independent temporal and spatial feature encoding in the premotor cortices versus integrated spatio-temporal encoding in contralateral primary motor cortex (Kornysheva and Diedrichsen 2014; Diedrichsen and Kornysheva 2015). Within the dynamical systems framework, this modularity would be related to the activation of several recurrent neocortical networks that are utilized to encode integrated spatio-temporal encoding in M1 and dedicated temporal encoding in premotor regions, the latter enabling the flexibility of the response independently of a spatial motor features, analogous to the temporal transfer observed behaviourally (Fig. 2a–d). In contrast, it is much less straightforward how such recurrent networks could be mapped to encode the spatial feature of sequences (e.g. finger order) independently of their exact temporal feature. If the encoding of movement sequences draws on consolidated multiunit trajectories of randomly recurrent units, the precise changes in multiunit space would be ignored, such that a certain cascade of states would be mapped onto the same spatial state (configuration of finger activations). The temporal evolution would then be specified at the stage when both are combined either by acting on integrated spatio-temporal M1 representations (Kornysheva and Diedrichsen 2014) or downstream in the case of direct cortico-spinal projections from the premotor cortex.

How do these regions interact with each other to achieve precise motor timing of skilled movements? Here only projections with a short latency ("online") transduction up to tens of milliseconds can be considered to exhibit control at time scales relevant to online motor control.

For a long time it has been assumed that the basal ganglia and the cerebellum operate in parallel to each other at the subcortical level, having separate thalamic relays to the neocortex (Bostan et al. 2013). However, in rodents (Ichinohe et al. 2000) and more recently in primates (Hoshi et al. 2005; Bostan et al. 2010) disynaptic connections from the DCN to the striatum have been established. The relay is located in the intralaminar nuclei of the thalamus which contain projections to the striatum. Recently, it has been determined that the propagation speed between DCN and the dorsolateral striatum can be as low as 10 ms (Chen et al. 2014). This suggests a rapid transmission of cerebellar output to striatal. High frequency and well-timed bursts in DCN neurons can modulate activity at the entry stage of the basal ganglia, thereby coordinating cerebellar output with the basal ganglia computations in real time. Interestingly when stimulation of the DCN was combined with concurrent cortico-striatal input, the cortico-striatal activation was potentiated (Chen et al. 2014). The cerebellar output signals which carry a high temporal resolution profile of a signal are therefore impacting the neocortical input at the level of the ramping activity of medium spiny neurons. At the same time the subthalamic nucleus to which striatal neurons project via the indirect pathway innervates the cerebellum via the pontine nuclei. The propagation speed of this connection is currently unknown.

As with the basal ganglia, the premotor cortex forms reciprocal disynaptic connections with the cerebellum (Kelly and Strick 2003). The DCN project to the neocortex via the ventrolateral nucleus of the thalamus and affect not only supragranular layers, but also directly layer V in M1 as shown by optogenetic stimulation of the cerebellar Purkinje cells (Proville et al. 2014). Importantly, the DCN inhibition is followed by a rebound excitation following the offset of Purkinje cell stimulation at around 60ms and in M1 40ms later. At the same time this study revealed a short-latency transmission between M1/S1 and the lateral cerebellar cortex, with onsets of Purkinje cell frequency modulation as early as 10ms after neocortical stimulation. Finally, non-invasive research in humans has shown that the latency of cerebellar inhibition of the cortex as measured by M1 triggered MEP is highest at 5 ms delay (Ugawa et al. 1991), confirming a rapid transmission between the cerebellum and the neocortex. In other words, it is likely that the (pre-) motor cortical networks relevant for temporal encoding receive a precisely timed (high resolution) signal from the cerebellum while the latter is modulated by neocortical input, with these interactions unfolding almost instantaneously.

Why do we need parallel timers in our brain operating in parallel and what is their specific contribution? A schematic model based on the current review is presented in Fig. 3 (cf. caption for details). At the current stage, any answers to this question will remain speculative. Most of the invasive electrophysiological recordings that could provide direct evidence for this report only from one region at a time. Yet, in an intact brain it is impossible to disentangle whether the activities reported relay the input of interconnected regions, or whether this activity originates and is causally involved in the production of well-timed movements. Even lesion studies (temporal inactivation, TMS, patients, etc.) are of limited use, as they cause

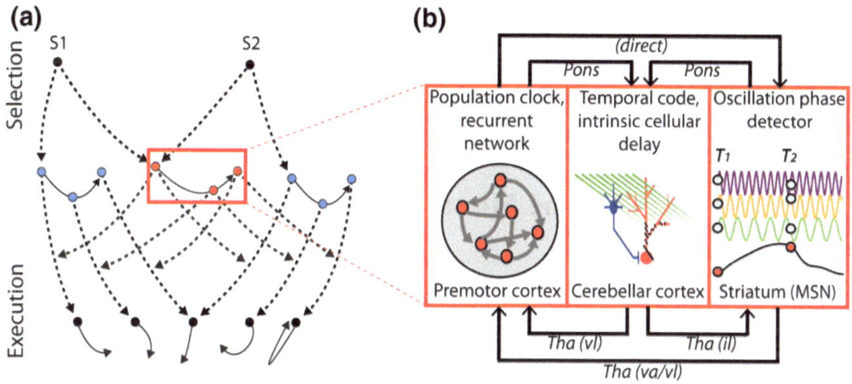


**Fig. 3** Temporal encoding for skilled spatiotemporal sequence production. **a** Modular representation of temporal (*red dots*, a longer and a shorter interval) and spatial (*blue dots*) sequence features. The temporal representation modulates the signal originating from two different spatial representations (*black broken arrows*) (Kornysheva et al. 2013). This allows two different sequences S1 and S2 to utilize the same learned temporal structure flexibly (adapted from Diedrichsen and Kornysheva 2015). **b** The premotor cortex, the cerebellar cortex and the striatum utilize different computational mechanisms that can be harnessed to learn and control motor timing—either independently of the movement in space as shown here or in an integrated spatio-temporal fashion (see main text). These regions are interconnected with each other by short-latency circuits via the thalamus and the pons, respectively. The following model of motor timing for skilled movement sequences is proposed in the current review: The neocortex produces sustained dynamic activity in a population of interconnected neurons which can be utilized for the duration of a whole sequence of movements (Buonomano and Laje 2010). This multi-unit activity is read out by the MSN in the striatum based on oscillation phase detection and chunked into a series of ramps that mark the interval between movement onsets or between an external stimulus and a motor response (Buhusi and Meck 2005). Crucially, the cortical and striatal activity is fed into the cerebellum, providing a sequential context signal for each movement unfolding in the seconds time range. This activity is transformed by the cerebellar cortex into a precise high temporal resolution output on a sub-seconds scale in the deep cerebellar nuclei for each sequence component. Through disynaptic projections, the latter modulates both the ramps in the striatum and the population clocks in the neocortex to achieve a more precisely timed representation of the sequence. Abbreviations: il—intralaminar; MSN—medium spiny neurons; S—sequence; T—time point; Tha—Thalamus; va—ventroanterior; vl—ventrolateral

reorganization in the network, that unless recorded, remains hidden and may impact conclusions with regard to behaviour. Short-lived local inactivation though muscimol, optogenetic stimulation (animal models) and transcranial magnetic stimulation (humans) whilst recording from the site to which the region that is disrupted projects are likely to provide more conclusive answers to this question. For instance, to assess the individual contribution of cortical and subcortical sites to learned timing, a pioneering study by Mauk and colleagues has been conducted to decompose the contributions of the neocortex versus the cerebellar nuclei to trace eye blink conditioning (Siegel and Mauk 2013). This task is known to rely not only on the cerebellum (in contrast to delay eyeblink conditioning), but also on the cortex and the hippocampus. Here it could be demonstrated that ramping activity observed in prefrontal cells, as well as the well-timed conditioned motor response is abolished when cerebellar output is inhibited, whereas the sustained activity during the duration of the CS remained intact. In the future similar studies need to be designed to directly probe the contribution of the premotor cortex, the striatum and the cerebellar cortex to skilled motor timing.

## Conclusions/Take Home Message

Precise motor timing of spatio-temporal skills is crucial for a variety of skilled movements. During the past decade there have been contradictory results with regard to how timing for spatio-temporal motor skills is represented in the brain. The encoding of motor timing is achieved either directly by measuring time intervals from movement onset or an external stimulus (dedicated timing) or indirectly via state-dependent encoding (emergent timing). Which mode is chosen depends on the characteristics of the motor task, such as the correlation of the temporal target with a state-dependent variable (e.g. position or velocity), the presence of temporal overlap across effectors requiring their coordination in time and the reliability of temporal versus state-dependent encoding for task success. The ability to transfer temporal features across different motor configurations in space indicate a modular representations of these features for the control of skilled motor sequences which can be found in the premotor as opposed to primary motor cortices. The idea that there is a localizable universal neural clock in the CNS, which is utilized across different domains, perceptual and motor, is an unlikely scenario. Partly this is evidenced by the fact that timing functions have been attributed to different areas across the brain. Instead, different neural mechanisms that operate in parallel—dynamical systems (random recurrent network), oscillation phase detection (ramps), patterned input and molecular delays at the cell level—constitute representations in neocortical motor areas, the striatum and the cerebellar cortex, respectively. These neural representations interact with each other in short-latency loops to produce well-timed behaviour.

**Acknowledgements** I thank Joern Diedrichsen for discussions on emergent versus dedicated motor timing, Freek Hoebeek for advice on the conductance speeds between the cerebellum and the neocortex for motor control, and Sundeep Teki for insightful feedback on the ideas presented in this review. This work was supported by the Sir Henry Wellcome Fellowship (098881/Z/12/Z).

## References

Ali F, Otchy T, Pehlevan C et al (2013) The basal ganglia is necessary for learning spectral, but not temporal, features of birdsong. Neuron 80:494–506. doi:10.1016/j.neuron.2013.07.049

Averbeck BB, Chafee MV, Crowe DA, Georgopoulos AP (2002) Parallel processing of serial movements in prefrontal cortex. Proc Natl Acad Sci USA 99:13172–13177. doi:10.1073/pnas.162485599

Baddeley A (2010) Working memory. Curr Biol 20:R136–R140. doi:10.1016/j.cub.2009.12.014

Bastian AJ, Martin TA, Keating JG, Thach WT (1996) Cerebellar ataxia: abnormal control of interaction torques across multiple joints. J Neurophysiol 76:492–509

Bostan A, Dum R, Strick P (2010) The basal ganglia communicate with the cerebellum. Proc Natl Acad Sci 107:8452–8456. doi:10.1073/pnas.1000496107

Bostan A, Dum R, Strick P (2013) Cerebellar networks with the cerebral cortex and basal ganglia. Trends Cogn Sci 17:241–254. doi:10.1016/j.tics.2013.03.003

Braitenberg V (1967) Is the cerebellar cortex a biological clock in the millisecond range? Prog Brain Res 25:334–346. doi:10.1016/S0079-6123(08)60971-1

Braun Janzen T, Thompson WF, Ammirante P, Ranvaud R (2014) Timing skills and expertise: discrete and continuous timed movements among musicians and athletes. Front Psychol 5:1482. doi:10.3389/fpsyg.2014.01482

Buhusi C, Meck W (2005) What makes us tick? Functional and neural mechanisms of interval timing. Nat Rev Neurosci 6:755–765. doi:10.1038/nrn1764

Buonomano DV, Laje R (2010) Population clocks: motor timing with neural dynamics. Trends Cogn Sci (Regul Ed) 14:520–527. doi:10.1016/j.tics.2010.09.002

Chen C, Fremont R, Arteaga-Bracho E, Khodakhah K (2014) Short latency cerebellar modulation of the basal ganglia. Nat Neurosci 17:1767–1775. doi:10.1038/nn.3868

Churchland MM, Santhanam G, Shenoy KV (2006) Preparatory activity in premotor and motor cortex reflects the speed of the upcoming reach. J Neurophysiol 96:3130–3146. doi:10.1152/jn.00307.2006

Conditt M, Mussa-Ivaldi F (1999) Central representation of time during motor learning. Proc Natl Acad Sci USA 96:11625–11630

d'Avella A, Saltiel P, Bizzi E (2003) Combinations of muscle synergies in the construction of a natural motor behavior. Nat Neurosci 6:300–308. doi:10.1038/nn1010

d'Avella A (2017) Modularity for motor control and motor learning. Adv Exp Med Biol 957:3–19

Dean P, Porrill J, Ekerot C-F, Jorntell H (2009) The cerebellar microcircuit as an adaptive filter: experimental and computational evidence. Nat Rev Neurosci 11:30–43. doi:10.1038/nrn2756

Dean P, Anderson S, Porrill J, Jorntell H (2013) An adaptive filter model of cerebellar zone C3 as a basis for safe limb control? J Physiol 591:5459–5474. doi:10.1113/jphysiol.2013.261545

Diedrichsen J, Kornysheva K (2015) Motor skill learning between selection and execution. Trends Cogn Sci (Regul Ed) 19:227–233. doi:10.1016/j.tics.2015.02.003

Diedrichsen J, Criscimagna-Hemminger S, Shadmehr R (2007) Dissociating timing and coordination as functions of the cerebellum. J Neurosci Official J Soc Neurosci 27:6291–6301. doi:10.1523/JNEUROSCI.0061-07.2007

Elliott MT, Welchman AE, Wing AM (2009) Being discrete helps keep to the beat. Exp Brain Res 192:731–737. doi:10.1007/s00221-008-1646-8

Evarts EV (1968) Relation of pyramidal tract activity to force exerted during voluntary movement. J Neurophysiol 31:14–27

Gao Z, Beugen B, De Zeeuw C (2012) Distributed synergistic plasticity and cerebellar learning. Nat Rev Neurosci 13:619–635. doi:10.1038/nrn3312
Georgopoulos AP, Kalaska JF, Caminiti R, Massey JT (1982) On the relations between the direction of two-dimensional arm movements and cell discharge in primate motor cortex. J Neurosci 2:1527–1537
Graybiel A (1998) The basal ganglia and chunking of action repertoires. Neurobiol Learn Mem 70:119–136. doi:10.1006/nlme.1998.3843
Heiney S, Kim J, Augustine G, Medina J (2014) Precise control of movement kinematics by optogenetic inhibition of Purkinje cell activity. J Neurosci 34:2321–2330. doi:10.1523/JNEUROSCI.4547-13.2014
Hikosaka O, Nakamura K, Sakai K, Nakahara H (2002) Central mechanisms of motor skill learning. Curr Opin Neurobiol 12:217–222. doi:10.1016/S0959-4388(02)00307-0
Hoshi E, Tremblay L, Féger J et al (2005) The cerebellum communicates with the basal ganglia. Nat Neurosci 8:1491–1493. doi:10.1038/nn1544
Huang C-C, Sugino K, Shima Y et al (2013) Convergence of pontine and proprioceptive streams onto multimodal cerebellar granule cells. eLife 2:e00400. doi:10.7554/eLife.00400
Ichinohe N, Mori F, Shoumura K (2000) A di-synaptic projection from the lateral cerebellar nucleus to the laterodorsal part of the striatum via the central lateral nucleus of the thalamus in the rat. Brain Res. doi:10.1016/S0006-8993(00)02744-X
Ishikawa T, Shimuta M, Häusser M (2015) Multimodal sensory integration in single cerebellar granule cells in vivo. eLife e12916
Ivry R, Schlerf J (2008) Dedicated and intrinsic models of time perception. Trends Cogn Sci 12:273–280. doi:10.1016/j.tics.2008.04.002
Ivry R, Spencer R (2004) The neural representation of time. Curr Opin Neurobiol 14:225–232. doi:10.1016/j.conb.2004.03.013
Jazayeri M, Shadlen MN (2015) A neural mechanism for sensing and reproducing a time interval. Curr Biol 25:2599–2609. doi:10.1016/j.cub.2015.08.038
Jirenhed D-A, Bengtsson F, Hesslow G (2007) Acquisition, extinction, and reacquisition of a cerebellar cortical memory trace. J Neurosci 27:2493–2502. doi:10.1523/JNEUROSCI.4202-06.2007
Johansson F, Jirenhed D-AA, Rasmussen A et al (2014) Memory trace and timing mechanism localized to cerebellar Purkinje cells. Proc Natl Acad Sci USA 111:14930–14934. doi:10.1073/pnas.1415371111
Johansson F, Carlsson HA, Rasmussen A et al (2015) Activation of a temporal memory in Purkinje cells by the mGluR7 receptor. Cell Rep 13:1741–1746. doi:10.1016/j.celrep.2015.10.047
Karniel A, Mussa-Ivaldi F (2003) Sequence, time, or state representation: how does the motor control system adapt to variable environments? Biol Cybern 89:10–21. doi:10.1007/s00422-003-0397-7
Kaufman MT, Churchland MM, Ryu SI, Shenoy KV (2015) Vacillation, indecision and hesitation in moment-by-moment decoding of monkey motor cortex. eLife. doi:10.7554/eLife.04677
Kelly R, Strick P (2003) Cerebellar loops with motor cortex and prefrontal cortex of a nonhuman primate. J Neurosci Official J Soc Neurosci 23:8432–8444
Koekkoek SK, Hulscher HC, Dortland BR et al (2003) Cerebellar LTD and learning-dependent timing of conditioned eyelid responses. Science 301:1736–1739. doi:10.1126/science.1088383
Kornysheva K, Diedrichsen J (2014) Human premotor regions parse sequences into their spatial and temporal features. eLife 3:e03043:e03043. doi:10.7554/eLife.03043
Kornysheva K, Sierk A, Diedrichsen J (2013) Interaction of temporal and ordinal representations in movement sequences. J Neurophysiol 109:1416–1424. doi:10.1152/jn.00509.2012
Laje R, Buonomano D (2013) Robust timing and motor patterns by taming chaos in recurrent neural networks. Nat Neurosci 16:925–933. doi:10.1038/nn.3405
Lewis PA, Miall RC (2003) Distinct systems for automatic and cognitively controlled time measurement: evidence from neuroimaging. Curr Opin Neurobiol 13:250–255

Matell MS, Meck WH (2004) Cortico-striatal circuits and interval timing: coincidence detection of oscillatory processes. Cognitive Brain Res 21(2): 139–70

Medina J, Mauk M (2000) Computer simulation of cerebellar information processing. Nat Neurosci 3(Suppl):1205–1211. doi:10.1038/81486

Medina J, Carey M, Lisberger S (2005) The representation of time for motor learning. Neuron 45:157–167. doi:10.1016/j.neuron.2004.12.017

Merchant H, Perez O, Zarco W, Gamez J (2013) Interval tuning in the primate medial premotor cortex as a general timing mechanism. J Neurosci 33:9082–9096. doi:10.1523/JNEUROSCI.5513-12.2013

O'Reilly J, McCarthy K, Capizzi M, Nobre A (2008) Acquisition of the temporal and ordinal structure of movement sequences in incidental learning. J Neurophysiol 99:2731–2735. doi:10.1152/jn.01141.2007

Overduin SA, d'Avella A, Roh J et al (2015) Representation of muscle synergies in the primate brain. J Neurosci 35:12615–12624. doi:10.1523/JNEUROSCI.4302-14.2015

Penhune V, Steele C (2012) Parallel contributions of cerebellar, striatal and M1 mechanisms to motor sequence learning. Behav Brain Res 226:579–591. doi:10.1016/j.bbr.2011.09.044

Perrett SP, Ruiz BP, Mauk MD (1993) Cerebellar cortex lesions disrupt learning-dependent timing of conditioned eyelid responses. J Neurosci 13:1708–1718

Proville RD, Spolidoro M, Guyon N, Dugué GP (2014) Cerebellum involvement in cortical sensorimotor circuits for the control of voluntary movements

Repp BH, Steinman SR (2010) Simultaneous event-based and emergent timing: synchronization, continuation, and phase correction. J Mot Behav 42:111–126. doi:10.1080/00222890903566418

Sakai K, Kitaguchi K, Hikosaka O (2003) Chunking during human visuomotor sequence learning. Exp Brain Res 152:229–242. doi:10.1007/s00221-003-1548-8

Sakai K, Hikosaka O, Nakamura K (2004) Emergence of rhythm during motor learning. Trends Cogn Sci. doi:10.1016/j.tics.2004.10.005

Shenoy K, Sahani M, Churchland M (2012) Cortical control of arm movements: a dynamical systems perspective. Neuroscience 36:337–359. doi:10.1146/annurev-neuro-062111-150509

Shima K, Isoda M, Mushiake H, Tanji J (2007) Categorization of behavioural sequences in the prefrontal cortex. Nature 445:315–318. doi:10.1038/nature05470

Shin JC, Ivry RB (2002) Concurrent learning of temporal and spatial sequences. J Exp Psychol Learn Mem Cogn 28:445–457

Shin JC, Ivry RB (2003) Spatial and temporal sequence learning in patients with Parkinson's disease or cerebellar lesions. J Cogn Neurosci 15:1232–1243. doi:10.1162/089892903322598175

Siegel J, Mauk M (2013) Persistent activity in prefrontal cortex during trace eyelid conditioning: dissociating responses that reflect cerebellar output from those that do not. J Neurosci 33:15272–15284. doi:10.1523/JNEUROSCI.1238-13.2013

Spencer RM, Ivry RB (2005) Comparison of patients with Parkinson's disease or cerebellar lesions in the production of periodic movements involving event-based or emergent timing. Brain Cogn 58:84–93. doi:10.1016/j.bandc.2004.09.010

Spencer RM, Zelaznik HN, Diedrichsen J, Ivry RB (2003) Disrupted timing of discontinuous but not continuous movements by cerebellar lesions. Science 300:1437–1439. doi:10.1126/science.1083661

Studenka BE, Zelaznik HN (2011) Circle drawing does not exhibit auditory-motor synchronization. J Mot Behav 43:185–191. doi:10.1080/00222895.2011.555796

Tanji J, Shima K (1994) Role for supplementary motor area cells in planning several movements ahead. Nature 371:413–416. doi:10.1038/371413a0

Teki S, Grube M, Griffiths T (2011) A unified model of time perception accounts for duration-based and beat-based timing mechanisms. Front Integr Neurosci 5:90. doi:10.3389/fnint.2011.00090

Ten Brinke MM, Boele H-JJ, Spankc JK et al (2015) Evolving models of pavlovian conditioning: cerebellar cortical dynamics in awake behaving mice. Cell Rep 13:1977–1988. doi:10.1016/j.celrep.2015.10.057

Thomson AM, Bannister PA (2003) Interlaminar connections in the neocortex. Cereb Cortex 13:5–14. doi:10.1093/cercor/13.1.5
Timmann D, Watts S, Hore J (1999) Failure of cerebellar patients to time finger opening precisely causes ball high-low inaccuracy in overarm throws
Ugawa Y, Day BL, Rothwell JC, Thompson PD, Merton PA, Marsden CD (1991) Modulation of motor cortical excitability by electrical stimulation over the cerebellum in man. J Phys 441(1):57–72
Ullen F, Bengtsson S (2003) Independent processing of the temporal and ordinal structure of movement sequences. J Neurophysiol 90:3725–3735. doi:10.1152/jn.00458.2003
Zeeuw C, Hoebeek F, Bosman L et al (2011) Spatiotemporal firing patterns in the cerebellum. Nat Rev Neurosci 12:327–344. doi:10.1038/nrn3011
Zelaznik HN, Rosenbaum DA (2010) Timing processes are correlated when tasks share a salient event. J Exp Psychol Hum Percept Perform 36:1565–1575. doi:10.1037/a0020380
Zelaznik HN, Spencer RM, Ivry RB et al (2005) Timing variability in circle drawing and tapping: probing the relationship between event and emergent timing. J Mot Behav 37:395–403. doi:10.3200/JMBR.37.5.395-403
Zhou H, Lin Z, Voges K et al (2014) Cerebellar modules operate at different frequencies. Elife 3:e02536. doi:10.7554/eLife.02536

# Predictability and Robustness in the Manipulation of Dynamically Complex Objects

**Dagmar Sternad and Christopher J. Hasson**

**Abstract** Manipulation of complex objects and tools is a hallmark of many activities of daily living, but how the human neuromotor control system interacts with such objects is not well understood. Even the seemingly simple task of transporting a cup of coffee without spilling creates complex interaction forces that humans need to compensate for. Predicting the behavior of an underactuated object with nonlinear fluid dynamics based on an internal model appears daunting. Hence, this research tests the hypothesis that humans learn strategies that make interactions predictable and robust to inaccuracies in neural representations of object dynamics. The task of moving a cup of coffee is modeled with a cart-and-pendulum system that is rendered in a virtual environment, where subjects interact with a virtual cup with a rolling ball inside using a robotic manipulandum. To gain insight into human control strategies, we operationalize predictability and robustness to permit quantitative theory-based assessment. Predictability is quantified by the mutual information between the applied force and the object dynamics; robustness is quantified by the energy margin away from failure. Three studies are reviewed that show how with practice subjects develop movement strategies that are predictable and robust. Alternative criteria, common for free movement, such as maximization of smoothness and minimization of force, do not account for the observed data. As manual dexterity is compromised in many individuals with neurological disorders, the experimental paradigm and its analyses are a promising platform to gain insights into neurological diseases, such as dystonia and multiple sclerosis, as well as healthy aging.

D. Sternad (✉) · C.J. Hasson
Department of Biology, Electrical and Computer Engineering, and Physics, Northeastern University, 134 Mugar Life Science Building, 360 Huntington Avenue, Boston, MA 02115, USA
e-mail: D.Sternad@neu.edu

C.J. Hasson
Department of Physical Therapy, Movement and Rehabilitation Sciences, Department of Biology, Northeastern University, Boston, MA, USA



J. Laczko and M.L. Latash (eds.), *Progress in Motor Control*,
Advances in Experimental Medicine and Biology 957,
DOI 10.1007/978-3-319-47313-0_4

**Keywords** Theory-based · Cart-and-pendulum · Energy margin · Experimental paradigm · Nonlinear fluid dynamics · Hapticmaster · Chaos · Cup-and-ball · Hand trajectory · Dystonia · Noise

## Introduction

Everyday life is full of actions that involve interaction with objects. Grasping and lifting a book involves manipulation of a free rigid object; turning a key in a keyhole involves moving a rigid object against a kinematic constraint. Functional interaction with objects—tool use—is ubiquitous in activities of daily living and the basis for our evolutionary advantage. Tools extend and augment fundamental human capabilities. Surprisingly, how humans interactively control objects or tools is still little understood. Manipulation requires sensing the mechanics and the geometry of the object and adjusting one's movements and forces accordingly to exploit object properties. Manipulation becomes particularly intriguing when the objects have internal degrees of freedom that add complex dynamics to the interaction. An exotic example is cracking a whip, where the flexible whip creates challenging dynamics (infinitely many degrees of freedom) that the hand has to interact with (Bernstein et al. 1958; Goriely and McMillen 2002; Hogan and Sternad 2012). A more mundane example is leading a cup of coffee to one's mouth to drink: the transporting hand applies a force not only to the cup, but also indirectly to the liquid, which in turn acts back onto the hand. These continuous forces require sensitive adjustments to avoid spilling the coffee (Hasson et al. 2012a; Mayer and Krechetnikov 2012; Hasson and Sternad 2014; Sauret et al. 2015). Humans are strikingly adept at interacting with a large variety of such objects, but most studies on object manipulation have been confined to either multi-digit grasping of a static object or grip forces needed for transporting solid objects (Flanagan et al. 1993; Flanagan and Wing 1997; Santello et al. 1998; Gao et al. 2005; Fu and Santello 2014). This chapter will focus on physical interactions with complex objects that are largely unchartered territory in motor neuroscience to date.

Over the last two decades motor neuroscience has made advances in understanding the control of simple movements, for example straight-line reaches in the horizontal plane including adaptation to external force fields or visual perturbations. This research has shed light on significant aspects of adaptation and control, such as error correction mechanisms and internal models (Shadmehr and Mussa-Ivaldi 1994; Scheidt et al. 2001). This paradigm has continued the long tradition of motor neuroscience examining elementary behaviors under strict experimental control. Seminal paradigms range from single-joint wrist movements in primates (Evarts 1968), to the speed-accuracy paradigm (Fitts 1954), to today's center-out reaching task for human and primate studies (Kalaska 2009). While these paradigms render manageable data for analysis and modeling, they are far removed from the richness of everyday actions and interactions. Unfortunately it is difficult, if not impossible, to extrapolate insights to more complex movements. For example, when extending multi-joint movements from 2D to 3D, non-commutative finite rotations introduce entirely new problems (Zatsiorsky 1998; Charles and Hogan 2011). Further and important for this line of study, physical contact with objects introduces bidirectional forces that pose a control challenge that is completely absent in free movements (Hogan 1985). Different from the sequential flow of information processing, physical interactions are fundamentally bidirectional—each system affects the other with mutual causality, an observation first expressed in Newton's third law.

## *Previous Research on Complex Object Manipulation*

Previous research on human control of dynamically complex objects has adopted a variety of theoretical perspectives that, as a whole, still present a rather disconnected picture. One line of studies examined balancing a pole, the classic control theoretical problem of stabilizing an inherently unstable system. Different control mechanisms were proposed, ranging from intermittent to continuous, predictive control, with forward or inverse models (Mehta and Schaal 2002; Gawthrop et al. 2013; Insperger et al. 2013). Nonlinear time-series analysis of the hand trajectory has probed the role of noise and delays to distinguish between continuous versus intermittent control (Cluff et al. 2009; Milton 2011; Milton et al. 2013) or the perceptual information used to stabilize the pole (Foo et al. 2000). Valero-Cuevas and colleagues examined the manual compression of a spring, modeling this dynamical system to include a subcritical pitchfork bifurcation to account for buckling (Venkadesan et al. 2007). Other studies have focused on the role of visual and haptic information to learn complex manipulation (Huang et al. 2002, 2007; Danion et al. 2012). Yet other research examined the displacement of a linear mass-spring object and proposed optimization criteria, such as generalized kinematic smoothness (Dingwell et al. 2004), accuracy and effort (Nagengast et al. 2009), and minimum acceleration with constraints on the center of mass (Leib and Karniel 2012). While interesting insights have been gained, most studies implicitly or explicitly assume that the human has, or has to learn an internal model of the manipulated object. As already hinted above, this may be daunting.

## *Hypothesis 1: Predictability*

When interacting with complex objects, instantaneous action and reaction is critical. Control models for artificial systems have posited internal models and inverse dynamics control plus feedback control, as they are largely devoid of long feedback delays and with relatively low levels of noise (Flanagan et al. 1999, 2003; Kawato 1999; Takahashi et al. 2001). In contrast, in humans feedback-based corrective control is virtually irrelevant due to trans-cortical or trans-cerebellar loop delays on the order of 100 ms or more, which requires exact extrapolation from current state estimates

(Pruszynski et al. 2011). This is difficult as variability and noise in the human system is high, with an approximate precision in timing of 9 ms (Faisal et al. 2008; Cohen and Sternad 2012). Instead, intrinsic musculo-skeletal properties augmented by spinal reflexes deliver essentially instantaneous reaction and can provide stabilization to counteract noise or instability (Colgate and Hogan 1988; Burdet et al. 2001; Franklin et al. 2003; Selen et al. 2009; Lee et al. 2014). While mechanical impedance is essential, dexterous control in the presence of delays nevertheless requires one to anticipate, preempt, and exploit the forces and motions of an object. Yet, prediction for continuous nonlinear objects with chaotic, i.e. unpredictable, behavior is challenging or impossible, even for artificial systems with short delays and low noise. Therefore, rather than expending the neural resources to learn a complex dynamics model, we suggest an alternative hypothesis: humans make the interactions with objects more predictable. This can be achieved by simplifying the interactive dynamics via linearization or avoidance of chaotic regimes.

### *Hypothesis 2: Robustness*

A precise internal dynamic model with complex nonlinear dynamics is difficult, if not impossible to learn. On the other hand, such complex models may not be necessary. For example, humans can proficiently control an automobile without knowing its full dynamical model or even understanding how the various mechanical components of a car work. To cope with such situations, the nervous system should select movement strategies that are robust to modeling errors. The branch of control theory called *robust control* is devoted to solving this problem, i.e. designing controllers that have good performance and stability in spite of modeling errors (Zhou and Doyle 1998). Note that such a controller may not have the same level of performance as one that has access to a perfect dynamics model, but choosing a suboptimal movement strategy, i.e. a "good enough" solution (Loeb 2012) may be an acceptable trade-off for increased robustness to modeling errors. Therefore, we hypothesize that rather than expending the neural resources to learn a complex dynamics model, humans learn a simpler model and select a robust control strategy that offers greater safety margins against failure.

## The Model Task: Moving a Cup of Coffee

To test the two hypotheses—humans select movement strategies that make interactions with complex objects predictable and robust—an appropriate test bed is needed. Transporting a cup of coffee is a good candidate as the cup filled with liquid has complex dynamics and there are clear consequences for failure, i.e. spilled coffee. However, transporting a cup with sloshing coffee is a complex problem in fluid dynamics (Mayer and Krechetnikov 2012; Sauret et al. 2015). Hence, the task

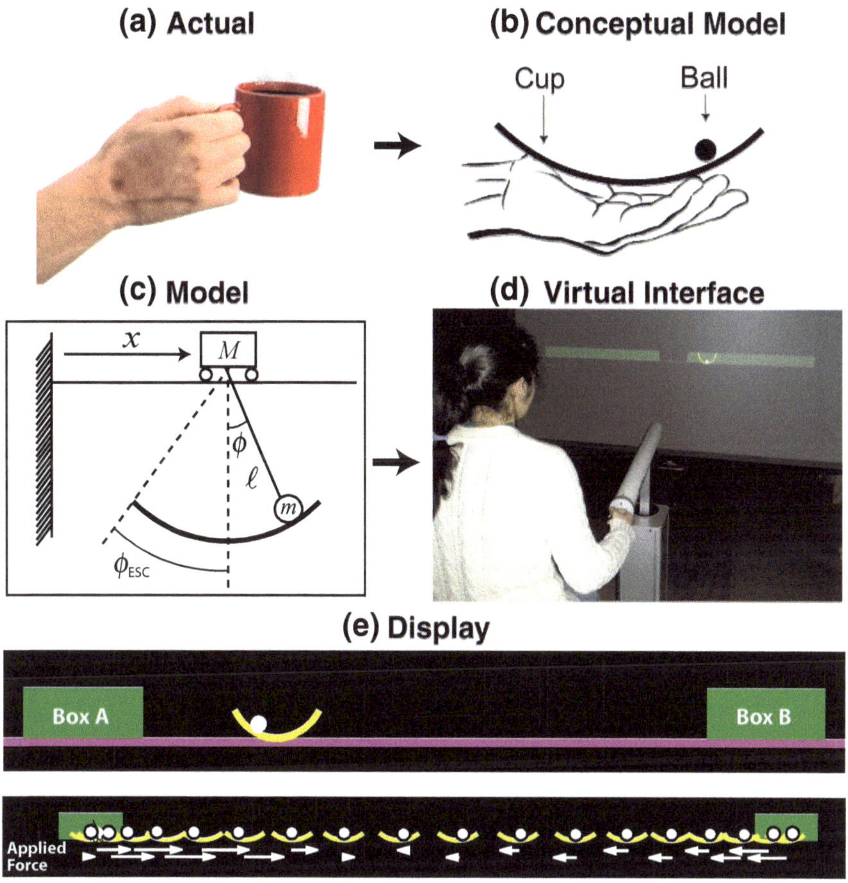


**Fig. 1** From the task to the experiment. **a** The actual task. **b** The conceptual model. **c** The cart-and-pendulum model underlying the displayed cup and ball. The cup is the arc of the circular pendulum path, the pendulum bob is the ball. **d** Virtual implementation with robot arm and visual and haptic interface. **e** The display with start and end box targets. The schematic below visualizes the applied force as arrows in accelerating and decelerating directions. Figure modified from Nasseroleslami et al. (2014) with permission under Creative Commons Attribution (CC BY) license

was simplified to that of moving a cup with a ball rolling inside, representing the complex dynamics of the coffee [Fig. 1a, b; (Hasson et al. 2012a)]. Implemented in a virtual environment the cup was visualized as an arc in 2D and modeled as a point mass moving along a horizontal axis. The ball's motion was modeled by a suspended pendulum; the arc of the cup corresponded to the ball's semi-circular path (Fig. 1c). This model system was implemented in a virtual environment, where subjects exert forces on the virtual cup via a robotic manipulandum (Fig. 1d shows the screen display and Fig. 1e (bottom panel) shows a movement of the cup and ball with the applied forces shown at different time points). Importantly, movements of the cup also accelerate the ball, which in turn acts back on the hand. Despite these simplifications, the model system retained essential elements of complexity: it is nonlinear and creates complex interaction forces between hand and object.

In this simplification, the equations of the cup-and-ball system are identical to the well-known cart-and-pendulum problem (Hinrichsen and Pritchard 2005; Ogata 2010). The cup is the cart with a point mass $M$ that moves horizontally; the pendulum comprises a point mass $m$ (the ball) attached to a mass-less rod of length $\ell$ with one angular degree of freedom $\theta$. Subjects control the ball indirectly by applying forces to the cup, and the ball can "escape", i.e. it can be lost from the cup when the angular distance to the rim is exceeded. The hand moving the cup is represented by an external applied horizontal force $F_{\mathrm{A}}$. The equations of the system dynamics are:

$$(m+M)\ddot{x} = m\ell\left(\ddot{\theta}\cos\theta + \dot{\theta}^2 \sin\theta\right) + F_{\mathrm{A}} \tag{1}$$

$$\ell\ddot{\theta} = \ddot{x}\cos\theta - g\sin\theta$$

where $\theta$, $\dot{\theta}$, and $\ddot{\theta}$ are the ball's angular position, velocity, and acceleration; $x$, $\dot{x}$ and $\ddot{x}$ are the cart/cup's position, velocity and acceleration; $g$ is gravitational acceleration; damping to pendulum and cart motion can also be added if desired.

To implement this cup-and-ball system in a virtual environment, the cart and the pendulum rod were hidden, but the pendulum bob (the ball) remained visible (Fig. 1e). Subjects manipulate the virtual cup-and-ball system via a robotic arm, which also exerts forces from the virtual object onto the hand [HapticMaster, Motek (van der Linde and Lammertse 2003)]. Using admittance control, the HapticMaster has three controllable degrees of freedom, but was constrained to motion on a horizontal line for the experiments. The pendulum's $\theta$ and $\dot{\theta}$ were computed using a 4th-order Runge-Kutta-integrator, and the force of the ball on the cup $F_{\mathrm{Ball}}$ was computed based on Eq. 1: $F_{\mathrm{Ball}} = m\ell\left(\ddot{\theta}\cos\theta + \dot{\theta}^2\sin\theta\right)$. This force, combined with any forces exerted by a human $F_{\mathrm{A}}$, accelerated the virtual mass $(m+M)$. The robot motors moved the manipulandum according to $\ddot{x}$ and the visual display was updated. For more details see (Hasson et al. 2012a).

This formalization and its virtual implementation has several advantages. (1) The focus is on the interaction forces between the hand and the object. Confining the physical interaction to a single "interaction port" via the robot handle avoids the complexity of grasp formation (Santello and Soechting 2000; Nowak and Hermsdörfer 2003). (2) Compared to real objects that have dozens of modes, this formalization reduces the object to two modes that facilitate analytical treatment (Hasson et al. 2012b). (3) The virtual implementation enables versatile manipulation of task parameters, including linear and nonlinear aspects. (4) The task involves "skill" and requires practice to arrive at smooth and stable execution. (5) The virtual implementation of the task is equivalent to the dynamic model. Hence, the measured human kinematics and kinetics lends itself to novel mathematical analyses to assess how humans sense and exploit the object's dynamic properties. In sum, the task has manageable but sufficient richness with multiple routes to increment complexity.

## Predictable and Unpredictable Interactions—Chaos

Most studies involving object manipulation have used linear systems, such as mass-springs (Dingwell et al. 2002; Svinin et al. 2006; Danion et al. 2012). By definition, such systems display predictable behavior. For example, if one were to oscillate a linear mass spring with the goal of attaining a given oscillation, the execution variables, the amplitude $A$ and frequency $f$ of the cup oscillation relate linearly to the applied forces and the resulting motion of the system: If the system is sinusoidally forced at 1 Hz, it will oscillate sinusoidally at 1 Hz. However, with a nonlinear system, such as the cup-and-ball, this mapping becomes non-trivial: the same forcing input may cause the system to oscillate at an array of frequencies with unpredictable and chaotic behavior.

To illustrate this chaotic behavior in the cup-and-ball system, we applied inverse dynamics to obtain the required force $F_A$ for a given oscillatory cup motion, specified by the scalar execution variables cup amplitude $A$ and cup frequency $f$, with initial ball angle $\theta_0$ and ball velocity $\dot{\theta}_0$. Shown in Fig. 2 are two simulated examples with the same sinusoidal cup movement $x$. The only difference is in the initial angle of the ball $\theta_0$, with $\dot{\theta}_0$ set to zero. In one case ($\theta_0 = 1.0\,\text{rad}$), the force required to produce this motion $x$ is periodic and predictable. In the other case ($\theta_0 = 0.4\,\text{rad}$), the force required to produce the same cup motion shows highly irregular fluctuations. To characterize the pattern of force profiles with respect to the cup dynamics, $F_A$ was strobed at every peak of the cup position $x$. The marginal distributions of strobed force values are plotted as a function of ball angle $\theta_0$ in the bottom panel (Fig. 2). This input-output relation reveals bifurcations with a pattern similar to the period-doubling behavior of chaotic systems. This feature has important implications for control: small changes in initial states can dramatically change the long-term behavior and lead to unpredictable solutions.

### *Quantifying Predictability*

We hypothesize that subjects seek solutions with predictable object behavior (*Hypothesis 1*). To quantitatively test this hypothesis, predictability must be operationalized. One possible measure is the mutual information (MI) between the applied force and the motion of the object, which characterizes the long-term predictability of the object's dynamics (Cover and Thomas 2006; Nasseroleslami et al. 2014). MI is a nonlinear correlation measure defined between two probability density distributions of two random variables and quantifies the information shared

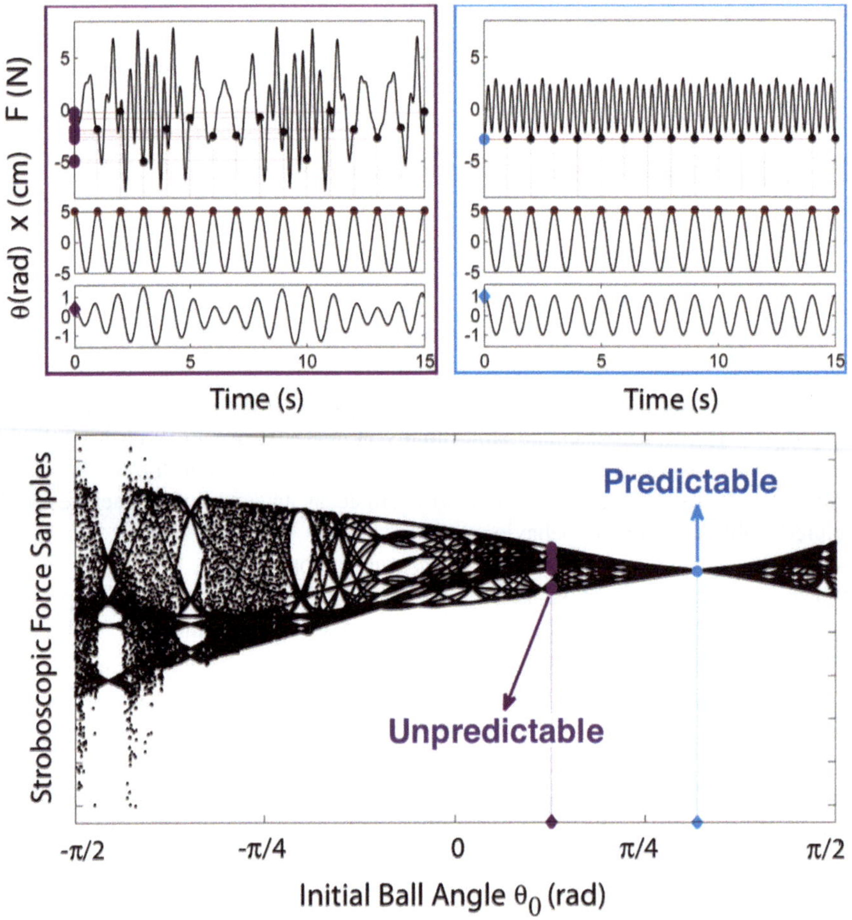


**Fig. 2** Simulated force profiles derived from inverse dynamics with specified cup and ball trajectories. The profiles are applied force, cup position, and ball angle (from *top* to *bottom*). The *left panel* was initiated with ball angle $\theta_0$ = 0.4 rad; the right panel with $\theta_0$ = 1.0 rad. The bifurcation diagram below shows the marginal distributions of force values strobed at all peak cup position (see dots in *upper panels*). The two initial ball angles are shown and marked as predictable and unpredictable. The diagram combines 1000 simulations with different initial ball angles $\theta_0$ in the range between $-\pi/2$ to $\pi/2$ rad. The force distributions plotted as a function of ball angle indicate chaotic dynamics. Figure modified from Nasseroleslami et al. (2014) with permission under Creative Commons Attribution (CC BY) license

between the two. MI is calculated between $F_A$ and the phase of the ball $\varphi$. This phase was calculated in phase space, spanned by ball angle and velocity:

$$MI(\varphi, F_A) = \iint p(\varphi, F_A) \log_e \frac{p(\varphi, F_A)}{p(\varphi)p(F_A)} d\varphi dF_A \tag{2}$$

where $p(.)$ denotes a probability density function. MI can also be calculated for the phase of the cup. MI presents a scalar measure of the performer's strategy that can be calculated for all amplitudes and frequencies of the cup and all initial conditions of the ball. MI can be summarized for each point of the 4D space of execution variables: $A, f, \theta_0$, and $\dot{\theta}_0$.

## *Experimental Insights*

Our recent study provided evidence that subjects increase the predictability of object dynamics with practice and favor predictable solutions over those that minimize expended force and smoothness, criteria that are widely supported criteria for free movements (Nasseroleslami et al. 2014). In this study, subjects (n = 8) oscillated the virtual cup between two targets with a robotic manipulandum, paced by a metronome at 1 Hz for 50 trials, each lasting 45 s. They were free to choose their movement amplitude and relative phase between the ball and cup.

The cup and ball oscillations were analyzed to determine how choices of movement amplitude and relative phase related to three result variables or costs: predictability, exerted force, and movement smoothness (Fig. 3). Figure 3a shows the result space for mutual information; lighter shading indicates that combinations of cup amplitude and ball angle render higher mutual information (higher predictability). The large point indicates the strategy with the highest mutual information. To compare potential alternative explanations, two other result measures, or commonly used costs, were derived for the same model: minimum force and maximum smoothness. The expended force was calculated by the square of the

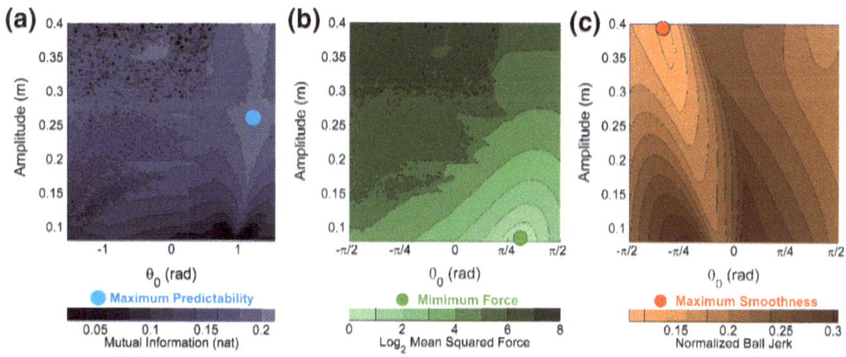


**Fig. 3** Result spaces that combine result variables or costs in the space spanned by the execution variables initial ball angle, cup amplitude, frequency (fixed at 1 Hz), and initial ball velocity (set to zero). **a** Mutual information. **b** Mean squared force (log transformed). **c** Mean squared jerk of the ball motion (normalized for amplitude); the large point in each graph indicates the location of maximum cost. Importantly, the minimum/maximum values are located in different parts of the map, providing different predictions. Figure modified from Nasseroleslami et al. (2014) with permission under Creative Commons Attribution (CC BY) license

mean integral over $F_A$ over the course of the trial, mean squared force MSF. Figure 3B shows the resulting pattern of force for the same space of execution variables; lighter shading refers to strategies requiring less force. The point highlights that the minimum force solution is obtained at the smallest allowable cup amplitudes. Lastly, smoothness or jerk was evaluated of the cup trajectory for each of the strategies defined in the execution space. Figure 3c shows smoothness of the ball movements for each strategy, with lighter shades denoting higher smoothness. The point shows that a strategy with high amplitude reaches maximum smoothness or minimum jerk. Importantly, the three maxima lie in different locations of the execution space. Therefore, by looking at which amplitude and relative phase subjects choose, we can infer which of the three costs are most important for subjects' movement control.

Following these simulations, equivalent measures for the execution variables $A, f, \theta_0$, and $\dot{\theta}_0$ had be derived from experimental data. However, the experimental trajectories were not fully determined by the initial values of ball states as variations could be due to online corrective changes. Therefore, to estimate the execution variables from the experimental trajectories, the initial values were extracted at each cycle $k$ (see Fig. 4). Peak excursions of the cup trajectory served as strobe points to estimate $A, f, \theta_0, \dot{\theta}_0$ and calculate trial averages $\bar{A}, \bar{f}, \bar{\theta}_0, \bar{\dot{\theta}}_0$ that served as correlates for the variables in the simulations. To exclude transients only the time window after 25 s was considered for analysis. To evaluate *Hypothesis 1*, that subjects seek predictable object interactions, *MI*, and the alternative costs mean squared force *MSF*, and mean squared jerk *MSJ* were calculated for each measured strategy $\bar{A}_k, \bar{f}_k, \bar{\theta}_k, \bar{\dot{\theta}}_k$. Calculation of *MI* followed the same procedure as in the simulated *MI*, except that probability density functions were estimated experimentally (for more details see (Nasseroleslami et al. 2014). To calculate *MSF*, the continuous force profile of each trial was squared and averaged, analogous to the simulated data. *MSJ* was calculated according to the standard equations (Hogan and Sternad 2009).

The main experimental results are summarized in Fig. 5; the mutual information plot is overlaid with contours of selected simulated force values (green). The figure shows how subjects gravitated towards areas with higher *MI*, i.e. strategies with more predictable interactions. In the left panel, each point represents the average strategy for each 45 s trials for all subjects; darker red indicates early practice and lighter red indicates late practice. The right panel shows the same data separated by subject: the red arrows mark how each subject's average strategy changed from early practice (mean of first 5 trials) to late practice (mean of last 5 trials). Both figures clearly show that all subjects increased their movement amplitude, associated with an increase in overall exerted force. The majority of subjects switched from low- to high-predictability regions in the result space. None of the subjects moved toward the minimum force strategy, nor towards a strategy with maximum smoothness. Analysis of *MSF* and *MSJ* over trials shows that indeed exerted force increased and smoothness decreased with practice, counter to findings in free unconstrained movements. Overall, the results rejected the two alternative criteria and were consistent with *Hypothesis 1*.

 

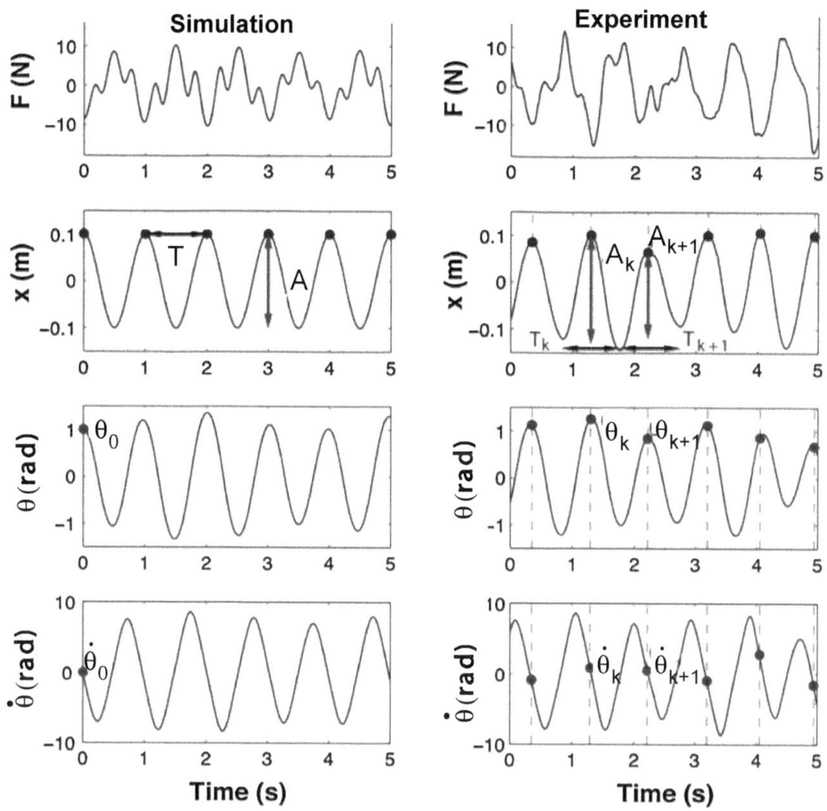


**Fig. 4** Exemplary profiles from inverse dynamics simulations and corresponding experimental data for applied force, cup position, ball angle and velocity. Estimates for the execution variables in the data were derived for each cycle as shown and then averaged across the trial to obtain scalar estimates for each trial. Figure modified from Nasseroleslami et al. (2014) with permission under Creative Commons Attribution (CC BY) license

## Robust Interactions

The reviewed results suggest that when there is a choice, humans select a strategy that increases the predictability of the human-object interaction. More predictable human-object interactions may lessen the control burden; however, errors in control undoubtedly exist, especially when only rough approximations of internal models of object dynamics are available. Thus, keeping interactions predictable may not be enough—a good strategy should also be robust to control errors. The cup-and-ball task lends itself to experimental investigation of robustness, as there is a well-defined threshold for failure, i.e. the ball escapes the cup—coffee is spilled. Note that in the previous experiment, the ball could not escape, but swung around following the circular path of the pendulum in situations of varying difficulty. By

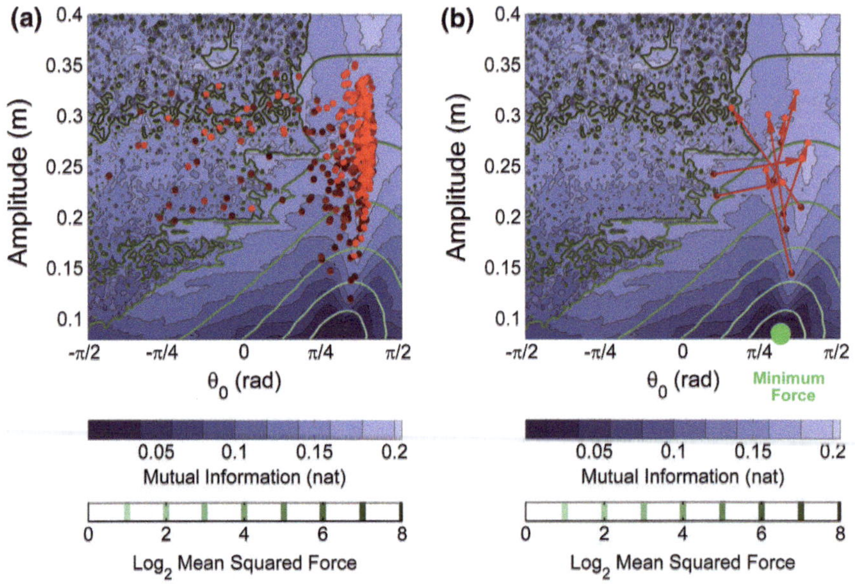


**Fig. 5** Main results in the result space for mutual information. The green contours denote different values of mean squared force superimposed onto the same space. **a** Data pooled from all subjects; each point is one trial. Darker red pertains to earlier trials than lighter red. The data show that in the course of practice, subjects shifted their movement strategies to the area of high mutual information (high predictability). **b** Averaged data from eight subjects; each arrow represents one subject, the tail of the arrow is the mean of the first 5 trials, the tip of the arrow is the mean of the last 5 trials. Notice that none of the subjects shifted down towards the area of minimum force, indicated by the green point. Figure modified from Nasseroleslami et al. (2014) with permission under Creative Commons Attribution (CC BY) license

introducing a "rim" and also using a shallower cup, we could probe the use of fragile and robust strategies.

We hypothesized that as a subject learns to manipulate the object, s/he should find strategies that are more robust to failure. In a risky strategy, the ball gets close to the rim of the cup and any small error may lead to loss of the ball. Therefore, a safety margin is critical and might present a sensitive measure distinguishing "fragile" from robust control. We hypothesized that this safety margin should increase with practice (*Hypothesis 2*). Further, we expected that the size of the safety margin depends on the performance variability. Individuals have different degrees of variability and those with more variable movements should seek larger safety margins (*Hypothesis 2a*). Further, if variability decreases with practice, then the safety margin should change accordingly (*Hypothesis 2b*). We will now review two studies that addressed these questions in young and also older healthy adults.

## *Quantifying Robustness*

To test these hypotheses, the safety margin needed to be defined. Safety margins have been most frequently characterized in gait and posture and are typically quantified by the degree of spatial and/or temporal difference between the body center of mass/center of pressure and the base of support boundary (Hof et al. 2005; Hasson et al. 2008). While useful, such measures can be difficult to generalize, because they are specific to upright stance and can depend on the physical attributes of the individual. Therefore, we developed a more general formulation, defined in terms of energy, i.e. an energy margin.

Most objects that we may interact with are initially at rest, and when we pick them up or handle them, we impart energy to them. For example, we push on a shopping cart to start moving it or pick up a cup of coffee to drink. If too much energy is imparted to such objects, an undesirable outcome may occur, such as overturning the shopping cart or spilling the coffee. We define the energy margin *EM* by the difference between the current energy to the energy level that causes failure [see (Hasson et al. 2012a) for more details].

Specifically for the cup-and-ball system, *EM* quantifies how close the total energy of the ball $TE_{\mathrm{BALL}}$ is to the energy level that would cause the ball to exceed the rim, i.e. the escape energy $E_{\mathrm{ESC}}$

$$EM = (EM_{\mathrm{ESC}} - TE_{\mathrm{BALL}})/E_{\mathrm{ESC}} \tag{3}$$

*EM* is normalized to $E_{\mathrm{ESC}}$ so that the maximum value is $EM = 1$ (maximum safety). Small values indicate a "dangerous" condition; if *EM* remains below zero the ball will escape from the cup unless a corrective action is taken. $TE_{\mathrm{BALL}}$ is given by

$$TE_{\mathrm{BALL}} = KE_{\mathrm{BALL}} + PE_{\mathrm{BALL}} + PSE_{\mathrm{BALL}}$$

$$= \left[\frac{1}{2}m\left(\ell\dot{\theta}\right)^2\right] + [mg\ell(1 - \cos\theta)] + \left[-m\ddot{x}\ell\sin\theta + m|\ddot{x}|\ell\right] \tag{4}$$

where $KE_{\mathrm{BALL}}$ is the kinetic energy of the ball, $PE_{\mathrm{BALL}}$ is the potential energy of the ball, and $PSE_{\mathrm{BALL}}$ is a pseudo-energy because the ball is in an accelerated reference frame relative to the cup. $E_{\mathrm{ESC}}$ is defined as

$$E_{\mathrm{ESC}} = mg\ell(1 - \cos\theta_{\mathrm{ESC}}) - m|\ddot{x}|\ell\sin\theta_{\mathrm{ESC}} + m|\ddot{x}|\ell \tag{5}$$

In these equations, there are only three time-varying quantities, the ball angle $\theta$, the ball angular velocity $\dot{\theta}$, and the cup acceleration $\ddot{x}$. These variables are measured and defined as the execution variables, which jointly determine the result variable *EM*. Essentially, *EM* takes the instantaneous state of the cup and ball, which includes inputs from the human hand, and extrapolates to determine whether the ball will escape, given the current value for $\ddot{x}$. At the very next instant in time, a new determination is made based on updated execution values of $\theta$, $\dot{\theta}$, and $\ddot{x}$, and so on for future time points. This analysis approach follows the same logic as for the rhythmic task described above: identify the execution variables that fully determinate the result variable. However, instead of mapping to a predictability measure, *MI* (alternative measures or *MSF* and *MSJ*), the execution variables are mapped into the energy margin *EM*. This same analysis strategy was previously applied to other tasks such as throwing and bouncing a ball (Dijkstra et al. 2004; Cohen and Sternad 2009; Sternad et al. 2014).

For any movement of the cup and ball, the energy margin fluctuates over the time of the movement, as shown in Fig. 6a for an exemplar point-to-point translation of the cup and ball. The normalization of *EM* to $E_{ESC}$ affords an assessment of the risk at any instant during an ongoing movement. When $EM > 0$ and the margin is large, any unexpected disturbance can easily be dealt with or "absorbed". However, when $EM \leq 0$, the ball will escape in a finite "time-to-escape" (red dotted lines in Fig. 6a), unless action is taken to increase the *EM* before the ball reaches the rim. The exemplary profile shows fluctuations that are concurrent, but not coincident with the ball excursions, as the applied force is also important. The same trial can also be plotted as a trajectory in 3D space spanned by the three execution variables $\theta$, $\dot{\theta}$, and $\ddot{x}$. (Figure 6b). The result variable is *EM*. The critical energy $E_{ESC}$ defines a closed two-dimensional manifold two oblique cones joined together;

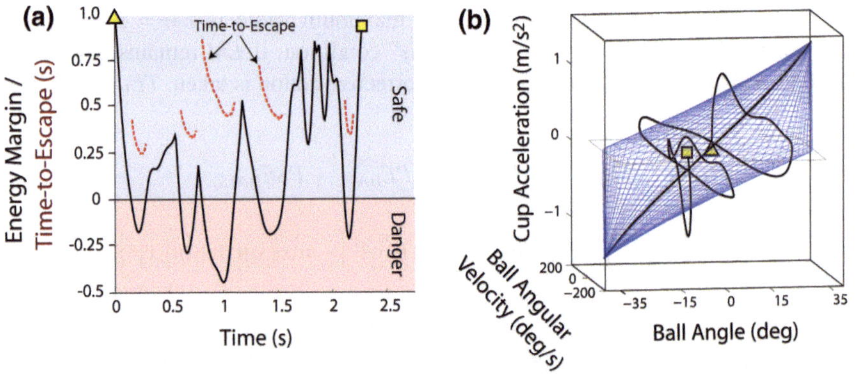


**Fig. 6** Exemplary profile of energy margin of one trial during early practice. **a** The energy margin *EM* as a function of time. With the initial high *EM*, the ball is at rest and is unlikely to escape from the cup, even when exposed to a disturbance. However, when the *EM* drops below zero the ball is in a state where it will escape from the cup in a finite time (shown as the red dotted "Time-to-Escape" lines). **b** For the same trial, the three variables that determine *EM*, ball angle and angular velocity and cup acceleration, are shown in a three dimensional execution space. The trial starts in the center (*yellow triangle*) and moves through the space as the trial progresses until the cup is stopped at the spatial target (*yellow square*). The blue mesh represents the escape energy threshold, $EM = 0$; whenever the trajectory breaches the manifold there is danger of ball-escape unless a corrective action is taken. Figure modified from Hasson and Sternad (2014) with permission under Creative Commons Attribution (CC BY) license

see blue mesh. If the trajectory stays within this manifold, $EM > 0$, the time-to-escape is infinite, and the ball is never in danger of escaping. If the manifold is breached, $EM < 0$ and the ball may escape. If the subject applies a corrective action to change $\ddot{x}$ in an appropriate way, then failure may be prevented. However, the available time to make such a correction is finite. If the correction takes too long, the ball will be lost. Note that the time-to-escape is computed at each instant in time, assuming constant $\ddot{x}$, but is then updated at the next instant in time when a new set of execution variables ($\theta$, $\dot{\theta}$, and $\ddot{x}$) is available.

## *Experimental Insights*

A prior study sought to test the hypothesis that humans seek robust movement strategies with appropriate safety margins (Hasson et al. 2012a). Subjects were asked to make a discrete point-to-point translation of the cup, and to complete the movement in a target time of 2 s without losing the ball from the cup. This completion time was comfortable and afforded selection among several strategies. For comparison, a separate group of subjects performed a minimum-time movement, translating the cup as fast as possible over the same distance. Both groups improved their performance, i.e. the timing error and movement time decreased for the target-time and minimum-time groups, respectively. As hypothesized, subjects in the target-time condition increased their energy margin over practice (Fig. 7a). In contrast, the energy margin decreased in the minimum-time task (Fig. 7b). Accordingly, the minimum-time group lost the ball about 10 times as often as the target-time group at the end of practice. These changes in the energy margin typically occurred throughout the entire movement profile, as highlighted by the shading in Fig. 7a, b, although some portions of the movement tended to show larger changes than others. These findings suggest that when urged to move as fast as possible, subjects "live dangerously" and use small energy margins. However, when multiple movement options are available humans prefer those that are more robust to errors in control. This result supported *Hypothesis 2a*.

For a different view on how the energy margin changed with practice, a number of trials from one representative subject are shown in execution space in Fig. 7c. The blue mesh again represents the $E_{ESC}$ manifold; two perspectives on the same data are shown for clarity. Early in practice, the movement trajectories are variable and frequently break through the $E_{ESC}$ manifold by a significant amount, often leading to loss of the ball. This happened mostly near the end of the movement when subjects tried to stop the cup (seen as high cup deceleration). However, after practicing the task, a clear structure becomes visible and the trajectories conform to the $E_{ESC}$ manifold. This "contraction" of the trajectories raises the energy margin, increasing robustness. As long as the trajectory is within the $E_{ESC}$ manifold there is no chance of the ball escaping from the cup. This could be advantageous, as minor

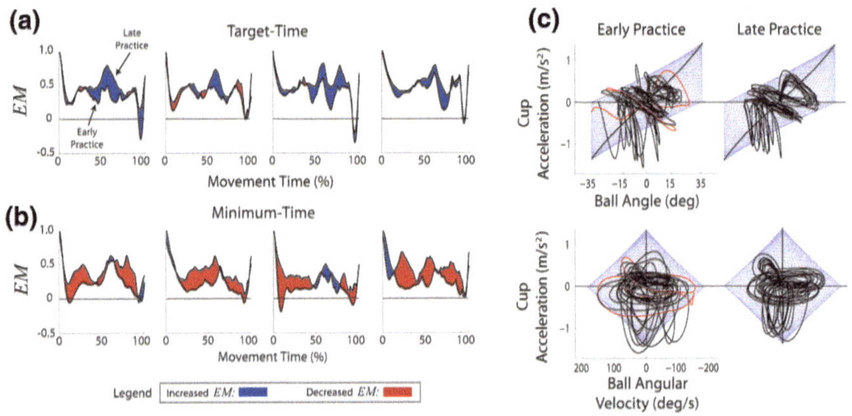


**Fig. 7** Changes in the energy margin *EM* with practice. **a** Difference in the energy margins between an early and a late trial in four subjects of the target-time group. **b** Difference in the energy margins between an early and a late trial in four subjects of the minimum-time group. **c** Examples of early and late practice trials for one subject. Trajectories are plotted in execution space, defined by ball angle and velocity and cup acceleration. Two different views of the three-dimensional execution space are shown. Trials in which the ball escaped are shown in red. Note that not all trials that exit the manifold result in failure. Figure modified from Hasson and Sternad (2014) with permission under Creative Commons Attribution (CC BY) license

errors in control would not cause failure, which could free up cognitive resources for higher–level movement planning operations.

Motivated by the robustness hypothesis, we also predicted that the size of the safety margin should depend on subjects' motor variability (*Hypothesis 2a*). This follows previous work suggesting that variability plays a central role in movement control such that the motor system optimizes movements to minimize the effects of variability on task goals (Harris and Wolpert 1998; Trommershäuser et al. 2005; Gepshtein et al. 2007; Cohen and Sternad 2009; Hudson et al. 2010; Sternad et al. 2011; Chu et al. 2013). Specifically, individuals with greater trial-to-trial variability should choose a larger energy margin, and vice versa. To test this hypothesis, the degree of correlation between the energy margin and trial-to-trial variability was assessed for both the target-time and minimum-time tasks. Consistent with *Hypothesis 2a*, results showed a positive correlation, i.e. subjects with high variability at the end of practice also had large safety margins at the end of practice (Fig. 8a). There was no correlation for the target-time task. This could be ascribed to the individual variations in strategies in the target-time group, while subjects in the minimum-time group displayed more similar strategies. When examining potential correlations *across practice* within each individual, there was a significant correlation for the target-time group. Consistent with *Hypothesis 2b*, subjects with large decreases in variability also changed their strategies to smaller energy margins, and vice versa (Fig. 8b). Those subjects who developed a consistent movement pattern may have been more confident in their ability, and therefore did not need large energy margins. Conversely, subjects with greater trial-to-trial variability

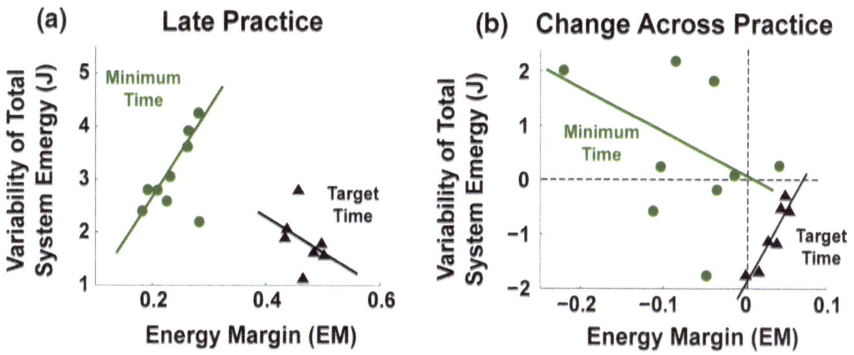


**Fig. 8** Changes in energy margin *EM* as a function of trial-to-trial variability and task condition. **a** Correlations between *EM* and trial-to-trial variability of the total ball-and-cup system energy $TE_{STD}$ over the last 30 trials for the Target-Time group (*black triangles*) and the Minimum-Time group (*green circles*). **b** Correlations between the change in *EM* and variability from early (first 30 trials) to late (last 30 trials) practice within each subject. Figure redrawn from Hasson et al. (2012a, 2012b)

chose a larger energy margin to accommodate the greater uncertainty. A connection between variability and safety margins was subsequently demonstrated in other recent studies (Chu et al. 2013, 2016; Hadjiosif and Smith 2015).

Robust control of behavior seems especially essential for individuals with diminished control abilities and who are fragile and prone to injury. One such population is frail older adults who may face catastrophic consequences in the event of an error in movement control, such as a fall. Paradoxically, even though older adults should utilize larger safety margins, in many cases the opposite has been shown. For example, when walking over obstacles or navigating stairs, older adults have smaller foot-obstacle clearances (Begg and Sparrow 2000; McFadyen and Prince 2002; Hamel et al. 2005). During quiet standing their postural sway measures show reduced spatiotemporal margins of stability (Slobounov et al. 1998; Van Wegen et al. 2002). We posited that such "high risk" strategies arise because older adults have more difficulty controlling complex whole body movements. Older adults may strive for high safety margins, but may be unsuccessful due to sensorimotor limitations. To explore this conjecture, we invited older adults to practice the cup and ball task. We tested the hypothesis that older adults have lower energy margins compared to younger adults (*Hypothesis 3a*), but as they learn to control the cup-and-ball dynamics, we expected their energy margins to increase significantly (*Hypothesis 3b*). Support for the latter hypothesis would show that they are indeed striving for larger energy margins as they gain better control of the object dynamics. We therefore asked them to perform the same discrete transport of the cup and ball, emphasizing that they should not lose the ball.

The results showed that with practice, both young and older adults improved their skill in the target-time task (decreased their timing error). Not surprisingly, the younger adults performed better and dropped the ball less often. When comparing

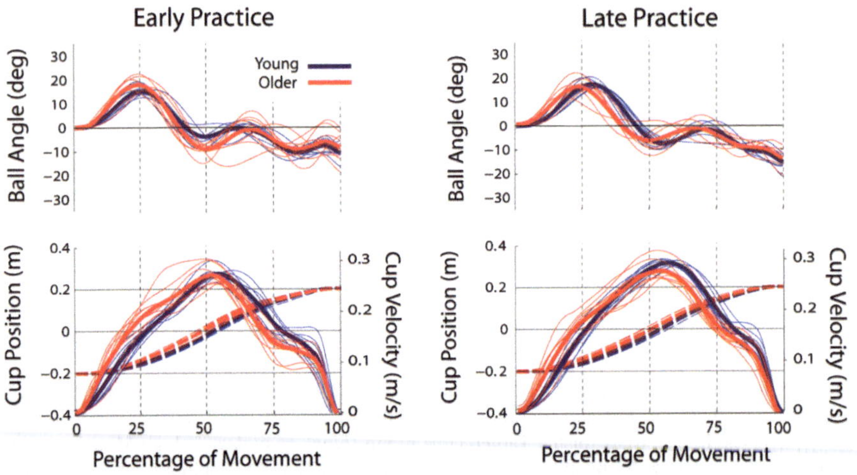


**Fig. 9** Cup and ball kinematics in early and late practice for young (*blue*) and older adults (*red*). Individual subjects are shown as *thin lines* and the group means are the *thicker lines*. Variability decreases in both groups, but no other evident differences are discernible. Figure modified from Hasson and Sternad (2014) with permission under Creative Commons Attribution (CC BY) license

the kinematic profiles, i.e. the position and velocity of the cup and ball, only minor differences between the two age groups were discernable (Fig. 9). It was only the energy margins that revealed the differences between the two groups: early in practice, the older adults performed with a significantly lower energy margin (Fig. 10a), supporting *Hypothesis 3a*. Nonetheless, the older adults were able to increase their energy margin with practice, although not to the level of the younger adults (Fig. 10b). This implies that as older adults learned to interact with the cup-and-ball dynamics, they were able to increase the robustness of their movement strategies and, consequently, lost the ball less frequently (Fig. 10c), supporting *Hypothesis 3b*. However, while the energy margins continued to increase in the younger adults, it plateaued in the older adults. This suggests that sensorimotor limitations in older adults limit their ability to keep the cup and ball in a regimen with high safety margins.

## A Task-Based Approach for Understanding Human-Object Interactions

How do humans successfully manipulate tools in daily life, an ability that has a long evolutionary history? Manipulation of complex dynamic objects presents daunting challenges, although more for the scientist than the human actor. Extrapolating our current understanding of human control of free movements to those involving object manipulation may not be an incremental process. For example, feedback

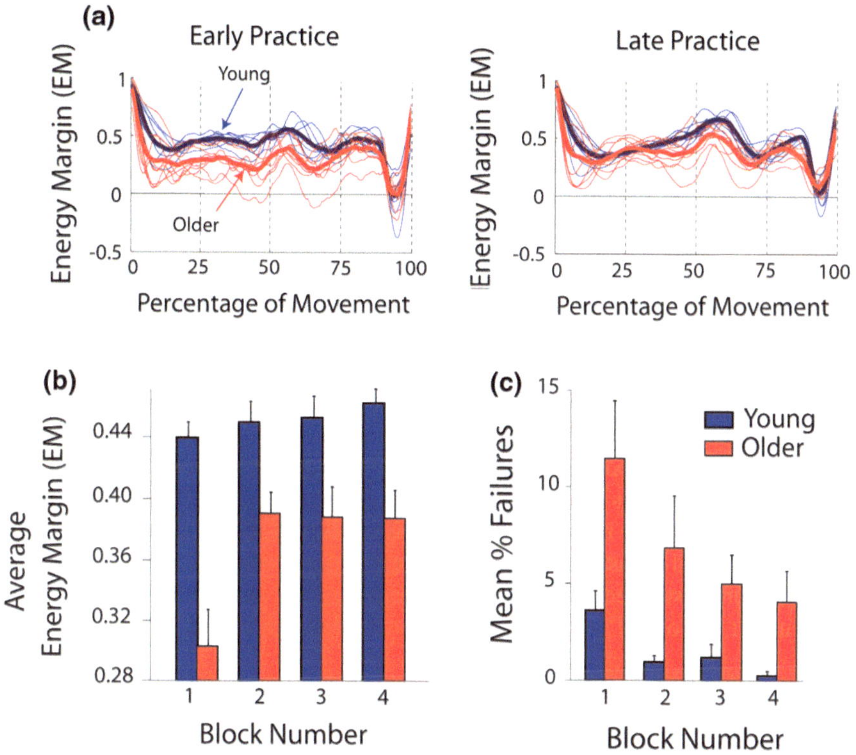


**Fig. 10** Changes in the energy margin *EM* and number of ball escapes with practice for young and older adults. **a** The energy margin as a function of normalized movement time in early and late practice: individual subjects are shown as *thin lines* and the group means are the *thicker lines*. Note that in early practice the older subjects had a lower energy margin for most of the movement. **b** Group average *EM* across four blocks of practice. Both young and older adult increase their energy margins, but older adults have significantly smaller energy margins. **c** Group average percentage of trials in which the ball was dropped across four practice blocks. Older adults show visibly more failures than young subjects, but they also improve with practice. Figure modified from Hasson and Sternad (2014) with permission under Creative Commons Attribution (CC BY) license

control based on internal models of the object dynamics appears problematic given the long delays and high levels of noise in the human neuromotor system. Void of knowing the control architecture, we adopted a task-based approach. We analyzed the task and derived the solution space with no assumptions about the human actor and control. Starting with a physical model of the object dynamics and the task, we first identified execution and result variables. Mapping execution to selected result variables rendered a space of solutions. Based on this understanding of the physics, we could formulate quantitative hypotheses about potential strategies and objective functions that humans might use. Implementing the task in an interactive virtual environment we then measured human performance and directly evaluated task performance in the result space. This task-based approach has also been successfully applied in other tasks (Sternad et al. 2014).

## Take Home Message

Unlike the body's own limbs, interactions with objects in the external world can be quite unpredictable. This is particularly true for objects with complex dynamics that cannot be directly controlled, such as a cup of coffee or a jostling baby carriage. Using the cup of coffee as a model task, we reviewed studies showing that with practice humans learn to control such objects by making the interaction both predictable and robust. These criteria are important for all individuals, however they may be of special importance to individuals with disabilities, where unpredictable and fragile interactions with the world incur marked effects on the quality of life. For these populations, it would be beneficial to develop interventions that promote predictability and robustness and thereby complement traditional movement criteria from free unconstrained movements such as movement smoothness and economy. The current ecological task may be a first step in this direction.

**Acknowledgments** This work was supported by the NIH-R01-HD045639, NIH-R01-HD081346, NIH-R21-DC013095, and NSF-EAGER-1548514 awarded to Dagmar Sternad and NIH-F32-AR061238 awarded to Christopher J. Hasson.

## References

Begg RK, Sparrow WA (2000) Gait characteristics of young and older individuals negotiating a raised surface: implications for the prevention of falls. J Gerontol Ser A Biol Sci Med Sci 55: M147–M154

Bernstein B, Hall DA, Trent HM (1958) On the dynamics of a bull whip. J Acoust Soc Am 30:1112–1115

Burdet E, Osu R, Franklin D, Milner T, Kawato M (2001) The central nervous system stabilizes unstable dynamics by learning optimal impedance. Nature 414:446–449

Charles SK, Hogan N (2011) Dynamics of wrist rotation. J Biomech 44:614–621

Chu WTV, Sternad D, Sanger TD (2013) Healthy and dystonic children are sensitive to their motor variability. J Neurophysiol 109:2169–2178

Chu WTV, Park S, Sanger TD, Sternad D (2016) Dystonic children can learn a novel motor skill: strategies that are tolerant to high variability. IEEE Trans Neural Syst Rehabil Eng 8:847–858

Cluff R, Riley MA, Balasubramanian R (2009) Dynamical structure of hand trajectories during pole balancing. Neurosci Lett 464:88–92

Cohen R, Sternad D (2009) Variability in motor learning: relocating, channeling and reducing noise. Exp Brain Res 193:69–83

Cohen RG, Sternad D (2012) State space analysis of intrinsic timing: exploiting task redundancy to reduce sensitivity to timing. J Neurophysiol 107:618–627

Colgate J, Hogan N (1988) Robust control of dynamically interacting systems. Int J Control 48:65–88
Cover TM, Thomas JA (2006) Elements of information theory. Wiley, Hoboken
Danion F, Diamond J, Flanagan J (2012) The role of haptic feedback when manipulating nonrigid objects. J Neurophysiol 107:433–441
Dijkstra TM, Katsumata H, De Rugy A, Sternad D (2004) The dialogue between data and model: passive stability and relaxation behavior in a ball bouncing task. Nonlinear Stud 11:319–344
Dingwell JB, Mah CD, Mussa-Ivaldi FA (2002) Manipulating objects with internal degrees of freedom: evidence for model-based control. J Neurophysiol 88:222–235
Dingwell J, Mah C, Mussa-Ivaldi F (2004) Experimentally confirmed mathematical model for human control of a non-rigid object. J Neurophysiol 91:1158–1170
Evarts E (1968) Relation of pyramidal tract activity to force exerted during voluntary movement. J Neurophysiol 31:14–27
Faisal AA, Selen LP, Wolpert DM (2008) Noise in the nervous system. Nat Rev Neurosci 9: 292–303
Fitts PM (1954) The information capacity of the human motor system in controlling the amplitude of movement. J Exp Psychol 47:381–391
Flanagan J, Wing A (1997) The role of internal models in motion planning and control: evidence from grip force adjustments during movements of hand-held loads. J Neurosci 17:1519–1528
Flanagan J, Tresilian J, Wing A (1993) Coupling of grip force and load force during arm movements with grasped objects. Neurosci Lett 152:53–56
Flanagan J, Nakano E, Imamizu H, Osu R, Yoshioka T, Kawato M (1999) Composition and decomposition of internal models in motor learning under altered kinematic and dynamic environments. J Neurosci 34:1–5
Flanagan J, Vetter P, Johansson R, Wolpert D (2003) Prediction precedes control in motor learning. Curr Biol 13:146–150
Foo P, Kelso J, de Guzman G (2000) Functional stabilization of unstable fixed points: human pole balancing using time-to-balance information. J Exp Psychol Human Percept Perform 26: 1281–1297
Franklin D, Osu R, Burdet E, Kawato M, Miner T (2003) Adaptation to stable and unstable dynamics achieved by combined impedance control and inverse dynamics model. J Neurophysiol 90:3270–3282
Fu Q, Santello M (2014) Coordination between digit forces and positions: interactions between anticipatory and feedback control. J Neurophysiol 111:1519–1528
Gao F, Latash ML, Zatsiorsky VM (2005) Internal forces during object manipulation. Exp Brain Res 165:69–83
Gawthrop P, Lee K-Y, Halaki M, O'Dwyer N (2013) Human stick balancing: an intermittent control explanation. Biol Cybern 107:637–652
Gepshtein S, Seydell A, Trommershäuser J (2007) Optimality of human movement under natural variations of visual–motor uncertainty. J Vision 7, 13. doi:10.1167/7.5.13
Goriely A, McMillen T (2002) Shape of a cracking whip. Phys Rev Lett 88:244301
Hadjiosif AM, Smith MA (2015) Flexible control of safety margins for action based on environmental variability. J Neurosci 35:9106–9121
Hamel KA, Okita N, Higginson JS, Cavanagh PR (2005) Foot clearance during stair descent: effects of age and illumination. Gait Posture 21:135–140
Harris CM, Wolpert DM (1998) Signal-dependent noise determines motor planning. Nature 394:780–784
Hasson C, Sternad D (2014) Safety margins in older adults increase with improved control of a dynamic object. Front Aging Neurosci 6, 158. doi:10.3389/fnagi.2014.00158
Hasson CJ, Van Emmerik RE, Caldwell GE (2008) Predicting dynamic postural instability using center of mass time-to-contact information. J Biomech 41:2121–2129
Hasson C, Shen T, Sternad D (2012a) Energy margins in dynamic object manipulation. J Neurophysiol 108:1349–1365

Hasson CJ, Hogan N, Sternad D (2012b) Human control of dynamically complex objects. In: 4th IEEE RAS & EMBS International Conference on Biomedical Robotics and Biomechatronics (BioRob). IEEE, pp 1235–1240

Hinrichsen D, Pritchard A (2005) Mathematical systems theory: modelling, state space analysis, stability and robustness. Springer, New York

Hof A, Gazendam M, Sinke W (2005) The condition for dynamic stability. J Biomech 38:1–8

Hogan N (1985) Impedance control: An approach to manipulation: Part II—implementation. J Dyn Syst 107:8–16

Hogan N, Sternad D (2009) Sensitivity of smoothness measures to movement duration, amplitude, and arrests. J Motor Behav 41:529–534

Hogan N, Sternad D (2012) Dynamic primitives of motor behavior. Biol Cybern 106:727–739

Huang FC, Gillespie RB, Kuo AD (2002) Haptic feedback and human performance in a dynamic task. In: 10th symposium on haptic interfaces for virtual environmental and teleoperating systems (Haptics '02). IEEE, Orlando, pp 24–31

Huang FC, Gillespie RB, Kuo AD (2007) Visual and haptic feedback contribute to tuning and online control during object manipulation. J Motor Behav 39:179–193

Hudson TE, Tassinari H, Landy MS (2010) Compensation for changing motor uncertainty. PLoS Comput Biol 6:e1000982

Insperger T, Milton J, Stepan G (2013) Acceleration feedback improves balancing against reflex delay. J R Soc Interface 10(79), 20120763

Kalaska J (2009) From intentions to actions: motor cortex and the control of reaching movements. In: Sternad D (ed) Progress in Motor Control V: a multidisciplinary approach. Springer, New York

Kawato M (1999) Internal models for motor control and trajectory planning. Curr Opin Neurobiol 9:718–727

Lee H, Krebs H, Hogan N (2014) Multivariable dynamic ankle mechanical impedance with active muscles. IEEE Trans Neural Syst Rehabil Eng 22:971–981

Leib R, Karniel A (2012) Minimum acceleration with constraints of center of mass: a unified model for arm movements and object manipulation. J Neurophysiol 108:1646–1655

Loeb GE (2012) Optimal isn't good enough. Biol Cybern 106:757–765

Mayer HC, Krechetnikov R (2012) Walking with coffee: why does it spill? Phys Rev E 85:046117

McFadyen BJ, Prince F (2002) Avoidance and accommodation of surface height changes by healthy, community-dwelling, young, and elderly men. J Gerontol Ser A Biol Sci Med Sci 57: B166–B174

Mehta B, Schaal S (2002) Forward models in visuomotor control. J Neurophysiol 88:942–953

Milton J (2011) The delayed and noisy nervous system: implications for neural control. J Neural Eng 8:065005

Milton J, Fuerte A, Bélair C, Lippai J, Kamimura A et al (2013) Delayed pursuit-escape as a model for virtual stick balancing. Nonlinear Theor Appl 4:129–137

Nagengast A, Braun D, Wolpert D (2009) Optimal control predicts human performance on objects with internal degrees of freedom. PLoS Comput Biol 5:e1000419

Nasseroleslami B, Hasson C, Sternad D (2014) Rhythmic manipulation of objects with complex dynamics: predictability over chaos. PLoS Comput Biol 10:e1003900

Nowak D, Hermsdörfer J (2003) Selective deficits of grip force control during object manipulation in patients with reduced sensibility of the grasping digits. Neurosci Res 47:65–72

Ogata K (2010) Modern control engineering. Prentice Hall, New York

Pruszynski JA, Kurtzer I, Scott SH (2011) The long-latency reflex is composed of at least two functionally independent processes. J Neurophysiol 106:449–459

Santello M, Soechting JF (2000) Force synergies for multi-fingered grasping. Exp Brain Res 133:457–467

Santello M, Flanders M, Soechting J (1998) Postural hand synergies for tool use. J Neurosci 18:10105–10115

Sauret A, Boulogne F, Cappello J, Dressaire E, Stone H (2015) Damping of liquid sloshing by foams. Phys Fluids 27:022103

Scheidt R, Dingwell J, Mussa-Ivaldi F (2001) Learning to move amid uncertainty. J Neurophysiol 86:971–985

Selen L, Franklin D, Wolpert D (2009) Impedance control reduces instability that arises from motor noise. J Neurosci 29:12606–12616

Shadmehr R, Mussa-Ivaldi FA (1994) Adaptive representation of dynamics during learning of a motor task. J Neurosci 14:3208–3224

Slobounov SM, Moss SA, Slobounova ES, Newell KM (1998) Aging and time to instability in posture. J Gerontol Ser A Biol Sci Med Sci 53:B71–B80

Sternad D, Abe MO, Hu X, Muller H (2011) Neuromotor noise, error tolerance and velocity-dependent costs in skilled performance. PLoS Comput Biol 7:e1002159

Sternad D, Huber ME, Kuznetsov N (2014) Acquisition of novel and complex motor skills: stable solutions where intrinsic noise matters less. Adv Exp Med Biol 826:101–124

Svinin M, Goncharenko I, Luo Z-W, Hosoe S (2006) Reaching movements in dynamic environments: how do we move flexible objects? IEEE Trans Robot 22:724–739

Takahashi C, Scheidt R, Reinkensmeyer D (2001) Impedance control and internal model formation when reaching in a randomly varying dynamical environment. J Neurophysiol 86:1047–1051

Trommershäuser J, Gepshtein S, Maloney LT, Landy MS, Banks MS (2005) Optimal compensation for changes in task-relevant movement variability. J Neurosci 25:7169–7178

van der Linde R, Lammertse P (2003) HapticMaster—a generic force controlled robot for human interaction. Ind Robot Int J 30:515–524

Van Wegen E, Van Emmerik R, Riccio G (2002) Postural orientation: age-related changes in variability and time-to-boundary. Human Mov Sci 21:61–84

Venkadesan M, Guckenheimer J, Valero-Cuevas F (2007) Manipulating the edge of instability. J Biomech 40:1653–1661

Zatsiorsky VM (1998) Kinematics of human motion. Human Kinetics, Champaign, IL

Zhou K, Doyle JC (1998) Essentials of robust control. Prentice Hall, Upper Saddle River, NJ

# Part II
# Fifty Years of the Equilibrium-Point Hypothesis

# Fifty Years of Physics of Living Systems

**Mark L. Latash**

**Abstract** The equilibrium-point hypothesis and its more recent version, the referent configuration hypothesis, represent the physical approach to the neural control of action. This hypothesis can be naturally combined with the idea of hierarchical control of movements and of synergic organization of the abundant systems involved in all actions. Any action starts with defining trajectories of a few referent coordinates for a handful of salient task-specific variables. Further, referent coordinates at hierarchically lower levels emerge down to thresholds of the tonic stretch reflex for the participating muscles. Stability of performance with respect to salient variables is reflected in the structure of inter-trial variance and phenomena of motor equivalence. Three lines of recent research within this framework are reviewed. First, synergic adjustments of the referent coordinate and apparent stiffness have been demonstrated during finger force production supporting the main idea of control with referent coordinates. Second, the notion of unintentional voluntary movements has been introduced reflecting unintentional drifts in referent coordinates. Two types of unintentional movements have been observed with different characteristic times. Third, this framework has been applied to studies of impaired movements in neurological patients. Overall, the physical approach searching for laws of nature underlying biological movement has been highly stimulating and productive.

**Keywords** Equilibrium-point hypothesis · Motor equivalence · Variance · Tonic stretch reflex · Control theory · Neural computation · Hierarchical control · Synergy · Selective stability · Back-coupling · Equifinality · Referent coordinates

---

M.L. Latash (✉)
Department of Kinesiology, The Pennsylvania State University, Rec. Hall-267,
University Park, PA 16802, USA
e-mail: mll11@psu.edu

 
J. Laczko and M.L. Latash (eds.), *Progress in Motor Control*,
Advances in Experimental Medicine and Biology 957,
DOI 10.1007/978-3-319-47313-0_5

## Recent History of Physics of Living Systems

The idea that biological systems should be studied with the same scientific method as all other material objects in the Universe is rather old. The great physicist, Erwin Schrödinger wrote over 70 years ago a seminal book "What is life? The physical aspect of the living cell" (Schrödinger 1948). About 15 years later, a rather unique Department named *Physics of Living Systems* was created in the Moscow Institute of Physics and Technology (MIPT, known also as Fiztekh). This was a very small program with fewer than 10 students graduating each year. It did, however, produce a visible group of researchers in the field of motor control including Sergei Adamovich, Alexander Frolov, Yuri Ivanenko, Yuri Levik, Konstantin Popov, and a few others including the author of this Chapter, and Anatol Feldman—the creator of the equilibrium-point (EP) hypothesis.

From its inception, the physical approach to biological systems was mostly limited to problems at the molecular and cellular levels. Applications of this approach to whole organisms, up to humans, were all but nonexistent, with the field of movement studies dominated by descriptions of behavior and a few formal models that had no explicit links to laws of nature. One notable exception was biomechanics, with examples of successful application of the apparatus of classical mechanics to animal movements. However, even those examples applied existing, well-known laws of physics (mechanics) to biological objects without trying to discover new laws that would be specific to biological motion (cf. Gelfand 1991; Gelfand and Latash 1998). It has been obvious to any external observer that motion of biological objects is special. For example, knowing the mechanical parameters, initial conditions, and external forces is sufficient to predict motion of an inanimate object. Biological objects, however, are much less predictable; in particular, they frequently move uphill, fly against the wind, and swim against the current.

At the time of its creation, the name of the mentioned Fiztekh Department was largely a promise that at some time in future physics of living systems would become reality. In the subjective opinion of the author, it became reality in 1965 with the publication of the first in the series of three papers (Asatryan and Feldman 1965; Feldman 1966a, b) that introduced the basics of the EP-hypothesis. The approach accepted in those studies amounted to searching for parameters (invariants) of unknown laws that linked physical variables such as muscle length and force. Based on the accumulated experimental foundation, Feldman suggested that muscle force and length were linked by the mechanism of the tonic stretch reflex with only one parameter that could be modified by the central nervous system, namely threshold ($\lambda$) of the tonic stretch reflex.

Currently, two approaches compete in the field of motor control, physical (based on laws of natural science) and computational (based on engineering and control theory). The former approach tries to discover laws of nature and their salient parameters that dictate the production of natural movements. The latter approach assumes that the brain is a computational device that solves the complex problems of interactions among the numerous elements within the body and between the body and the environment (e.g., in the form of internal models, Wolpert et al. 1998; Kawato 1999; Shadmehr and Wise 2005). The postulate of neural computation is rejected by the physical approach as incompatible with natural science. On the other hand, the importance of the computational approach for such fields as robotics may be viewed as well established.

## Living Systems and Parameters of Relevant Physical Laws

Within the physical approach, one may define a *living system* as a system able to: (1) unite basic physical laws, those common across animate and inanimate object, into chains and clusters leading to new stable and pervasive relations among physical variables—*new physical laws*—involving new parameters; and (2) modify these parameters in a purposeful way. In other words, inanimate systems are slaves of the basic physical laws, while living systems create new physical laws and then modify parameters of those new laws to achieve their goals. Within this definition, the EP-hypothesis was revolutionary in its formulation of a new law of physics specific for a living muscle with its reflex connections and identification of a salient parameter for that law, $\lambda$.

It took 50 years for the EP-hypothesis to become recognized as one of the (or arguably the only) hypothesis in the field of motor control that makes it a subfield of physics of living systems. Over this time, it had to deal with numerous criticisms. It was at some point fashionable to criticize or even disprove the EP-hypothesis (Gottlieb 1998; Popescu and Rymer 2000; Hinder and Milner 2003; for review see Feldman and Latash 2005; Feldman 2015). All these criticisms were based on misinterpretations of some of the basic postulates of the EP-hypothesis and/or accepting its simplified (and hence incorrect) versions, making predictions, and then showing them to be false. The fashion to disprove the EP-hypothesis has fortunately passed. Now trying to prove or disprove this hypothesis may be safely viewed as waste of time. The true task is developing the hypothesis and there are plenty of actively explored directions of such developments including generalization of the EP-hypothesis to natural multi-muscle movements, linking it to perception, applying it to disordered movements, and, possibly most importantly, linking it to neurophysiological mechanisms.

Generalization of the EP-hypothesis to the control of natural multi-muscle movements has led to the emergence and development of the idea of control with referent coordinates (RC) sometimes addressed as the RC-hypothesis (reviewed in Feldman 2015). This is a quickly developing field, and the following brief and simplified description may be viewed as a snapshot of the current view on the control of natural movements.

Any goal-directed action starts with the formulation of a goal, which may be viewed as a relatively low-dimensional set of salient coordinates (Task in Fig. 1). For example, a pointing movement to a target involving numerous muscles may be associated with setting only three parameters (RCs) for the pointing effector (e.g., the tip of the index finger) corresponding to the three coordinates of the target in space. Implementation of the action requires setting RCs for individual joints and muscles leading to a series of few-to-many transformations (the upper level and lower level transformations only are shown in Fig. 1). Such transformations have been commonly addressed as redundant (Prablanc et al. 2003; Karniel 2011), while recently, a different term has been suggested—abundant (Latash 2012)—reflecting a particular attitude to the problem of motor redundancy (Bernstein 1967). According to the principle of motor abundance, apparently redundant transformations are not the sources of computational problems for the CNS (solved, for example, with optimization approaches, Prilutsky and Zatsiorsky 2002) but useful, rich designs that allow ensuring stability of salient performance variables in the presence of unpredictable changes in both intrinsic body states and external forces, and necessity to perform secondary tasks using the same sets of elements (reviewed in Latash et al. 2007; Latash 2008).

This chapter addresses the recent progress in some of the areas within this general scheme of control. First, I will review the links of the general scheme of control with RCs to such notions as hierarchical control, theory of synergies, and

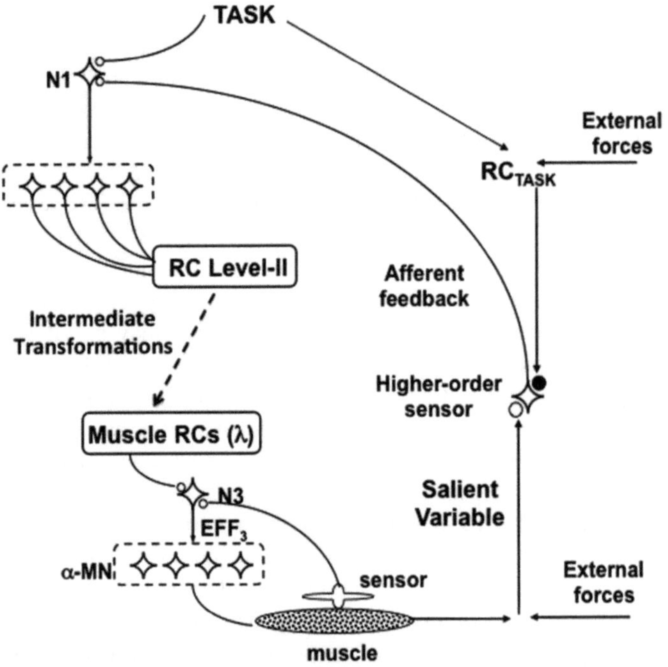


**Fig. 1** A multi-level scheme for the production of a natural movement with several few-to-many mapping organized in a synergic way. Action starts with defining a low-dimensional set of referent coordinates for task-specific salient variables, $RC_{TASK}$. It leads to sets of λs for the involved muscles compatible with the $RC_{TASK}$. At each step, the feedback may be organized similarly to the central back-coupling scheme in Latash et al. (2005)

uncontrolled manifold (UCM) hypothesis. Then, a few recent results are reviewed related to such issues as synergies within spaces of hypothetical control variables (such as *r*- and *c*-commands within the EP-hypothesis, Feldman 1980, 1986), the nature of unintentional movements and drifts in force observed when a person stops receiving visual information on an ongoing action (e.g., Slifkin et al. 2000; Vaillancourt and Russell 2002; Shapkova et al. 2008; Ambike et al. 2015a), and attempts at translating the recent progress into clinical studies (reviewed in Latash et al. 2010; Latash and Huang 2015).

## Hierarchical Control of Actions by Abundant Systems

The hierarchical scheme in Fig. 1 is readily compatible with the idea of synergies as neural organizations that lead to stabilization of salient performance variables by covaried adjustment within an abundant set of elemental variables, i.e., the outputs of elements (reviewed in Latash et al. 2002, 2007). This is not the only meaning of the word *synergy* in movement science. At least two more meanings are used broadly in the literature. Among clinicians, *synergy* frequently has a negative connotation; it implies stereotypical patterns of muscle activations (forces and joint rotations) interfering with purposeful movements that are commonly seen in patients after stroke (DeWald et al. 1995). In motor control literature, *synergy* frequently means something like "a set of variables showing parallel changes" (d'Avella et al. 2003; Ivanenko et al. 2004; Ting and Macpherson 2005). Uniting elemental variables into such synergies has been viewed as a means of alleviating the problem of motor redundancy by decreasing the number of variables manipulated by the CNS.

One of the crucial features of natural movements, which allow them to be efficient in the changing environment, is their stability. This term has been used in the motor control literature in different meanings. Here, under stability, I imply ability of a time-varying (dynamic) system to return to a state or trajectory following a small, transient perturbation. Gregor Schöner (1995) introduced the important notion of *task-specific stability*, which is a characteristic of biological systems but not of inanimate objects. This notion implies that a multi-element system is able to reorganize interactions among elements leading to stability of some performance variables but not others, depending on the task and intention of the actor. This notion is applicable to abundant systems only.

The introduced definition of stability suggests that there is only one method to assess stability directly, that is, to apply a small, transient perturbation and quantify behavior of the system. While this method seems straightforward, its application to the human motor system is far from trivial because of the poorly controlled phasic reflexes and reflex-like reactions to perturbations (Hammond 1955; Tatton et al. 1978; reviewed in Shemmell et al. 2010). Two indirect methods have been developed. The first is based on the idea that, over a set of repetitive trials, a dynamic system deviates from average trajectory more in directions of low stability and less in deviations of high stability. As a result, computing variance across trials at comparable phases is expected to show lower values in relatively stable directions. If a particular performance variable is selectively stabilized, variance within a subspace that leads to no change in that variable (its uncontrolled manifold, UCM, Scholz and Schöner 1999) is expected to be relatively large as compared to variance in the orthogonal to the UCM directions (ORT). The inequality $V_{UCM} > V_{ORT}$ has been used as a signature of a synergy stabilizing that performance variable, sometimes reduced to a single metric, an index of synergy $\Delta V = (V_{UCM} - V_{ORT})/V_{TOTAL}$, where all variance indices are computed per dimension in the corresponding spaces.

Another indirect method of assessing stability in different directions of a multi-element system is based on the notion of quick corrective actions as brief perturbations into the system. If a person is asked to perform a quick corrective action, much of the motion is expected to be in directions of lower stability. For example, imagine that you hold a small metal spring from a typical pen between the thumb and the index finger. If you now try to squeeze the spring quickly, very likely it will buckle and jump away, i.e., move not in the desired direction but in an orthogonal direction (cf. Venkadesan et al. 2007; Lawrence et al. 2014). This is due to the fact that the spring is relatively stable along its main axis and is relatively unstable in orthogonal directions. If the CNS stabilizes a performance variable selectively, an attempt to change this variable quickly is expected to lead to large deviations in directions of low stability, i.e., within the UCM for that variable. Such deviations have been addressed as *motor equivalent* (ME). A number of studies have documented very large amounts of ME action during quick corrections of actions despite the fact that ME action by definition is inefficient in correcting a deviation of the performance variable (Mattos et al. 2011, 2013, 2015). Note that both ΔV and ME indices are proxies of stability.

Mechanisms of selective stability (mechanisms of synergies) are basically unknown. A number of schemes have been suggested including both computational (such as optimal feedback control, Todorov and Jordan 2002) and neurophysiological (such as central back-coupling, Latash et al. 2005) schemes. A scheme uniting the ideas of back-coupling and of control with RCs has been suggested (Martin et al. 2009). Indeed, irrespective of specific realization, the idea of synergic control is naturally compatible with hierarchical control with RCs: If an action is initiated with specifying a low-dimensional task-specific set of RCs at a hierarchically high level (with some, relatively low, inter-trial variance), trajectories of corresponding performance variables are expected to be stabilized against intrinsic noise (e.g., associated with deviations of RCs at lower levels, see Fig. 1) and changes in external forces. The latter are due to the length- and velocity-dependence of muscle force (which is universal, not task-specific) and may get contributions from other, learned, feedback mechanisms.

While action stability is vital, too much stability may be detrimental, for example, if a person tries to produce a very quick change in a performance variable on the background of a steady-state action. Recent studies have shown that humans possess a feed-forward mechanism that allows changing stability of salient performance variables (as reflected in indices of corresponding multi-element synergies) in preparation to a quick action or reaction, *anticipatory synergy adjustments* (ASAs, Olafsdottir et al. 2005; Shim et al. 2005). ASAs are seen as a drop in the index of synergy stabilizing a performance variable 200–300 ms prior to a planned quick change in that variable. This purposeful destabilization of a variable facilitates its quick change in any direction, even in situations when the direction of change in not known in advance (Zhou et al. 2013).

## Synergies in the Space of Referent Coordinates

Since the original publication (Scholz and Schöner 1999), the UCM hypothesis and the associated method of analysis of the structure of variance have been used actively to explore synergies in spaces of elemental variables such as joint rotations, digit forces, and muscle groups (M-modes). The formulation of the problems in those studies has always been deficient: They searched for synergies in spaces of elemental performance variables, not control variables. Merging the ideas of control with RCs with the UCM-based analysis was elusive until recently. This was partly due to the fact that measuring RCs (e.g., λs for individual muscles) is by itself a challenging task. Attempts at reconstructing time profiles of RCs required multiple trials and were built on simplified mechanical models of the moving effectors (Latash and Gottlieb 1991; Latash 1992; Ambike et al. 2015b), which were criticized as inadequate (Gribble et al. 1998). Recently, however, a method has been developed that promises to overcome this limitation (Ambike et al. 2016b).

Consider the control of a joint spanned by two muscles acting in opposite directions, an agonist–antagonist pair (Fig. 2a). The neural control of each muscle may be described with a single parameter, λ, threshold of the tonic stretch reflex. For a given λ, there is a dependence of active muscle torque on joint angle (shown as curved lines in Fig. 2). Note that λ defines neither muscle force, nor its length, nor its activation level: All three change along the muscle characteristic. Joint behavior will be defined by its torque-angle characteristic, which is the algebraic sum of the two muscle characteristics. Actual joint torque and position in equilibrium will be defined by interactions of the joint with the external load. The pair of control variables $\{\lambda_{AG}, \lambda_{ANT}\}$ may be substituted with an equivalent pair $\{r; c\}$ (Feldman 1980). The *r*-command (reciprocal command) corresponds to unidirectional shifts in the two λs; such shifts favor activation of one muscle within the pair compared to the other muscle. The *c*-command (coactivation command) leads to counter-directional changes in the λs leading to parallel changes, an increase or a decrease, in the activation of both muscles.

The $\{r; c\}$ pair of control variables can be generalized for any effector. For example, the neural control of action by a fingertip along one dimension can be described with two variables, *R* and *C*, which are analogous to the *r*- and *c*-commands but represent the action of all the muscles that contribute to the fingertip action (Fig. 2b). One of the commands, *R*, defines a referent coordinate for the

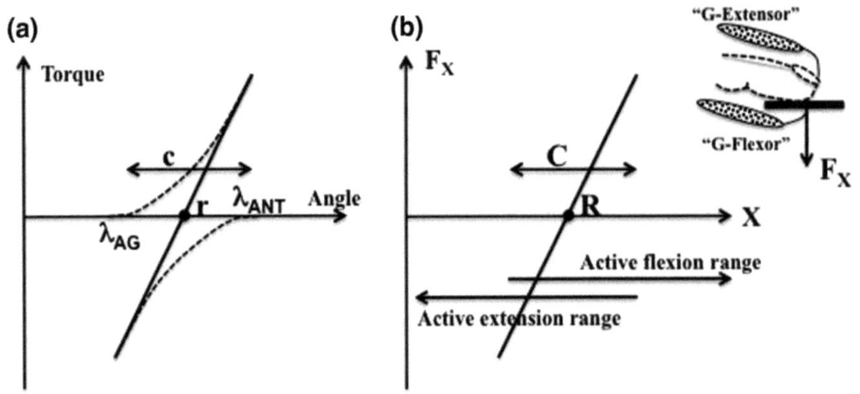


**Fig. 2** **a** The torque-angle characteristics for two muscles, agonist and antagonist, acting at a joint. Their control can be described with two λs, $\lambda_{AG}$ and $\lambda_{ANT}$ or with two equivalent commands, *r*-command and *c*-command; **b** The control of a fingertip can be described with two equivalent commands, $RC_{FT}$ (referent coordinate) and $k_C$ (apparent stiffness)

fingertip, i.e., a spatial coordinate, at which it produces zero net force on the environment. If the fingertip is kept away from its RC (for example, by a stop), it would produce a net force on the stop, and the force magnitude ($F_{FT}$), in a linear approximation, would depend on the difference between the fingertip actual coordinate ($AC_{FT}$) and $RC_{FT}$ with a scaling coefficient $k_C$, apparent stiffness of the fingertip: $F_{FT} = k_C(RC_{FT} - AC_{FT})$. The coefficient $k_C$ reflects the other basic command, *C*-command.

Figure 3a illustrates the task of accurate force production. When a person is asked to produce a particular force level by pressing on a stop with the fingertip, an infinite number of $\{RC_{FT}; k_C\}$ combinations can be used to produce the required force (panel A in Fig. 3). So, the apparently nonredundant task of fingertip force production is in fact redundant (abundant!) at the control level. There is a hyperbolic line corresponding to a required force level (Fig. 3b). This line is the UCM for this task in the space of elemental control variables or, to be more exact, in the space of mechanical variables that may be viewed as direct reflections of the hypothetical neural variables.

Now one can formulate a question: Do humans use the abundance afforded by the presence of two commands, *R* and *C*, during single-finger accurate force production tasks? Are there $\{R; C\}$ synergies stabilizing the fingertip force reflected in the $\{RC_{FT}; k_C\}$ synergies? This is not a trivial question. Indeed, deviations from the prescribed force level depend only on deviations from the UCM in Fig. 3 and not on deviations along the UCM. So, one can perform the task very accurately, i.e., with small deviations from the UCM with equal deviations along the UCM, larger deviations, or smaller deviations. This is illustrated in the right panel of Fig. 3, which shows three data distributions with about the same dispersion orthogonal to the UCM (about the same accuracy of performance). They are associated with

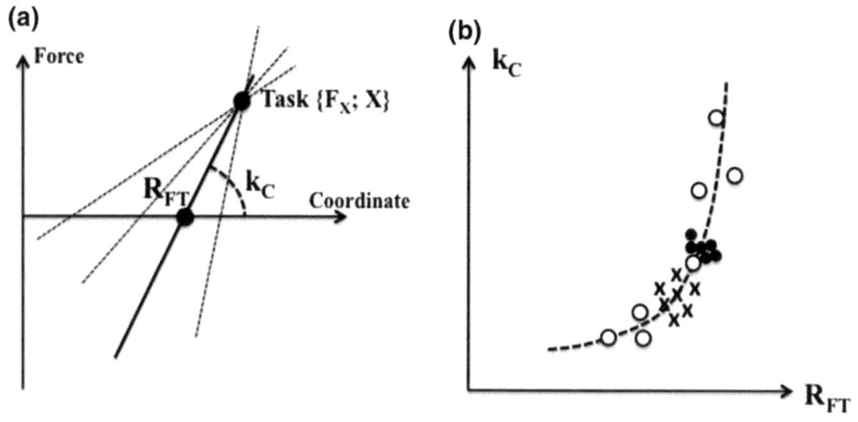


**Fig. 3** **a** The solution space (UCM) for the task of accurate fingertip force production on the $\{RC_{FT}; k_C\}$ plane; **b** A close-up showing possible data distributions compatible with a particular level of accuracy in the task. Note that deviations from the UCM are about the same for the three data distributions, while deviations along the UCM may be smaller (*crosses*), about the same (*open circles*), or larger (*filled circles*) that deviations from the UCM. Distributions similar to the one illustrated with filled circles were observed in experiments

smaller (filled circles), about equal (crosses), and much larger (open circles) dispersion along the UCM.

So far, only one experimental study explored $\{RC_{FT}; k_C\}$ synergies during fingertip force production (Ambike et al. 2016b). In that study, the subjects pressed on a force sensor and produced fingertip force to match a visual target. Then, the "inverse piano" (Martin et al. 2011) device was used to lift the finger smoothly over a small amplitude (1 cm at 2 cm/s), while the subject was instructed not to interfere with possible fingertip force changes. Visual feedback was turned off to help the subject not to correct the force changes that emerged naturally when the finger was lifted. Finger force increased during the lift in proportion to the change in the fingertip coordinate. Note that, according to the idea of control with RCs, in the initial state, a combination of $RC_{FT}$ and $k_C$ led to the actual force generated by the fingertip and measured by the force sensor. Assuming that the subject did not change the neural commands, the finger lift led to a change in the difference between the $RC_{FT}$ and $AC_{FT}$ and, as a result, to a proportional change in the force. Using linear regression between the force and coordinate allows reconstructing $RC_{FT}$ and $k_C$ as the intercept and the slope of the regression line. Of course, the method is based on several assumptions including the linear relation between the fingertip force and coordinate (supported by the observed linear regressions with $R > 0.9$) and the questionable extension of the regression line beyond the range of observations.

Experiments have shown that subjects indeed use relatively broad ranges of both $RC_{FT}$ and $k_C$, and the two are nearly perfectly covarying to keep force deviations from the desired value low (as in the distribution with filled circles in Fig. 3b). The covariation was demonstrated using a randomization method (cf. Müller and Sternad 2003) when values of the two variables were selected randomly from different trials, hence no covariation existed in such surrogate data sets, and variance of total force from the surrogate data sets was computed. It was two orders of magnitude larger than force variance in the original data set demonstrating that the randomization removed very strong covariation between RC and $k_C$ directed at keeping force variance low.

To our knowledge, this is the first demonstration of synergies in the space of elemental variables that are close to the hypothetical neural variables within the hypothesis on control with RCs. Of course, the method is crude and relies on assumed linear behavior of the effector within a small range of perturbations. However, this method is a promising first step towards developing analysis of synergies within spaces of control variables.

## Unintentional Movements as Drifts of Referent Coordinates

According to the EP-hypothesis, transient changes in external forces should not lead to changes in the final steady state of a moving effector as long as the subject does not introduce corrections and the muscle force-generating properties remain unchanged (reviewed in Feldman 2015). This prediction has been addressed as *equifinality*. Equifinality was confirmed in several studies with transient force changes (Bizzi et al. 1976; Kelso and Holt 1980; Schmidt and McGown 1980; Latash and Gottlieb 1990), while other studies reported violations of equifinality during movements in unusual environments such as during arm movements performed while sitting in a rotating centrifuge or moving in an artificial force field with negative damping (Lackner and DiZio 1994; DiZio and Lackner 1995; Hinder and Milner 2003). Within the framework of the EP-hypothesis, violations of equifinality suggest that the subject was unable to follow the instruction "do not intervene" and unintentionally shifted RCs for the moving effector (Feldman and Latash 2005). In particular, Archambault et al. (2005) suggested that humans were unable to tolerate perturbations leading to muscle stretch and adjusted commands ($\lambda$) to those muscles unintentionally. This interpretation has been developed recently based on examples of violations of equifinality in more typical experimental conditions, without destabilizing force fields.

In fact, unintentional drifts in motor performance have been known for some time. Imagine a person who is asked to produce force against a stop (e.g., to press with a finger on a force sensor) and match a target force level using visual feedback. If the visual feedback is turned off and the subject is instructed to keep producing the same force level, the actual force typically shows a slow decline with the time constant of between 10 and 20 s (Vaillancourt and Russell 2002; Ambike et al. 2015a). The subject is unaware of the force drift, even if it leads to a substantial force drop, >30 % from the original level. A similar drop in grip force was seen in experiments when subjects moved a handle up and down under anesthesia of cutaneous receptors in the hand (Augurelle et al. 2003) and when the handle grasped by the subject expanded and contracted very slowly (Ambike et al. 2014). The drop in force means that there is a drift in RC for the fingertip and/or a drift in the apparent stiffness $k_C$. This force drop also illustrates the limitations of the somatosensory system in providing feedback on kinetic variables such as fingertip force.

The instruction "continue doing what you have been doing" implies that the subject should not change the intentional input into the hierarchically lower structures involved in the task performance. Removing visual feedback is another method to avoid corrective changes in this descending input. Spontaneous drift in performance under such conditions is a reflection of natural behavior of the physical system (including both neural and muscular components) that participates in the performance. The aforementioned unintentional force drift suggests that this physical system drifts toward a state with minimum of its potential energy (as all physical systems do).

More examples of unintentional changes in RCs leading to violations of equifinality were observed in experiments with external transient perturbations applied on the background of static tasks, such as holding a hand position against a force vector (Zhou et al. 2014, 2015a, b) and producing a constant force by the four fingers of a hand (Wilhelm et al. 2013; Reschechtko et al. 2014, 2015). These examples form two groups. First, a transient change in the external force applied during the positional task leads to relative equifinality in the final hand position and orientation, but equifinality is violated at the level of joint configurations (Zhou et al. 2014). In other words, if the external force increases and then decreases smoothly, the hand comes to a new position, which is more or less the same as the initial hand position. In contrast, the joint configuration shows large variance, which is compatible with the unchanged hand position and orientation. In the multi-finger tasks, lifting and lower a finger (a transient positional perturbation) during the steady-state phase of the task leads to relatively low variance in the total force while sharing of the force among the four fingers shows large variability.

Taken together, these studies show that application of external transient perturbations leads to relative equifinality at the task level but not at the level of elemental variables. This is not a completely unexpected result given the existence of synergies (in a sense $V_{UCM} > V_{ORT}$) stabilizing the task-specific variables, such as hand position and orientation in the kinematic task and total finger force in the kinetic task (cf. Wilhelm et al. 2013; Zhou et al. 2014). Such synergies reflect lower stability of the involved elemental variables in directions leading to no change in the task-specific performance variables, i.e., along the corresponding UCM. Naturally, a perturbation is expected to lead to larger deviations of the elemental variables along the directions of low stability; this is true for both segments of a transient perturbation, its application, and removal. Hence, these results confirm directly (i.e., with perturbations) one of the main postulates of the UCM-based method, i.e., that the inequality $V_{UCM} > V_{ORT}$ indeed reflects different stability of the system along the UCM and ORT subspaces.

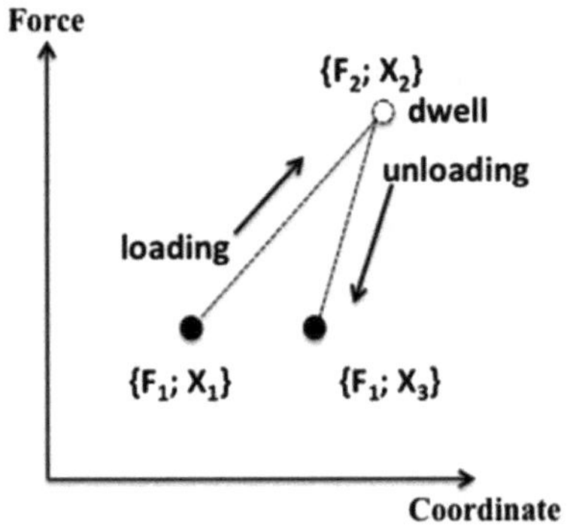


**Fig. 4** Schematic illustration of the main results in an experiment with transient perturbations interrupted by a brief dwell time. In the initial state, the subject produces active force against an external load ($F_1$) at a certain coordinate ($X_1$). An increase in the load to $F_2$ leads the hand to a new coordinate ($X_2$) where the system is again at equilibrium. After a few seconds, the load returns to $F_1$, the hand moves toward the initial coordinate but stops short of $X_1$ ($X_3$). Note the different slopes of the trajectories during the application and removal of the force perturbation

Another example of violations of equifinality was observed at the task level. Such violations take place if a transient perturbation is interrupted by a dwell time interval between the perturbation application and removal (Zhou et al. 2014, 2015a, b). This process is much faster than the slow force drift observed without perturbations, with a time constant of about 1 s. Such effects were observed both in multi-joint positional tasks with the application of force perturbations and in multi-finger pressing tasks under the application of positional perturbations with the help of "inverse piano." These processes show that a perturbation triggers a drift of the RC at the task level toward the new actual coordinate of the effector. The drift is never complete such that, when the perturbation is removed, the effector moves back to its initial state but with an undershoot that may cover up to 50 % of the displacement (force change) of the effector caused by the perturbation (Fig. 4).

## Voluntary, Involuntary, and Unintentional Movements

Within the EP (RC) hypothesis, there are two types of movements, voluntary and involuntary. The former are associated with shifts of the RC at the task level (λ at the single-muscle level); their mechanical effects and muscle activation changes depend on both the RC shift and the external force field. For example, the same shift of λ can lead to a movement in isotonic condition, to force generation in isometric conditions, and to both in intermediate loading conditions. Involuntary movements are associated with changes in the external load without a change in RC. Typical examples include joint motion in response to unloading as in the classical experiments of Feldman (1966a).

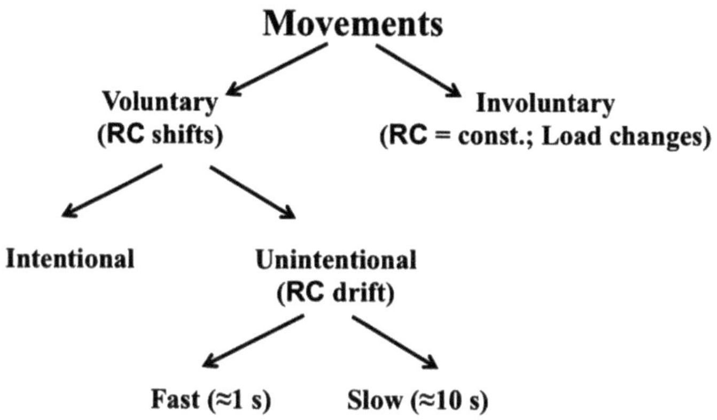


**Fig. 5** Within the EP (RC) hypothesis, changes in RC lead to voluntary movements (and/or active force production). If external forces (load) change while RC stays unchanged, movements are also observed (involuntary movements). Recent studies have suggested the existence of unintentional voluntary movements, i.e. those associated with a change in RC, which happens without subject's intention. There are two types of such movements, slow (with characteristic times of ≈ 10–20 s) and fast (with characteristic times of ≈ 1 s)

The aforementioned examples of RC drifts that happen when the subject is trying not to change the RC suggest that there is one more class of movements. They may be called voluntary because they are associated with RC changes, but they are unintentional (Fig. 5). There are two classes of such movements that differ rather dramatically in the characteristic times, fast ($\tau \approx 1$ s) and slow ($\tau \approx 10$–20 s). If we assume that these drifts reflect natural behavior of the physical (including physiological) system involved in the task, they suggest the existence of two relaxation processes with such characteristic times.

This hypothesis maps well on the idea of two subspaces, UCM and ORT, with different characteristic stability and, correspondingly, different characteristic times of processes within those subspaces. Indeed, processes within the UCM are expected to be slow (low stability) while processes within ORT are expected to be fast (high stability). Note that even a seemingly nonredundant task (for example, force production by a single finger) is abundant at the "hidden" levels such as, for example, the level of the involved joints, muscles, motor units, etc. So, the notions of UCM and ORT are applicable to such tasks. During steady states, drifts are mostly limited to the UCM and, hence, are slow. In contrast, perturbations move the system within the ORT space and trigger fast unintentional movements.

Both unintentional drifts, fast and slow, have been observed for task-related performance variables, i.e., within the ORT space. The idea that the slow drift originates within the UCM suggests that the two subspaces, UCM and ORT, are coupled such that a drift in one leads to a drift in the other at a comparable rate. This hypothesis has not been tested experimentally.

There are several observations that keep the offered description incomplete. For example, force drift during low initial forces was observed in the opposite direction, i.e., toward higher forces (Ambike et al. 2015a). When a subject is asked to perform not a steady-state task but a cyclical task, turning visual feedback off reveals both drifts: A slow drift in the midpoint of the force cycle to lower values and a fast drift of the amplitude of the cyclical force changes towards higher values (Ambike et al. 2016a). When a perturbation moves the system not away from its RC but towards the RC, no visible drift is seen (Qiao et al. 2015). Taken together, these observations suggest that unintentional movement characteristics are defined by a number of factors that may include, in particular, properties of the somatosensory system, preferred (maybe optimal in some sense) magnitude of performance, and reactions of the muscle to lengthening and shortening perturbations (cf. Archambault et al. 2005). This is an exciting field for future experimental and theoretical studies.

## Translation into Clinics: Muscle Tone and Spasticity

The EP (RC) hypothesis offers a fruitful and consistent framework based on natural science to analyze movement disorders in a variety of neurological patients. One of the most challenging and important tasks of researchers is to define terms used by clinicians in a way that allows incorporating them into a general scheme of the neural control of movement. For example, such commonly used terms as increased or decreased *muscle tone*, impaired *joint coordination*, impaired *postural or movement stability*, and others should be defined explicitly and operationally; the last word means that methods of objective quantitative assessment of the corresponding phenomena have to be developed.

One of the commonly used notions in clinical movement studies is *muscle tone*. A detailed discussion of this notion has been published recently (Latash and Zatsiorsky 2016). Here I would only like to mention the definition proposed by Bernstein and Kots (1963) who emphasized the importance of reflexes for muscle tone and linked the notion of muscle tone to tuning the neuromotor apparatus to tasks of active postural or movement control. In contrast, consider how muscle tone is measured during clinical examinations (the person is asked to relax and the effector is moved over its range of motion) or tools that are claimed to measure muscle tone (applying deformation orthogonal to the surface of the body and measuring the resistive force and deformation). None of these methods addresses the Bernstein understanding of muscle tone as a contributor to active movement and postural tasks, since the subject is always asked to relax and do nothing.

Consider Fig. 6, which illustrates the behavior of a muscle during its passive stretch. If in the original state the muscle is relaxed at a particular length $L_0$, its threshold of the tonic stretch reflex value ($\lambda$) is undefined. We only know that $\lambda > L_0$. For $\lambda_1 \approx L_0$, any motion stretching the muscle would trigger its activation and relatively strong resistance. For $\lambda_3$, which is beyond the anatomical range of muscle length, motion over the whole range would lead to only weak resistance due to properties of passive tissues. For an intermediate $\lambda_2$, the resistance would be weak until muscle length reaches $\lambda_2$ and then will increase substantially. Should one

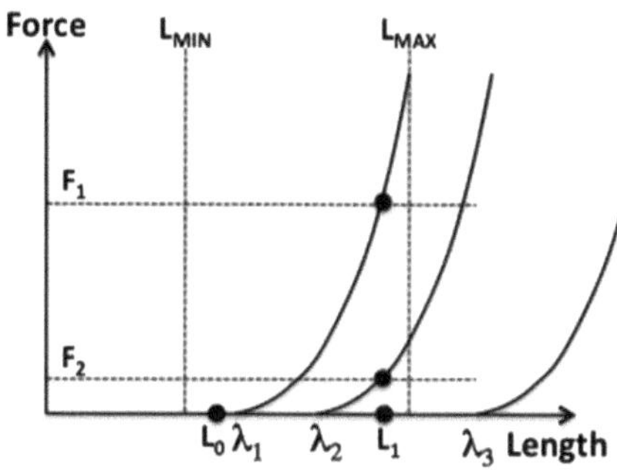


**Fig. 6** A scheme of the relation between muscle active force and length. If the muscle is relaxed, its tonic stretch reflex threshold (λ) is longer than its actual length ($L_0$). Passive stretching the muscle to $L_1$ may lead to its different resistance depending on the distance between λ and $L_0$ $F_1$ for $\lambda_1$, $F_2$ for $\lambda_2$, and $F = 0$ for $\lambda_3$). Muscle tone may be associated with the distance between $L_0$ and λ. Passive length-dependent resistance of the tissues is ignored

call the case with $\lambda_1$ hypertonic, the case with $\lambda_3$ hypotonic, and the case with $\lambda_2$—normal tone? This does not seem to make sense because all three subjects in this experiment were healthy and simply interpreted the instruction "to relax" differently. Does it make sense to associate the notion of muscle tone with the distance from actual muscle length to λ when the subject is instructed to relax? This question is open to research and discussion among clinicians.

The framework of the EP-hypothesis has been effectively used in the analysis of spasticity, a common motor disorder following spinal cord injury, stroke, multiple sclerosis, and other neurological impairments. Traditionally, spasticity has been associated with so-called positive signs (increased muscle tone and uncontrolled contractions) and negative signs (weakness up to complete paralysis) (Landau 1974). A number of more recent studies have suggested, however, that the positive and negative signs interact with each other. For example, effective suppression of uncontrolled contractions, e.g., with intrathecal baclofen, is associated with unmasking of better voluntary movements (Corcos et al. 1986; Latash et al. 1990; Latash and Penn 1996).

These and other observations led to a scheme that considers spasticity as impairment in the control of control of RCs within the whole natural range (Levin and Feldman 1994; Musampa et al. 2007; Calota and Levin 2009). At the muscle level, this means impaired control of shifts of λs of affected muscles. A healthy person is able to change λ beyond the anatomical range of muscle length values, $\{L_{MIN}; L_{MAX}\}$. This allows producing high forces at short muscle length values ($\lambda < L_{MIN}$) and relaxing the muscle when it is stretched to its maximal length ($\lambda > L_{MAX}$) (Fig. 7). Patients with spasticity are assumed to have a limited range of λ changes, $\{L_-; L_+\}$. This limits the voluntary control of the muscle to the narrow range of length values; at shorter muscle length ($L < L_-$), the muscle is paralyzed, while at longer muscle length ($L > L_+$), the muscle shows involuntary contraction. This scheme leads to multiple questions regarding, for example, changes in the

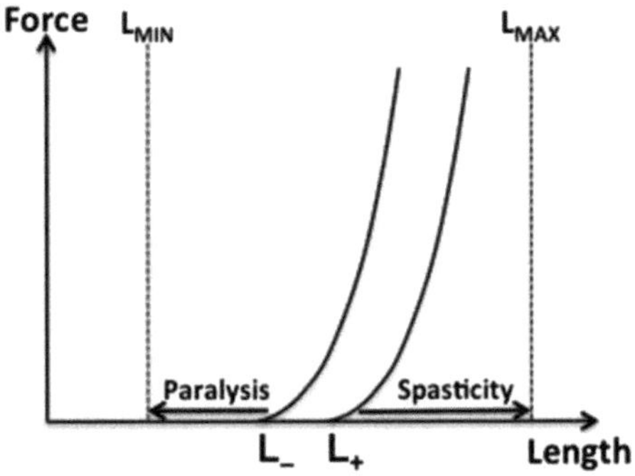


**Fig. 7** An illustration of a spastic muscle. The subject lost his/her ability to move the tonic stretch reflex threshold (λ) over its full range. If the range of λ is limited by $L_-$ and $L_+$ it is impossible to activate voluntarily the muscle when its length is shorter than $L_-$. It also becomes impossible to relax the muscle when its length is longer than $L_+$. The biomechanical range of muscle length changes in shown with vertical dashed lines, $L_{MIN}$ to $L_{MAX}$

effects of stimulation of somatosensory receptors on λ shifts in spasticity (leading, for example, to the withdrawal reaction), interactions leading to cyclical changes in unintentional muscle activity (such as in clonus), etc.

## Translation into Clinics: Impaired Control of Stability

Another aspect of movement disorders that has recently been studied actively within the physical approach is disorders of movement stability (reviewed in Latash and Huang 2015). Note that stability of natural actions is crucial for their success given the changing unpredictable environment and intrinsic states of the body. As a result, loss of stability of motor patterns leads to major impairments in the motor function. Well-known examples include ataxia in patients with cerebellar disorders and loss of postural stability in Parkinson's disease (reviewed in Fahn and Jankovic 2007).

The framework of the UCM hypothesis has been used to quantify task-specific stability in a number of neurological disorders across a variety of motor tasks, from multi-finger accurate pressing and manipulation (Park et al. 2012, 2013) to multi-joint reaching (Reisman and Scholz 2003) and multi-muscle whole-body actions (Falaki et al. 2016). Overall, these studies revealed strikingly different changes in stability of motor patterns in patients with cortical and subcortical disorders.

Patients with subcortical disorders including those with Parkinson's disease (Park et al. 2012; Jo et al. 2015), multi-system atrophy (Park et al. 2013), and multiple sclerosis affecting subcortical loops (Jo et al. 2016a) showed significantly decreased indices of motor synergies during steady-state phases and significantly reduced and delayed anticipatory synergy adjustments (ASAs) in preparation to quick actions. These two aspects of impaired synergic control (impaired control of stability, ICS) have been referred to as impaired stability and impaired agility, respectively (Latash and Huang 2015).

ICS could be seen in patients with Parkinson's disease at early stages. In particular, in patients at stage-I of Hoehn-Yahr (Hoehn and Yahr 1967), which is characterized by clinical signs of PD limited to only one side of the body, changes in the synergy index and ASAs during multi-finger tasks were seen in both hands (Park et al. 2012, 2014). Patients at stage-II, defined as the stage with no clinical signs of postural instability, show decreased indices of multi-muscle synergies and decreased ASAs seen during whole-body postural tasks (Falaki et al. 2016). Taken together, these observations suggest that ICS, as reflected by changes in synergic control, can be seen in tasks and body parts that show no clinically detectable signs of the disease. This is a potentially important set of results promising a sensitive behavioral biomarker of Parkinson's disease. It is also important that indices of impaired synergic control are sensitive to dopamine-replacement treatment of Parkinson's disease (Park et al. 2014) and to deep brain stimulation (Hang Jin Jo et al., unpublished).

In contrast to the pronounced ICS in patients with subcortical disorders, the very first studies of multi-joint reaching in persons after cortical stroke led to an unexpected result (Reisman and Scholz 2003). The motor patterns during movements of the contralesional (strongly impaired) arm were clearly different compared to those of the ipsilesional (relatively unimpaired) arm. However, the structure of inter-trial variance quantified in the joint configuration space with respect to the endpoint trajectory was similar between the two arms. So, the impairment of motor patterns was not accompanied by impaired stability of the movement. A later study confirmed these results for multi-finger accurate force production tasks (Jo et al. 2016b). Surprisingly, ASAs were changed (delayed and reduced) after cortical stroke in both ipsilesional and contralesional hands (Fig. 8). So, stroke patients show only one aspect of ICS, namely impaired agility, while stability of their movement patterns by synergies does not seem to be affected.

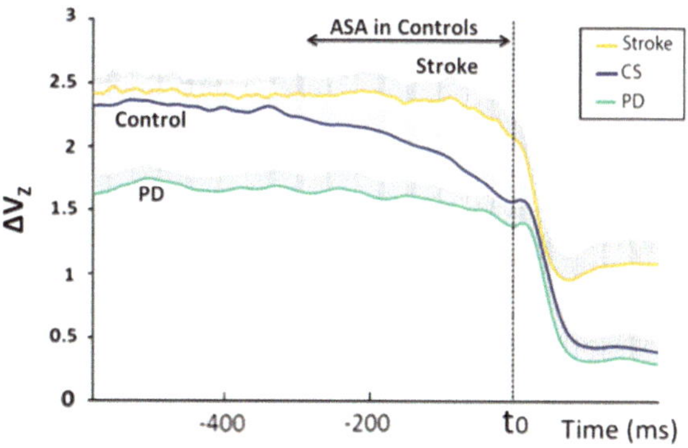


**Fig. 8** Illustration of the synergy index ($\Delta V_Z$, log-transformed) in patients with Parkinson's disease (PD), stroke (ST), and healthy controls (HC). The subjects performed the task of accurate force production with four fingers pressing on individual force sensors followed by a force pulse into the target under visual feedback initiated at time t0

Let us try to interpret these observations within the EP (RC) hypothesis framework. Recent studies using transcranial magnetic stimulation (TMS) during single-joint motor tasks have suggested that the corticospinal tract signals are encoding not forces, coordinates, or muscle activation levels, but changes in λs for the involved muscles (Raptis et al. 2010; Sangani et al. 2011). It is reasonable to generalize these findings to natural movements and assume that corticospinal signals encode RC values for the task-specific variables. Multi-element synergies stabilizing salient performance variables are associated with the assumed mapping of the task-specific RCs on RCs at lower hierarchical levels, e.g., as in the central back-coupling hypothesis (see Fig. 1; Latash et al. 2005). Feedback loops involved in such mappings have adjustable gains leading to different indices of synergies within the abundant spaces of elemental variables stabilizing the corresponding salient performance variables. Within this scheme, there are two major groups of neural variables specified at the hierarchically higher control level. Changes in one of the two groups, NV1 in Fig. 9, lead to changes in salient performance variables, while changes in the other group, NV2 in Fig. 9, lead to changes in stability of those performance variables as ensured by synergies at the hierarchically lower level.

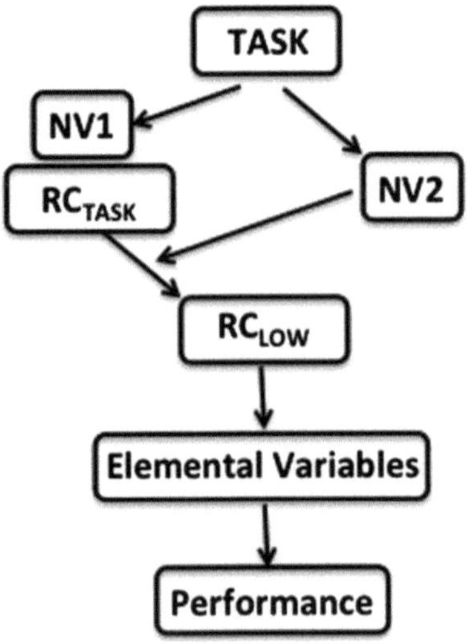


**Fig. 9** Two groups of neural variables are specified at the hierarchically higher control level, NV1 and NV2. A change in NV1 leads to changes in the referent coordinates at the task level ($RC_{TASK}$) for salient performance variables, while a change in NV2 leads to changes in stability of those variables (transformations from $RC_{TASK}$ to RCs at lower hierarchical levels, $RC_{LOW}$). Clinical studies suggest that both NV1 and NV2 are generated with a crucial role played by cortical structures. An impairment in NV1 leads to overall impaired performance including weakness and low accuracy as seen in averaged across trials patterns. An impairment in NV2 leads to decreased and delayed ASAs

The contrasting results in different patient groups suggest that both NV1 and NV2 are generated with a crucial role played by cortical structures. An impairment in NV1 leads to overall impaired performance including weakness and low accuracy as seen in averaged across trials patterns. An impairment in NV2 leads to decreased and delayed ASAs. In contrast, the circuitry involved in the few-to-many transformations seems to depend crucially on subcortical loops. This interpretation, while obviously simplistic, maps well on the idea of distributed processing modules based on subcortical loops involving the cerebellum and the basal ganglia (Houk 2005).

## Take Home Messages

The EP (RC) hypothesis is unique in its longevity and influence. Fifty years ago it signified the first step toward turning motor control into a subfield of physics of living systems. Over the recent years, the EP-hypothesis has been developing rapidly. In particular, the main idea of control with referent coordinates merged with the ideas of hierarchical control, synergic control, and naturally incorporated the apparatus of the uncontrolled manifold hypothesis. This approach has been productive not only in the field of the neural control of movement but also with important implications for kinesthetic perception, clinical studies, and effects of practice (reviewed in Feldman 2015). From the subjective optimistic view of the author, we are moving toward turning motor control into a subfield of natural science, physics of living systems.

**Acknowledgments** The author would like to thank his younger colleagues who played central roles in the reviewed studies, Satyajit Ambike, Ali Falaki, Hang Jin Jo, Daniela Mattos, Florent Paclet, Jaebum Park, Mu Qiao, Sasha Reschechtko, Stanislaw Solnik, Luke Wilhelm, and Tao Zhou. Preparation of this paper was in part supported by a grant R01 NS035032 from the National Institutes of Health, USA.

## References

Ambike S, Paclet F, Zatsiorsky VM, Latash ML (2014) Factors affecting grip force: anatomy, mechanics, and referent configurations. Exp Brain Res 232:1219–1231

Ambike S, Zatsiorsky VM, Latash ML (2015a) Processes underlying unintentional finger force changes in the absence of visual feedback. Exp Brain Res 233:711–721

Ambike S, Zhou T, Zatsiorsky VM, Latash ML (2015b) Moving a hand-held object: reconstruction of referent coordinate and apparent stiffness trajectories. Neurosci 298:336–356

Ambike S, Mattos D, Zatsiorsky VM, Latash ML (2016a) The nature of constant and cyclic force production: unintentional force-drift characteristics. Exp Brain Res 234:197–208

Ambike S, Mattos D, Zatsiorsky VM, Latash ML (2016b) Synergies in the space of control variables within the equilibrium-point hypothesis. Neurosci 315:150–161

Archambault PS, Mihaltchev P, Levin MF, Feldman AG (2005) Basic elements of arm postural control analyzed by unloading. Exp Brain Res 164:225–241

Asatryan DG, Feldman AG (1965) Functional tuning of the nervous system with control of movements or maintenance of a steady posture. I. Mechanographic analysis of the work of the limb on execution of a postural task. Biophysics 10:925–935

Augurelle AS, Smith AM, Lejeune T, Thonnard JL (2003) Importance of cutaneous feedback in maintaining a secure grip during manipulation of hand-held objects. J Neurophysiol 89:665–671

Bernstein NA (1967) The co-ordination and regulation of movements. Pergamon Press, Oxford

Bernstein NA, Kots YM (1963) Tone. Grand Medical Encyclopaedia, vol 32. State Encyclopaedia, Moscow, pp 418–422

Bizzi E, Polit A, Morasso P (1976) Mechanisms underlying achievement of final head position. J Neurophysiol 39:435–444

Calota A, Levin MF (2009) Tonic stretch reflex threshold as a measure of spasticity: implications for clinical practice. Top Stroke Rehab 16:177–188

Corcos DM, Gottlieb GL, Penn RD, Myklebust B, Agarwal GC (1986) Movement deficits caused by hyperexcitable stretch reflexes in spastic humans. Brain 109:1043–1058

d'Avella A, Saltiel P, Bizzi E (2003) Combinations of muscle synergies in the construction of a natural motor behavior. Nat Neurosci 6:300–308

DeWald JP, Pope PS, Given JD, Buchanan TS, Rymer WZ (1995) Abnormal muscle coactivation patterns during isometric torque generation at the elbow and shoulder in hemiparetic subjects. Brain 118:495–510

DiZio P, Lackner JR (1995) Motor adaptation to Coriolis force perturbations of reaching movements: endpoint but not trajectory adaptation transfers to the nonexposed arm. J Neurophysiol 74:1787–1792.

Fahn S, Jankovic J (2007) Principles and practice of movement disorders. Churchill Livingstone Elsevier, Philadelphia, PA, USA

Falaki A, Huang X, Lewis MM, Latash ML (2016) Impaired synergic control of posture in Parkinson's patients without postural instability. Gait Posture 44:209–215

Feldman AG (1966a) Functional tuning of the nervous system with control of movement or maintenance of a steady posture. II Controllable parameters of the muscle. Biophysics 11:565–578

Feldman AG (1966b) Functional tuning of the nervous system with control of movement or maintenance of a steady posture. III. Mechanographic analysis of execution by man of the simplest motor task. Biophysics 11:667–675

Feldman AG (1980) Superposition of motor programs. I Rhythmic forearm movements in man. Neurosci 5:81–90

Feldman AG (1986) Once more on the equilibrium-point hypothesis (λ-model) for motor control. J Mot Behav 18:17–54

Feldman AG (2015) Referent control of action and perception: challenging conventional theories in behavioral science. Springer, NY

Feldman AG, Latash ML (2005) Testing hypotheses and the advancement of science: recent attempts to falsify the equilibrium-point hypothesis. Exp Brain Res 161:91–103

Gelfand IM (1991) Two archetypes in the psychology of man. Nonlinear Sci Today 1:11–16

Gelfand IM, Latash ML (1998) On the problem of adequate language in movement science. Mot Control 2:306–313

Gottlieb GL (1998) Rejecting the equilibrium-point hypothesis. Mot Control 2:10–12

Gribble PL, Ostry DJ, Sanguineti V, Laboissiere R (1998) Are complex control signals required for human arm movements? J Neurophysiol 79:1409–1424

Hammond PH (1955) Involuntary activity in biceps following the sudden application of velocity to the abducted forearm. J Physiol 127:23P–25P

Hinder MR, Milner TE (2003) The case for an internal dynamics model versus equilibrium point control in human movement. J Physiol 549:953–963

Hoehn MM, Yahr MD (1967) Parkinsonism: onset, progression and mortality. Neurology 17:427–442

Houk JC (2005) Agents of the mind. Biol Cybern 92:427–437

Ivanenko YP, Poppele RE, Lacquaniti F (2004) Five basic muscle activation patterns account for muscle activity during human locomotion. J Physiol 556:267–282

Jo HJ, Park J, Lewis MM, Huang X, Latash ML (2015) Prehension synergies and hand function in early-stage Parkinson's disease. Exp Brain Res 233:425–440

Jo HJ, Mattos D, Lukassen EB, Huang X, Latash ML (2016a) Changes in multi-digit synergies and their feed-forward adjustments in multiple sclerosis. J Mot Behav (in press) e-pub: PMID: 27715488

Jo HJ, Maenza C, Good DC, Huang X, Park J, Sainburg RL, Latash ML (2016b) Effects of unilateral stroke on multi-finger synergies and their feed-forward adjustments. Neurosci 319:194–205

Karniel A (2011) Open questions in computational motor control. J Integr Neurosci 10:385–411

Kawato M (1999) Internal models for motor control and trajectory planning. Curr Opin Neurobiol 9:718–727

Kelso JA, Holt KG (1980) Exploring a vibratory systems analysis of human movement production. J Neurophysiol 43:1183–1196

Lackner JR, DiZio P (1994) Rapid adaptation to Coriolis force perturbations of arm trajectory. J Neurophysiol 72:1–15

Landau WM (1974) Spasticity: the fable of a neurological demon and the emperor's new therapy. Arch Neurol 31:217–219

Latash ML (1992) Virtual trajectories, joint stiffness, and changes in natural frequency during single-joint oscillatory movements. Neurosci 49:209–220

Latash ML (2008) Synergy. Oxford University Press, New York

Latash ML (2012) The bliss (not the problem) of motor abundance (not redundancy). Exp Brain Res 217:1–5

Latash ML, Gottlieb GL (1990) Compliant characteristics of single joints: preservation of equifinality with phasic reactions. Biol Cybern 62:331–336

Latash ML, Gottlieb GL (1991) Reconstruction of elbow joint compliant characteristics during fast and slow voluntary movements. Neurosci 43:697–712

Latash ML, Huang X (2015) Neural control of movement stability: lessons from studies of neurological patients. Neurosci 301:39–48

Latash ML, Levin MF, Scholz JP, Schöner G (2010) Motor control theories and their applications. Medicina 46:382–392

Latash ML, Penn RD (1996) Changes in voluntary motor control induced by intrathecal baclofen. Physiother Res Intern 1:229–246

Latash ML, Scholz JP, Schöner G (2002) Motor control strategies revealed in the structure of motor variability. Exerc Sport Sci Rev 30:26–31

Latash ML, Scholz JP, Schöner G (2007) Toward a new theory of motor synergies. Mot Control 11:276–308

Latash ML, Penn RD, Corcos DM, Gottlieb GL (1990) Effects of intrathecal baclofen on voluntary motor control in spastic paresis. J Neurosurg 72:388–392

Latash ML, Shim JK, Smilga AV, Zatsiorsky V (2005) A central back-coupling hypothesis on the organization of motor synergies: a physical metaphor and a neural model. Biol Cybern 92: 186–191

Latash ML, Zatsiorsky VM (2016) Biomechanics and motor control: defining central concepts. Academic Press, NY

Lawrence EL, Fassola I, Werner I, Leclercq C, Valero-Cuevas FJ. (2014) Quantification of dexterity as the dynamical regulation of instabilities: comparisons across gender, age, and disease. Front Neurol 5:53

Levin MF, Feldman AG (1994) The role of stretch reflex threshold regulation in normal and impaired motor control. Brain Res 657:23–30

Martin V, Scholz JP, Schöner G (2009) Redundancy, self-motion, and motor control. Neural Comput 21:1371–1414
Martin JR, Zatsiorsky VM, Latash ML (2011) Multi-finger interaction during involuntary and voluntary single finger force changes. Exp Brain Res 208:423–435
Mattos D, Latash ML, Park E, Kuhl J, Scholz JP (2011) Unpredictable elbow joint perturbation during reaching results in multijoint motor equivalence. J Neurophysiol 106:1424–1436
Mattos D, Kuhl J, Scholz JP, Latash ML (2013) Motor equivalence (ME) during reaching: Is ME observable at the muscle level? Mot Control 17:145–175
Mattos D, Schöner G, Zatsiorsky VM, Latash ML (2015) Task-specific stability of abundant systems: structure of variance and motor equivalence. Neurosci 310:600–615
Müller H, Sternad D (2003) A randomization method for the calculation of covariation in multiple nonlinear relations: illustrated with the example of goal-directed movements. Biol Cybern 89:22–33
Musampa NK, Mathieu PA, Levin MF (2007) Relationship between stretch reflex thresholds and voluntary arm muscle activation in patients with spasticity. Exp Brain Res 181:579–593
Olafsdottir H, Yoshida N, Zatsiorsky VM, Latash ML (2005) Anticipatory covariation of finger forces during self-paced and reaction time force production. Neurosci Lett 381:92–96
Park J, Wu Y-H, Lewis MM, Huang X, Latash ML (2012) Changes in multi-finger interaction and coordination in Parkinson's disease. J Neurophysiol 108:915–924
Park J, Lewis MM, Huang X, Latash ML (2013) Effects of olivo-ponto-cerebellar atrophy (OPCA) on finger interaction and coordination. Clin Neurophysiol 124:991–998
Park J, Lewis MM, Huang X, Latash ML (2014) Dopaminergic modulation of motor coordination in Parkinson's disease. Parkinson Rel Disord 20:64–68
Popescu FC, Rymer WZ (2000) End points of planar reaching movements are disrupted by small force pulses: an evaluation of the hypothesis of equifinality. J Neurophysiol 84:2670–2679
Prablanc C, Desmurget M, Gréa H (2003) Neural control of on-line guidance of hand reaching movements. Prog Brain Res 142:155–170
Prilutsky BI, Zatsiorsky VM (2002) Optimization-based models of muscle coordination. Exer Sport Sci Rev 30:32–38
Qiao M, Zhou T, Latash ML (2015) Positional errors introduced by transient perturbations applied to a multi-joint limb. Neurosci Lett 595:104–107
Raptis H, Burtet L, Forget R, Feldman AG (2010) Control of wrist position and muscle relaxation by shifting spatial frames of reference for motoneuronal recruitment: possible involvement of corticospinal pathways. J Physiol 588:1551–1570
Reisman D, Scholz JP (2003) Aspects of joint coordination are preserved during pointing in persons with post-stroke hemiparesis. Brain 126:2510–2527
Reschechtko S, Zatsiorsky VM, Latash ML (2014) Stability of multi-finger action in different spaces. J Neurophysiol 112:3209–3218
Reschechtko S, Zatsiorsky VM, Latash ML (2015) Task-specific stability of multi-finger steady-state action. J Mot Behav 47:365–377
Sangani SG, Raptis HA, Feldman AG (2011) Subthreshold corticospinal control of anticipatory actions in humans. Behav Brain Res 224:145–154
Schmidt RA, McGown C (1980) Terminal accuracy of unexpected loaded rapid movements: evidence for a mass-spring mechanism in programming. J Mot Behav 12:149–161
Scholz JP, Schöner G (1999) The uncontrolled manifold concept: identifying control variables for a functional task. Exp Brain Res 126:289–306
Schöner G (1995) Recent developments and problems in human movement science and their conceptual implications. Ecol Psychol 8:291–314
Schrödinger E (1948) What is life? The physical aspect of the living cell. Cambridge University Press, Cambridge
Shadmehr R, Wise SP (2005) The computational neurobiology of reaching and pointing. MIT Press, Cambridge, MA

Shapkova EYu, Shapkova AL, Goodman SR, Zatsiorsky VM, Latash ML (2008) Do synergies decrease force variability? A study of single-finger and multi-finger force production. Exp Brain Res 188:411–425
Shemmell J, Krutky MA, Perreault EJ (2010) Stretch sensitive reflexes as an adaptive mechanism for maintaining limb stability. Clin Neurophysiol 121:1680–1689
Shim JK, Olafsdottir H, Zatsiorsky VM, Latash ML (2005) The emergence and disappearance of multi-digit synergies during force production tasks. Exp Brain Res 164:260–270
Slifkin AB, Vaillancourt DE, Newell KM (2000) Intermittency in the control of continuous force production. J Neurophysiol 84:1708–1718
Tatton WG, Bawa P, Bruce IC, Lee RG (1978) Long loop reflexes in monkeys: an interpretive base for human reflexes. Prog Clin Neurophysiol 4:229–245
Ting LH, Macpherson JM (2005) A limited set of muscle synergies for force control during a postural task. J Neurophysiol 93:609–613
Todorov E, Jordan MI (2002) Optimal feedback control as a theory of motor coordination. Nat Neurosci 5:1226–1235
Vaillancourt DE, Russell DM (2002) Temporal capacity of short-term visuomotor memory in continuous force production. Exp Brain Res 145:275–285
Venkadesan M, Guckenheimer J, Valero-Cuevas FJ (2007) Manipulating the edge of instability. J Biomech 40(8):1653–1661
Wilhelm L, Zatsiorsky VM, Latash ML (2013) Equifinality and its violations in a redundant system: multi-finger accurate force production. J Neurophysiol 110:1965–1973
Wolpert DM, Miall RC, Kawato M (1998) Internal models in the cerebellum. Trends Cogn Sci 2:338–347
Zhou T, Wu Y-H, Bartsch A, Cuadra C, Zatsiorsky VM, Latash ML (2013) Anticipatory synergy adjustments: preparing a quick action in an unknown direction. Exp Brain Res 226:565–573
Zhou T, Solnik S, Wu Y-H, Latash ML (2014) Unintentional movements produced by back-coupling between the actual and referent body configurations: violations of equifinality in multi-joint positional tasks. Exp Brain Res 232:3847–3859
Zhou T, Zatsiorsky VM, Latash ML (2015a) Unintentional changes in the apparent stiffness of the multi-joint limb. Exp Brain Res 233:2989–3004
Zhou T, Zhang L, Latash ML (2015b) Characteristics of unintentional movements by a multi-joint effector. J Mot Behav 47:352–361

# The Relationship Between Postural and Movement Stability

Anatol G. Feldman

**Abstract** Postural stabilization is provided by stretch reflexes, intermuscular reflexes, and intrinsic muscle properties. Taken together, these posture-stabilizing mechanisms resist deflections from the posture at which balance of muscle and external forces is maintained. Empirical findings suggest that for each muscle, these mechanisms become functional at a specific, spatial threshold—the muscle length or respective joint angle at which motor units begin to be recruited. Empirical data suggest that spinal and supraspinal centers can shift the spatial thresholds for a group of muscles that stabilized the initial posture. As a consequence, the same stabilizing mechanisms, instead of resisting motion from the initial posture, drive the body to another stable posture. In other words by shifting spatial thresholds, the nervous system converts movement resisting to movement-producing mechanisms. It is illustrated that, contrary to conventional view, this control strategy allows the system to transfer body balance to produce locomotion and other actions without loosing stability at any point of them. It also helps orient posture and movement with the direction of gravity. It is concluded that postural and movement stability is provided by a common mechanism.

**Keywords** Reflexes · Spatial threshold · Joint angles · Reafference · Mechanical reductionism · Equilibrium point (EP) hypothesis · Referent body · Frame of reference (FR) · Motoneurons (MN) · Coactivation · Posture-movement problem

A.G. Feldman (✉)
Department of Neuroscience and Institute of Biomedical Engineering,
University of Montreal, Montreal, QC, Canada
e-mail: anatol.g.feldman@umontreal.ca

A.G. Feldman
Center for Interdisciplinary Research in Rehabilitation of Greater Montreal (CRIR),
Institut de réadaptation Gingras-Lindsay-de-Montréal, CIUSSS CSMTL, Montreal,
QC, Canada

A.G. Feldman
Jewish Rehabilitation Hospital, CISSS de Laval, Laval, Canada



J. Laczko and M.L. Latash (eds.), *Progress in Motor Control*,
Advances in Experimental Medicine and Biology 957,
DOI 10.1007/978-3-319-47313-0_6

## Introduction

Basic ideas on how the brain controls posture and movement have been originated about two centuries ago (Steinbuch 1811; von Uexküll 1926; von Helmholtz 1866) and in a most clear form were presented in seminal papers by von Holst (von Holst and Mittelstaedt 1950; von Holst 1954). Neurophysiological data were comparatively limited and many scientists focused on the mechanical output of motor actions. It seemed natural to assume that the brain is able to preprogram kinematic and kinetic characteristics and directly specify the requisite motor commands to muscles (i.e., EMG patterns or forces) to get the desired motor outcome. I call this approach to motor control mechanical reductionism (Feldman 2015). Many researchers still take for granted this approach despite indications that it runs into several problems in the explanation of essential features of motor behavior and despite experimental findings refuting this approach 50 years ago (Asatryan and Feldman 1965). Among the problems unresolved in the framework of mechanical reductionism is the posture-movement problem of how intentional movement from a stable posture can be produced without evoking resistance of posture-stabilizing mechanisms. In papers listed above, von Holst shared the common view that the brain preprograms motor commands and concluded that postural mechanisms are incompatible with mechanisms responsible for intentional motion. He further concluded that posture-stabilizing mechanisms are transiently neutralized when intentional movements are made and that this is achieved by suppressing movement-related afferent feedback (reafference). He additionally proposed that the nervous system uses a copy of predicted motor commands (efference copy) to suppress the reafference. Since then the notion of efference copy and similar notion of corollary discharge introduced by Sperry (1950) were broadly used in interpretations of motor behavior.

After discovery of coactivation of α- and γ-motoneurons (MNs) during active movements (see Granit 1955), it would be appropriate to conclude that the activity of agonist and antagonist muscles during intentional movement is supported due to activation of muscle spindle afferents by γ-MNs. This means reafference is enhanced, rather than suppressed during such movements, thus refuting von Holst's proposal. This, however, has not been noticed in many studies in which his proposal was accepted without any critical analysis (e.g., Gallistel 1980; McCloskey 1978; Cullen 2004; Proske and Gandevia 2009).

The inability to solve the posture-movement problem in the framework of mechanical reductionism led to another problem—the failure to clarify the relationship between postural and movement stability.

More progressive attempts to consider the control of posture and movement in the framework of the theory of dynamic systems. However, based on the same idea of preprogramming of kinematics and motor commands, these attempts run into the problems characteristic of traditional approaches to motor control (see however, Shoner Mattos et al. 2015). Applications of computational methods developed in robotics (Hollerbach 1982; Wolpert and Kawato 1998) did not leave the tenets of mechanical reductionism since these methods also relied on the notion of preprogramming of motor commands.

The tenets of mechanical reductionism were broken with the experimental identification of physiological parameters underlying the control of posture and movement in humans (Asatryan and Feldman 1965). This result motivated the equilibrium point (EP) hypothesis, which after systematic testing and development for about 50 years, has evolved into the theory of referent control of action and perception (Feldman 1986, 2011, 2015). The theory retained the original notion of the EP hypothesis that the brain controls motor actions by changing specific physiological parameters without the necessity to preprogram mechanical variables and motor commands. It has been shown that the parameters that the nervous system is able to change are threshold positions of the body segments at which muscles begin to be activated or deactivated (Asatryan and Feldman 1965). By changing such parameters the system shifts the equilibrium state in the interaction between the organism and the environment. As a consequence, changes in mechanical characteristics and motor commands emerge, without preprogramming, due to a natural cooperative tendency of neuromuscular elements to reach equilibrium. If the emerging motor action does not meet the task demand, the system adjusts the spatial activation thresholds to minimize the movement error. This control strategy is used not only during everyday life but also during learning novel motor skills.

There are different hierarchically organized neural levels and respective forms of threshold position control (Feldman 2015). In particular, the referent body configuration delimits possible positions of body segments or configurations of the entire body. The referent body orientation delimits possible body positions with respect to the direction of gravity. The referent body localization delimits possible places of the body in the environment. The referent coactivation zone delimits the range of body configurations at which agonist and antagonist muscles can be coactivated.

In addressing the stability topics, I will primarily focus on some of those forms of referent control. Each forms can be considered as the origin points of respective spatial frame of reference (FR) in which motor actions are produced. In this case, the EP is a secondary, emergent property of referent control. However, the EP and its shifts characterize an important feature of the organism-environment interaction and remain an integral part of the referent control theory.

The knowledge of physiological mechanisms underlying the organization of motoneuronal pool of a single muscle is a departure point of our understanding of how posture and movement are controlled. Therefore, before addressing the question of stability, I will describe how the motoneuronal pool of a single muscle is controlled. After that I clarify the EP concept in relation to stability of posture and movement. Finally, I will use different forms of referent control to clarify the relationship between postural and movement stability in several motor tasks. The reader may note that, in several important aspects, the present description of how the motoneuronal pool is organized and controlled is different from that in standard textbooks on motor control.

## Referent Control of the Motoneuronal Pool

According to the size principle established by Henneman et al. (1965), motoneurons (MNs) of a single muscle are recruited in a certain order. Recruitment usually starts from small MNs that induce comparatively low muscle forces that can be maintained for a long time (low-fatigable motor units) and ending with large MNs that can initiate stronger muscle forces but during short time (fast-fatigable motor units). Once recruited, MNs can go to the state (called plateau potential) in which they can generate spikes at a certain frequency increasing with increasing muscle stretching and force. To explain how the nervous system controls the motoneuronal pool, we need to consider MNs in the context of afferent and central influences received by MNs. Although this seems complicated, experimental data are sufficient to describe simple principles underlying neural control of MNs. These principles have been integrated in the theory of referent control of action and perception.

Based on experimental data, the following condition of muscle activation in normal conditions has been formulated (Fig. 1a). The muscle is activated if

$$x - \lambda^* > 0, \qquad (1)$$

where $x$ is muscle length and $\lambda^*$ is the threshold muscle length, i.e., the length at which MN recruitment is initiated. In dynamics, the threshold length depends on the speed of muscle lengthening ($v > 0$) or shortening ($v < 0$). It also depends on intermuscular interaction and history-dependent state of MNs ($\rho$):

$$\lambda^* = \lambda - \mu v + \rho \qquad (2)$$

Coefficient $\mu$ represents the dynamic sensitivity of the MN pool, presumably defined by the activity of dynamic $\gamma$-MNs that innervate muscle velocity-sensitive intrafusal fibers of spindle receptors (e.g., Matthews 1981). In a supra-threshold state, the activity of muscle (the number of recruited of MNs and firing of already recruited MNs) increases with the increasing difference between the actual and threshold muscle length. MNs are recruited in the order defined by their individual spatial thresholds, which is consistent with the size principle. The recruitment thresholds are velocity-dependent. The individual sensitivity ($\mu$) is probably higher for fast-fatigable than for fatigue-resistible motor units. As a consequence, with the increasing stretch velocity, the recruitment threshold may become lower for former than for latter motor units, such that the recruitment order can be reversed in some cases (Fig. 1c), as observed in some studies (e.g., Kanda et al. 1977).

The inequality (1) refers to any way of muscle activation or deactivation. The initially inactive muscle can be activated by stretching (increasing $x$) or centrally, by decreasing $\lambda$. If muscle is initially active, it can be deactivated by muscle shortening or by increasing $\lambda$. In particular, by shifting $\lambda$ beyond the upper limit of biomechanical range of muscle length, the muscle can be relaxed and be insensitive to

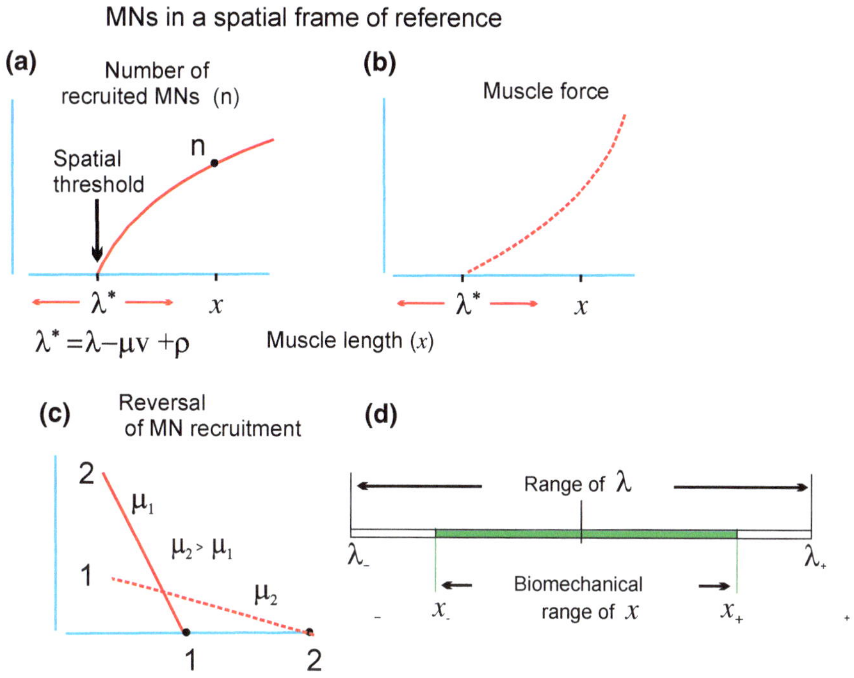


**Fig. 1** Basic principles underlying the organization and control of the motoneuronal pool. See text

muscle stretching within the entire biomechanical range of muscle length (unless the stretch speed is very high, as during the knee reflex).

Threshold defines where, in the biomechanical range, the muscles begins to be activated and generates nonlinear force-length characteristic (Fig. 1b). Defined by fixed value of $\lambda^*$, this characteristic is called invariant one (IC). This notion does not imply that the shape of characteristic is the same—it may be different for different activation thresholds. The ability to relax and activate the muscle at any length within the biomechanical range implies that the range of regulation of $\lambda$ exceeds the biomechanical range of changes in the muscle length, $x$ (Fig. 1d).

One can say that by specifying the threshold, the nervous system predetermines the spatial range within which the intrinsic muscle and reflex properties manifest position and velocity-dependent changes in the muscle force. These properties are essential for stability of posture and movement.

Not less essential is that, by changing $\lambda$, the system not only predetermines the position at which the muscle can balance the external load, but also transfer stability properties to that position. This notion is fundamental for the understanding of why intentional motion from a stable position does not evoke resistance of posture-stabilizing mechanisms as would occur if motion were produced by direct specification of motor commands to muscles—the posture-movement problem identified but not resolved by von Holst (von Holst and Mittelstaedt 1950; von Holst 1954). In other words, in the framework of referent (threshold) control the problem does not exist.

Several important conclusions can additionally be derived from the above description.

1. MNs of a single muscle function in one-dimensional spatial FR with the muscle length ($x$) as the only coordinate of this FR. There is the origin (referent) point in this FR: $\lambda^*$. MNs begin to be recruited at this point and the activity of MNs increases depending on $x - \lambda^*$. The origin point can be shifted by different spinal and supraspinal systems. It also depends on the speed of changes in the muscle length as well as on the intermuscular interaction.
2. The notion that MNs function in a spatial FR can be referred to many other neurons and the latter can be classified according to the spatial FR frames they function. Each FR is defined by the type of position-dependent information received by the respective neurons. Control levels target neurons associated with the FR that is most appropriate for the motor task and predetermine the spatial boundaries in this FR in which MNs can work. EMG patterns (motor commands to muscles) are not a matter of concern of neural control levels. These patterns emerge automatically depending on the deflection of the actual position of the body from the centrally specified spatial boundaries. Control levels determine whether the emergent motor action is adequate for reaching the goal and, if not, they adjust the spatial boundaries or change the type of FR to eventually reach the goal. This control strategy is equally applicable to everyday actions and motor learning.
3. The $\lambda^*$ can be considered as the dynamic threshold of the stretch reflex and thus the muscle is controlled by shifting the spatial threshold of this reflex.
4. By increasing the threshold, the nervous system can deactivate the muscle but this does not mean that the stretch reflex is blocked as such. Unless the threshold is increased to fully relax the muscle, functional properties of the reflex are fully preserved but in another part of the biomechanical range. This means that gating of reflexes usually considered in rhythmic movements results from modulation of the spatial reflex thresholds. In addition, the absence of reflex reactions should not be considered as resulting from blocking of reflexes. Threshold control implies that reflexes are not suppressed as such but remains fully functional in another part of biomechanical range. Similar reservations should be made for the notion of reflex gain modulation usually considered as a way the nervous system controls reflexes. In contrast, gain modulation should be considered as an emerging, not directly controlled by the nervous system property of threshold control. By focusing on gain modulation and reflex gating one considers effects, rather than causes of motor control.

## Dynamic Aspects of the EP Concept

Additional aspects of stability of posture and movement are associated with the EP concept. Mechanically, each EP represents a specific combination of the positions of body segments at which muscle and external torques are balanced. Since EP characterizes the interaction between the organism and environment, the nervous system can only influence but cannot fully predetermine the EP: its shifts can also result from changes in external forces, eliciting involuntary motion, as occurs during the unloading reflex. This reflex is illustrated by asking a subject to hold a book on the palm of the arm at certain position. When the book is suddenly removed by an assistant, the arm automatically moves to another EP.

The EP is a dynamic concept in the sense that it also implies certain reactions of muscle and external torques to deflections of the system from the EP: the latter can be stable (an attractor), semi-stable, or unstable (repellent) (Fig. 2). In some cases, the threshold control allows the nervous system to convert an unstable or semi-stable to a stable EP (Fig. 2). The neural circuitry and the neuromuscular properties allow the nervous system to mostly operate in the space of stable EPs, which means that deflections from an EP are met with resistance increasing with increasing deflection and its speed from the EP. If perturbations are transient, these reactions may be sufficient to regain the same EP (equifinality). However, the system may not tolerate strong or destabilizing perturbation even if they are transient and involves additional reactions to regain stability by transition to a new, stable EP. For example, during quiet standing, position- and velocity-dependent resistance might be insufficient to prevent body falling in response to sudden tilting of the platform on which the subject stands. In response to such a destabilizing perturbation, a step can be made to bring the body to a new, stable posture. In this case, step should be considered as a part of posture-stabilizing mechanisms, which further emphasizes the dynamic content of the EP concept. It also shows that while predicting the possibility of equifinality, the EP hypothesis does not predict that equifinality should be observed in response to any transient perturbation. Because of misunderstanding of this aspect, many researchers falsely rejected the EP hypothesis based on observations of equifinality violation in responses to perturbations (e.g., Lackner and Dizio 1994).

The EP concept is associated with an important physical principle not mentioned in standard textbooks on neural control of motor actions. This principle originated from analysis of stability in the theory of dynamics systems (Glansdorff and Prigogine 1971). Consider pointing arm motion from one stable EP to another. Muscle and external torques are balanced at the initial and final EPs, but, according to the physical principle, torques or any variables related to them (like forces, EMG activity) are not responsible for the choice of the arm position at which the balance is achieved. This statement becomes obvious when changes in the arm position are produced in isotonic conditions, when torques are the same at the initial and final EPs, implying that the choice between these EPs is not predetermined by torques. In contrast, according to the physical principle, the choice is predetermined by specific

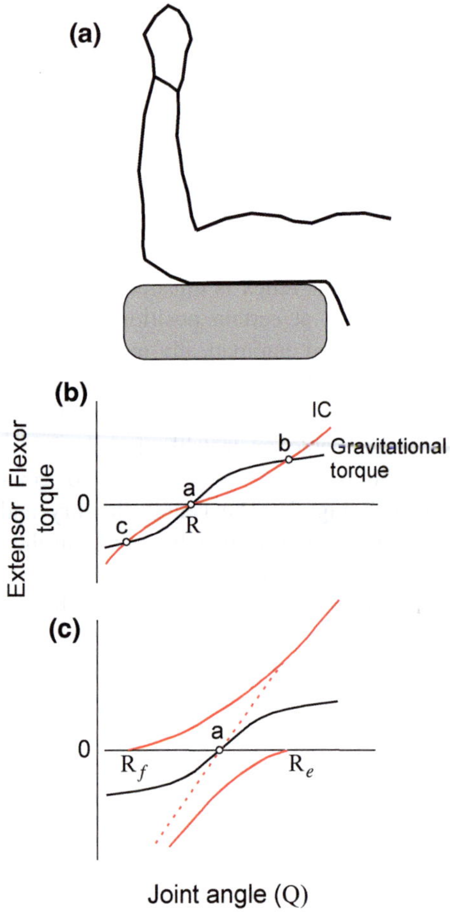


**Fig. 2** Stabilizing equilibrium points by combining R and C commands. When the human forearm is oriented vertically while shoulder lays on a horizontal surface (**a**), the gravitational torque at the elbow joint is zero but increases (in terms of the absolute value) with a deviation of the forearm from this position. To balance the zero gravitational torque the net muscle torque should also be zero. This condition can be met by setting the common threshold (R) for activation of elbow flexors and extensors at this position of the forearm. **b** Equilibrium point for this arm position will be unstable, since the gradient (*slope*) of the gravitational torque at this position is higher than muscle stiffness (*slope*) of invariant torque-angle characteristic, IC, at this position. In other words, the muscle torque generated in response to perturbations will be insufficient in overcoming the increasing gravitational torque to return the arm to the point *a*. When perturbed, the forearm will arrive at stable equilibrium point *b* or *c*, depending on the direction of perturbation. **c** To stabilize equilibrium point *a*, the system can facilitate flexor and extensor MNs to create a zone coactivation (between $R_f$ and $R_e$), thus increasing the net muscle stiffness. Reproduced with permission from Feldman (2015), Copyright Springer

neurophysiological parameters that are independent of any biomechanical variable characterizing the motor outcome. In other words, the choice is independent not only of muscle torques but also of motor commands (EMG signals) to muscles related to torques. Such parameters are missing in traditional approaches to control of motor actions based on the idea of direct preprogramming of kinematics and motor commands. Such parameters are also missing in computational approaches in which the motor output is preprogrammed based on some optimality criteria defined by a chosen cost function with the help of internal models of the system interacting with the environment. The inconsistency with the physical principle means that the computational approaches inadvertently rearrange cause and effect in the control of motor actions and, as a consequence, has no validity in applications to behavioral neuroscience. As illustrated later in this chapter, the cause–effect rearrangement results in several problems in action and perception that do not exist in the referent control theory. This explains why the referent control theory breaks with the traditional belief in direct preprogramming of kinematic and kinetic variables or EMG patterns.

## The Relationship Between Postural and Movement Stability

Parametric (referent) control is not taken into account in typical approaches to postural and movement stability, limiting the understanding of how stability is provided in healthy subjects and in subjects with neurological motor deficits.

Referent control suggests that to elicit motion, the nervous system shifts referent variables and thus converts posture-stabilizing to movement-producing mechanisms. In this case, posture-stabilizing mechanisms do not produce resistance to motion and the posture-movement problem does not exist. In other words, there is no conflict between postural and movement control mechanisms. Moreover, due to this control strategy, stability is transferred to the posture to which the motion is made, such that motion is finished without any tendency to move backward, to the initially stabilized posture. As mentioned above, in the case of a strong perturbation during standing, falling is often prevented by taking a step. Therefore, a step is part of the posture-stabilizing mechanisms. A more general rule stemming from referent control is that posture and movement are stabilized by common mechanisms.

I will clarify this rule by considering several motor actions.

## Referent Coactivation Zone and Agonist-Antagonist Coactivation During Active Motion

Coactivation of agonist and antagonist muscles is usually considered as a way of increasing position- and velocity-dependent resistance to perturbations, thus enhancing stability of posture and movements. Coefficients characterizing the position- and velocity-dependent sensitivity to perturbations are usually called stiffness and damping, respectively. We can coactivate muscles of the arm while preserving the arm position. We can also feel that we coactivate agonist and antagonist muscles while rapidly moving the arm from one position to another and that the coactivation (muscle stiffness) increases with increasing the speed of arm motion.

It seems strange that coactivation that does not produce motion in the postural task is used in motion and even increases with the increasing movement speed. Moreover, one can think that by coactivating muscles during active motion, we would stiffen the arm and increase resistance to motion. The puzzles are solved by considering two basic physiological phenomena in the context of referent control: reciprocal activation and coactivation of agonist and antagonist muscles. The system can specify a common threshold (R) for activation or deactivation of agonist and antagonist muscles. To shift the R, the system can facilitate agonist MNs and reciprocally de-facilitate antagonist MNs (the R command). The system can increase the speed of motion by increasing the rate of changes in the R command (Pilon and Feldman 2006; Zhang et al. 2016).

On the other hand, by facilitating agonist MNs in isolation or in combination with antagonist MNs, the nervous system can create a spatial zone in which agonist and antagonist muscles become coactivated (the referent C command). Outside this zone, only one muscle group is active (Fig. 3). The extent of this zone is defined by the degree of facilitation of MNs, independently of the R command. However, there is a certain hierarchy in the relationship between the R and C commands. A coactivation zone is created by shifting activation thresholds from those established by the R command. Therefore, the localization of the coactivation zone in the biomechanical range is predetermined by the R command. When it is necessary to increase the movement speed, a referent C command is used. Since the localization of the C zone in the biomechanical range is defined by the R command, the C zone moves with it. As a result, the increased muscle stiffness resulting from the C command is transferred to the position to which the arm moves and contributes to movement acceleration. Computer modeling showed (Pilon and Feldman 2006) that due to this control strategy, the C command initially benefits agonist muscles by increasing movement acceleration and then antagonist muscles by increasing movement deceleration such that terminal oscillations are minimized. Indeed, with the end of motion all muscles become coactivated and, being unnecessary at the final position, the coactivation gradually recedes while the final position is maintained. Naturally, when it is necessary to enhance the movement speed, a stronger C command is used.

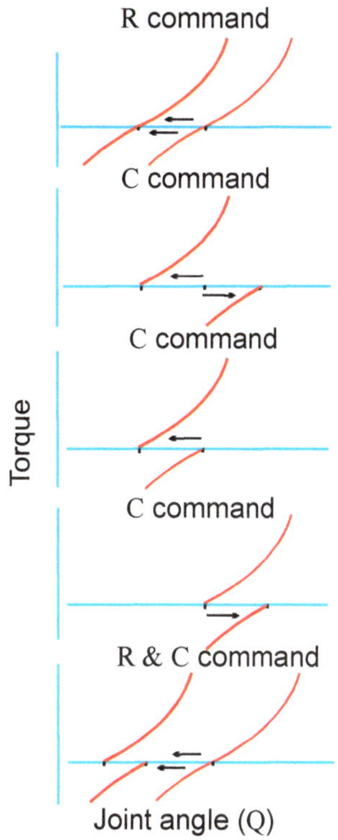


**Fig. 3** Referent R and C commands. Note that when the R command is shifted to produce motion, the coactivation zone is shifted with it (*lower panel*)

Thus, the C command is used to increase movement speed beyond the limit defined by maximal rate of shift in the R command. The system converts posture stabilizing to movement-producing mechanisms even in the cases when muscle coactivation is used. In contrast, if MNs were activated directly, muscle coactivation would increase resistance to motion.

This example also illustrates that with referent control, mechanisms that stabilize the initial posture are used to stabilize motion and the final posture.

## Control and Stabilization of Body Posture During Standing

Consider the case when the platform at which a subject quietly stands is suddenly tilted (toes up) by about 10° and remained tilted after that. After the transitional period, the subject continues to stand on the tilted platform. With the onset of tilting, ankle extensors are stretched. The stretch reflex tends to rotate the body backward and the subject is forced to do something to prevent falling. Naturally, with repetition of the perturbation, the stretch reflex is attenuated (Horak et al. 1989; Nashner 1976; Schieppati and Nardone 1995). One question has not been addressed in the cited studies. After adaptation to perturbation, the stretch reflex was suppressed. Was the reflex continued to be suppressed during standing on the tilted platform? The stretch reflex should be restored to stabilize posture during standing on the tilted platform, but how this is possible if the stretch reflex was suppressed with the onset of platform tilting after several trials? The notion of referent control is helpful in answering these questions. A after several trials subjects were able to increase the stretch reflex threshold of ankle extensors in response to tilting the platform to prevent stretch responses of these muscles. During the platform tilting, they continued to increase the threshold to minimize stretch responses. With the end of platform tilting, the threshold shifts were ceased such that the stretch reflex became fully functional during standing at the tilted platform (Fig. 4a). This example illustrates the general rule that by resetting the activation thresholds, the nervous system does not suppress the stretch reflex as such but transfer its properties to the final posture. Indeed, the validity of this description in terms of referent control can be tested by perturbation methods.

Consider how intentional leaning of the body can be produced. Referent body orientation indicates the position of the body with respect to the direction of gravity. At this position all skeletal muscles reach activation thresholds or, in the presence of C command, are activated but generate net zero torques. Activity of muscles changes with deflection of the body from the referent orientation.

Using the concept of referent body orientation, consider the case when a subject stands on a horizontal surface. One can assume that during quiet standing the referent body orientation coincides with the direction of gravity. The COM is located in front of referent line and the torque of body weight leans the body forward until this torque becomes balanced by the torque of activated ankle extensors. This is the initial situation during quiet standing before intentional leaning of the body.

Now consider how the subject can lean the body forward. The gravitational torque at the new body position is higher than at the initial body posture and, to balance the torque, extensor muscle torque should be enhanced. However, increasing the muscle torque at the initial posture would deflect the body backward, not forward. Again, referent control is helpful in the explanation of how the system performs the task. The system leans the referent body orientation forward thus initially diminishing the deflection of the actual body orientation from the new referent orientation. As a consequence, the activity of ankle extensors will initially drop, letting the gravitational torque lean the body forward. During leaning, ankle extensor muscles will be restretched and reactivated. The leaning will cease at a new body position at which the increasing gravitational torque becomes balanced by joint active and passive extensor torques (Fig. 4b). Again, the validity of the proposed referent control of body posture can be tested by perturbation methods.

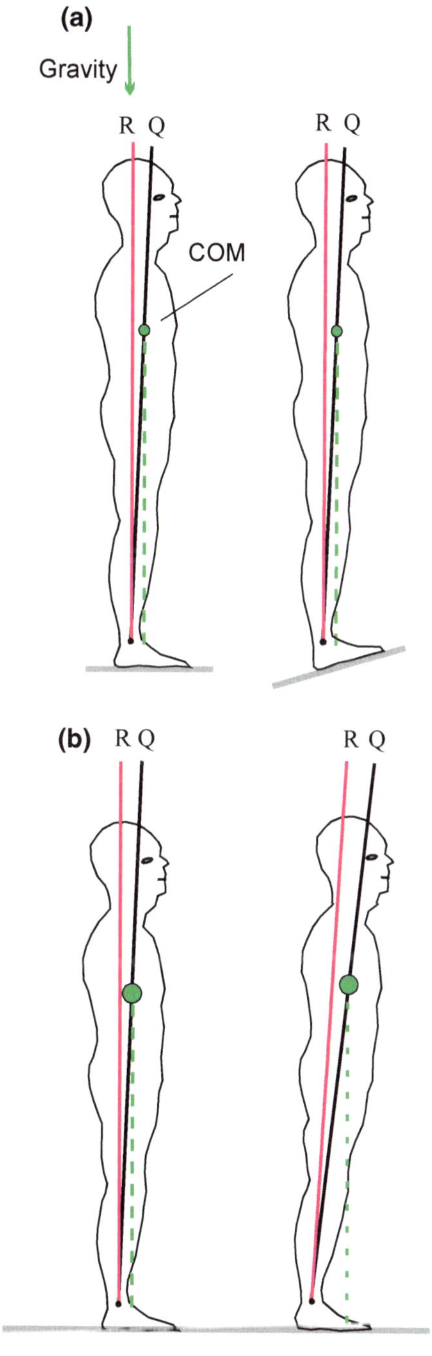


**Fig. 4** Referent control of posture during tilting the platform on which subject stand (**a**) or during self-initiated leaning forward (**b**). In **a**, the subject preserves the referent (R) and, as a consequence, the actual (Q) body orientation by changing the referent ankle joint angle. In contrast, to lean the body forward (**b**), the subject leans the referent body orientation so that the activity of anti-gravitational muscles decreases. The body begins to move under the influence of the gravitational torque. The body leans until stretched anti-gravitational muscles begin to balance the increased gravitational torque

## A Single Step and Gait as a Result of Referent Transfer of Body Balance in the Environment

The notion that posture and movement are stabilized by common mechanisms is helpful in the explanation of how, in principle, locomotion is initiated, maintained, and terminated. This analysis will brings us to the conclusion that the traditional claim that locomotion is associated with falling and catching of the body in each cycle is misleading.

Consider neurons that receive sensory inputs depending on the current localization of the body in the environment. Then according to the general rule, independent central inputs predetermine the referent (threshold) body localization at which these neurons begin to be recruited. Like in MNs, these neurons will generate activity depending on the deflection of the current from the referent body localization. Transmitted to MNs, this activity will affect threshold muscle lengths. It is assumed that these projections are organized such that the elicited changes in lambdas will elicit muscle activity tending to diminish the deflection of the body from its referent localization. These previous explanations will help us in the description of how a step or continuous locomotion is produced.

During quiet standing, the localization of the body coincides with the referent body localization. By sifting the referent body localization, the nervous system transfers body balance to a new place in the environment. Muscles are activated and tend to diminish the deflection of the body from the place to which the body balance is shifted. If the shift in the body localization is small, a single step will be sufficient to relocate the body to the new place. The speed of the step will depend on the rate of referent shift. With prolongation of the referent shifts, a continuous locomotion (gait or running, depending on the rate of referent shifts) will be performed until the referent shifts are ceased and the body reaches a new place in the environment.

In standard mechanical descriptions of locomotion, it is usually emphasized that when a step is made the COM moves outside the base of support (BOS) and the body begins to fall. In the framework of referent control this is not a problem: While leaving the initial BOS, the COM moves toward the new BOS to which the body balance was transferred. In non-perturbed locomotion, the speed of motion of the COM is always sufficient to move the COM toward the shifted BOS and reach the final BOS when referent shifts are ceased.

Some consequences of referent control of locomotion have been formulated and tested (Feldman et al. 2011; Krasovsky et al. 2014). Transient perturbations may temporarily slowdown the motion of COM such that the referent BOS will go too far to be reached. To prevent body falling, the system can shortly diminish the rate or even stop the referent shifts of the body localization such that the BOS will be reached. After that regular referent shifts can be resumed but the phase of steps will permanently be changed (phase resetting). This prediction was confirmed in the studies cited above.

To summarize, in addition to solving several problems in the explanations of motor actions, the notion of referent control clarifies the relationship between stability of posture and movement with the conclusion that posture and movements are stabilized by common mechanisms.

## Take Home Message

For a long time, many researchers were puzzled by the question of how intentional movements can be produced without resistance of reflexes that tend to minimize deflections from an initial posture (the classical posture-movement problem). By shifting the spatial domain in which reflexes function, the nervous system forces them to produce movements without any posture-movement problem.

**Acknowledgments** Supported by NSERC (Canada).

## References

Asatryan DG, Feldman AG (1965) Functional tuning of the nervous system with control of movement or maintenance of a steady posture: I. Mechanographic analysis of the work of the joint on execution of a postural task. Biophysics 10:925–934

Cullen KE (2004) Sensory signals during active versus passive movement. Curr Opin Neurobiol 14: 698–706

Feldman AG (1986) Once more on the equilibrium-point hypothesis (lambda model) for motor control. J Mot Behav 18:17–54

Feldman AG (2011) Space and time in the context of equilibrium-point theory. Wiley Interdiscip Rev Cogn Sci 2(3):287–304

Feldman AG (2015) Referent control of action and perception: challenging conventional theories in behavioral neuroscience. Springer, New York

Feldman AG, Krasovsky T, Banina MC, Lamontagne A, Levin MF (2011) Changes in the referent body location and configuration may underlie human gait, as confirmed by findings of multi-muscle activity minimizations and phase resetting. Exp Brain Res 210:91–115

Gallistel CR (1980) The organization of action: a new synthesis. Lawrence Erlbaum Associates, Inc, Hillsdale, p 432

Glansdorff P, Prigogine I (1971) Thermodynamics theory of structure, stability and fluctuations. Wiley-Interscience, London

Granit R (1955) Receptors and sensory perception. Yale University Press, New Haven

Henneman E, Somjen G, Carpenter DO (1965) Functional significance of cell size in spinal motoneurons. J Neurophysiol 28:560–580

Hollerbach JM (1982) Computers, brains, and the control of movement. Trends Neurosci 5: 189–192

Horak F, Diener H, Nashner L (1989) Influence of central set on human postural responses. J Neurophysiol 62:841–853

Kanda K, Burke RE, Walmsley B (1977) Differential control of fast and slow twitch motor units in the decerebrate cat. Exp Brain Res 8:57–74

Krasovsky T, Lamontagne A, Feldman AG, Levin MF (2014) Effects of walking speed on gait stability and interlimb coordination in younger and older adults. Gait Posture 39:378–385

Lackner JR, Dizio P (1994) Rapid adaptation to Coriolis force perturbations of arm trajectory. J Neurophysiol 72:299–313

Matthews PB (1981) Evolving views on the internal operation and functional role of the muscle spindle. J Physiol 320:1–30

Mattos D, Schoner G, Zatsiorsky VM, Latash ML (2015) Motor equivalence during multi-finger accurate force production. Exp Brain Res 233:487–502

McCloskey DI (1978) Corollary discharges: motor commands and perception. In: Brooks V (ed) Handbook of physiology, motor control, vol 2. American Physiological Society, Bethesda, MD, pp 1415–47

Nashner LM (1976) Adapting reflexes controlling the human posture. Exp Brain Res 26:59–72

Pilon JF, Feldman AG (2006) Threshold control of motor actions prevents destabilizing effects of proprioceptive delays. Exp Brain Res 174:229–239

Proske U, Gandevia SC (2009) The kinaesthetic senses. J Physiol 587(Pt 17):4139–4146

Schieppati M, Nardone A (1995) Time course of 'set'-related changes in muscle responses to stance perturbation in humans. J Physiol 487:787–796

Sperry RW (1950) Neural basis of the spontaneous optokinetic response produced by visual inversion. J Comp Physiol Psychol 43:482–489

Steinbuch J (1811). Beytrag Zur Physiologie Der Sinne (Contribution to the Physiology of Senses). Verlag, Nuremberg, p 408

von Helmholtz H (1866) Handbuch der Physiologischen Optik. Voss, Leipzig

Von Holst H (1954) Relations between the central nervous system and the peripheral organs. Br J Anim Behav 2:89–94

Von Holst E, Mittelstaedt H (1950/1973) Das reafferezprincip. Wechselwirkungen zwischen Zentralnerven-system und Peripherie, Naturwissenschaften 37: 467–476 (The reafference principle. In: Martin R (Trans), The behavioral physiology of animals and man. The collected papers of Erich von Holst. University of Miami Press, Coral Gables, FL, pp 139–173, 176–209)

von Uexküll J (1926) Theoretical Biology. Harcourt, Brace & Co, New York

Wolpert DM, Kawato M (1998) Multiple paired forward and inverse models for motor control. Neural Netw 11(7–8):1317–1329

Zhang L, Straube A, Eggert T (2016) Under threshold position control, peripheral mechanisms compensate efficiently for small perturbations of arm movements. Mot Control 20:87–108

# Principles of Motor Recovery After Neurological Injury Based on a Motor Control Theory

**Mindy F. Levin**

**Abstract** Problems of neurological rehabilitation are considered based on two levels of the International Classification of Functioning (ICF)—Body Structures and Function level and Activity level—and modulating factors related to the individual and the environment. Specifically, at the Body Structures and Function level, problems addressed include spasticity, muscle weakness, disordered muscle activation patterns and disruptions in coordinated movement. At the Activity level, deficits in multi-joint and multi-segment upper limb reaching movements are reviewed. We address how physiologically well established principles in the control of actions, Threshold Control and Referent Control as outlined in the Equilibrium-Point theory can help advance the understanding of underlying deficits that may limit recovery at each level.

**Keywords** Joint angle · International Classification of Function (ICF) · Stroke · Spasticity · Hand dysfunction · Trunk restraint · Virtual reality

## Introduction

In this chapter, problems of neurological rehabilitation are considered based on the International Classification of Functioning (ICF) to address how physiologically well established principles in the control of actions, such as those of the Equilibrium-Point theory can help advance the understanding and methods of treatment of neurological deficits.

One of the major goals of neurological rehabilitation is the identification of motor relearning techniques for facilitating recovery of functional arm and leg movements. Most of the work in this area has been done in patients with unilateral brain lesions due to stroke. Stroke is a major cause of long-term disability world-

M.F. Levin (✉)
School of Physical and Occupational Therapy, McGill University,
3654 Promenade Sir William Osler, Montreal, QC H3G 1Y5, Canada
e-mail: mindy.levin@mcgill.ca



J. Laczko and M.L. Latash (eds.), *Progress in Motor Control*,
Advances in Experimental Medicine and Biology 957,
DOI 10.1007/978-3-319-47313-0_7

wide (Kwakkel et al. 2003). The most widely recognized problem in survivors of stroke is motor impairment that restricts voluntary, well-coordinated, and effective movements predominantly on one side of the body (hemiparesis). In particular, long-term impairment of the upper limb on the side of the body contralateral to the brain lesion affects up to 75 % of people with stroke (Clarke et al. 2002), which limits the ability of an individual to perform the full range of daily life tasks and restricts their level of participation in society. Despite the large amount of research focusing on identifying underlying post-stroke sensorimotor deficits, it is still unclear which deficit or combination of deficits limits motor recovery. Consequently, there is a paucity of good evidence linking motor retraining approaches to effective upper limb functional recovery (Teasell et al. 2006).

Descriptions of post-stroke sensorimotor deficits are related to different levels of the International Classification of Functioning (ICF; Fig. 1) developed by the World Health Organization (WHO 2001). The ICF is an established globally-accepted framework to describe, assess, and compare function and disability which has been used to guide research and clinical practice. Based on the health condition, the ICF framework categorizes the spectrum of disability into three levels: Body Structures and Function, Activity, and Participation. It also identifies interactions between these levels associated with personal and environmental factors.

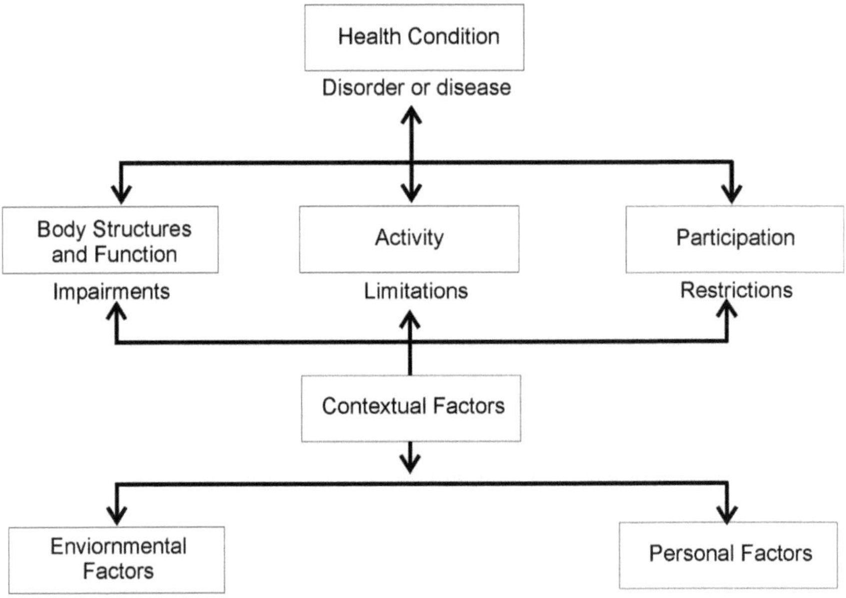


**Fig. 1** The International Classification of Function as a framework for the description, assessment, and comparison of function and disability, in which three levels are identified: (1) Body Structures and Function, (2) Activity, and (3) Participation, also termed, respectively, as impairments, limitations, and restrictions. These levels are modified by contextual factors with respect to the environment and the person. Modified from http://www.who.int/classifications/icf/en/

Recovery of functional movement is also related to behavioral, emotional, and cognitive processes (Hellstrom et al. 2003; Mayo et al. 2009; Robinson-Smith 2002; Rochat et al. 2013; Skidmore et al. 2010; Cisek 2007). Cognitive deficits in particular, are common after stroke affecting 32–56 % of patients and predict rehabilitation outcome, including motor recovery (Heruti et al. 2002; Hochstenbach et al. 1998; Skidmore et al. 2010; Toglia et al. 2011; Wagle et al. 2011). In addition to focal deficits, cognitive functions most commonly impaired after stroke are attention, working memory, processing speed, and executive function (Ballard et al. 2003; Jaillard et al. 2009). These have been linked to difficulties in initiating activities, solving motor problems and maintaining response consistently. Deficits in attention and executive function could exacerbate problems of planning and adapting movement trajectories in unpredictable environments, inhibiting and switching to another plan when unexpected obstacles arise, and using sensory feedback in subsequent motor actions. For example, in people with stroke training in a physical environment, better motor performance outcomes for reaching tasks were related to fewer deficits in memory and problem-solving (Cirstea et al. 2006; Dancause et al. 2002). Similarly, better success rates and response times for upper limb tasks were related to fewer deficits in cognitive flexibility and attention, respectively, for patients practicing reaching tasks with feedback on specific motor elements in a virtual reality environment (Kizony et al. 2004). Stroke severity has also been linked to disruption in implicit learning (Boyd et al. 2007), which is essential to neuroplasticity (Doyon and Benali 2005; Hallett et al. 1996) and involves skill learning without being aware of the rules that govern the specific performance or how the skill is actually learned.

From a behavioral and emotional perspective, depression and diminished self-efficacy also affect post-stroke recovery (De Ryck et al. 2014; Hacke 2003), quality of life (Lai et al. 2006; Robinson-Smith et al. 2000; Salbach et al. 2005) and post-rehabilitation improvement (Hellstrom et al. 2003). In particular, post-stroke subjects with depression were less able to use extrinsic feedback to improve motor performance (Subramanian and Levin 2013) and low upper limb self-efficacy was related to less arm use in a cohort of patients in the chronic phase of stroke (Baniña et al. 2016).

The construct of self-efficacy provides the theoretical basis for assessing the level of confidence of an individual to accomplish a given task (Bandura 1986). It is a situation-specific form of self-confidence and a central mechanism of social cognitive theory that regulates choice of actions (Bandura 1997). Decreased confidence may interfere with independent task accomplishment (Gage et al. 1997), especially, if it impacts the ability to solve the problem around in unexpected difficulties. Thus, patients with good motor skills but low self-efficacy are likely to perform below their true capabilities. Low self-efficacy may contribute to the discrepancy between upper limb activity capacity and arm use in everyday activities based on accelerometry monitoring (e.g., Urbin et al. 2015). The ideal match is when perceived self-efficacy is slightly higher than actual skills (Bandura 1986).

Deficits at the first two levels of the ICF model (Body Structures and Function, Activity) are of interest when discussing the basic motor control factors that may influence motor recovery.

## Body Structures and Function Level

Measures characterizing motor behaviors at the ICF *Body Structures and Function (impairment) level* include spasticity, decreased range of motion, muscle weakness, and impaired ability to isolate movements of individual joints and to perform coordinated movement of multiple joints (e.g., Brunnstrom 1966; Fugl-Meyer et al. 1975; Gowland et al. 1993).

Spasticity, defined as a motor disorder characterized by a velocity-dependent increase in tonic stretch reflexes ("muscle tone") with exaggerated tendon jerks (Lance 1980), develops gradually after the acute phase of stroke and persists into the chronic phase of recovery. The prevalence of post-stroke spasticity is highly variable, ranging from 17 to 42.6 % in chronic stroke (Wissel et al. 2013). The presence of spasticity may leads to secondary complications such as pain and muscle contractures as well as to difficulties in performing activities of daily living and to a reduced quality of life (Zorowitz et al. 2013). In standard clinical assessment, spasticity is characterized as the amount of resistance felt by the examiner when the muscle is stretched from a shortened to a lengthened position, regardless of the speed of stretch. Measured in this way, however, few studies have related the amount of spasticity to the functional motor deficit in the upper limb. However, for the lower limb, one study in post-stroke patients showed that velocity-sensitive stretch responses in the calf were correlated with low plantarflexor activation levels during the stance phase of gait. These responses were positively related ($r = 0.47$, $p < 0.01$; $r = 0.57$, $p < 0.001$) to the amount of muscle resistance evoked by stretch of the plantarflexors at rest, and negatively related to gait speed ($r = -0.47$ to $-0.53$, $p < 0.01$; Lamontagne et al. 2001). Although described behaviorally, the precise relationship between hypertonicity, hyperactive reflexes, weakness, and the production of voluntary movement has not been elucidated. Since disorders of muscle tone and movement deficits have been traditionally been considered as separate phenomena, the controversy is likely related to the lack of a coherent view of this relationship within the context of a motor control theory (see below).

During voluntary movement production, individuals with hemiparesis have varying degrees of spatial and temporal coordination deficits between movements of adjacent joints even for simple arm movements (elbow/shoulder, Beer et al. 2000; Cirstea et al. 2003a, b; Dewald et al. 1995; Levin 1996), when synchronizing hand orientation with hand opening/closing during reach-to-grasp movements to stationary targets (Michaelsen et al. 2004, 2006; Roby-Brami et al. 2003), and in modulation of precision grip forces according to task demands (Hermsdorfer et al. 1999; Wenzelburger et al. 2005). An example of how simple reaching movement to

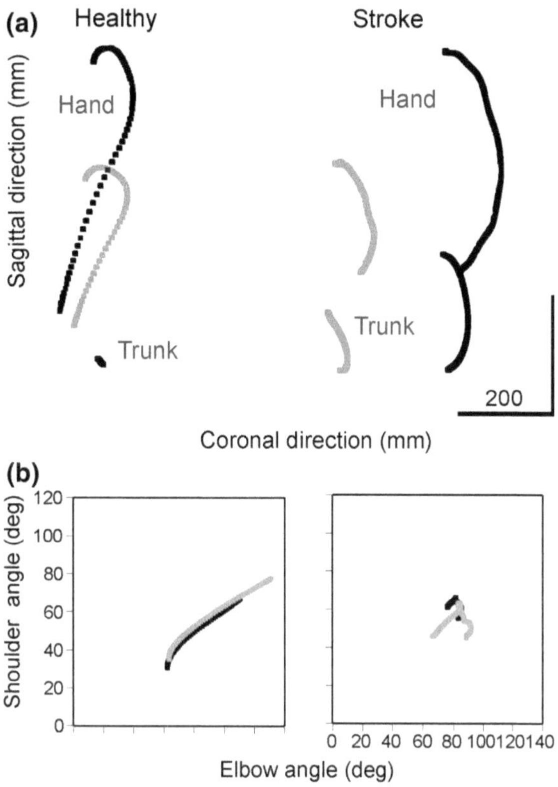


**Fig. 2** Examples of reaching trajectories of the hand for targets located in the sagittal midline at 2/3 arm's length (*grey lines*) and at arm's length (*black lines*) for one healthy individual (*left*), and one patient with moderate post-stroke hemiparesis (*right*). Traces for the close and far targets are separated in the *right* panel for clarity. **a** Excessive trunk movement is evident for each of the reaches in the patient. **b** Shoulder/elbow interjoint coordination patterns for reaches to close (*black lines*) and far (*grey lines*) targets for each of the reaches shown in (**a**). Figure shows disruption of interjoint coordination in the patient with stroke

a midline target is disrupted in patients with stroke is shown in Fig. 2. The figure illustrates two aspects of disrupted reaching in the spatial domain in a subject with moderately severe hemiparesis: excessive use of the trunk for reaching targets located both close and far from the body (Fig. 2a) and a deficit in the spatial coordination between movements of adjacent shoulder and elbow joints (Fig. 2b; Michaelsen et al. 2001). Temporal relationships between elbow and shoulder movements are also disrupted during rapid alternating pointing movements with both joints being recruited simultaneously and stereotypically compared to a more temporally sequential recruitment in age-matched healthy control subjects (Rodrigues et al. 2015).

## *Explanation of Motor Deficits at the Body Structures and Function Level Based on Equilibrium-Point Theory*

One mechanism that may underlie problems in movement production of the upper limb is disruption in the ability of the central nervous system to regulate stretch reflex thresholds of muscles during single (Levin et al. 2000) and multi-joint movements (Musampa et al. 2007). According to the principle of Threshold Control based on the Equilibrium-Point theory (Feldman 1966, 2015; Feldman et al. 2007), the healthy nervous system regulates muscle activation of a single muscle or group of muscles by shifting the spatial thresholds associated with the position of body segments at which the muscles begin to be recruited. The threshold regulation occurs through descending and peripheral influences on alpha motoneurons in the spinal cord. Through excitation or inhibition, afferent systems bring motoneurons further from or closer to their activation thresholds. The Equilibrium-Point theory describes a mechanism by which electrical thresholds are converted into the spatial domain so that a motoneuronal activation threshold is expressed as a joint angle at which the muscle begins to be active. Once initiated, muscle activation proceeds according to the biomechanics of the musculoskeletal system and its physical interactions with forces encountered in the external world, including gravity. In the healthy nervous system, the threshold of motoneuronal activation is regulated throughout and beyond the equivalent biomechanical limitations of the joint in order for muscle forces to be produced to balance external loads at any joint angle. This is illustrated in Fig. 3. According to the principle of Threshold Control, in the healthy nervous system, in order to relax a muscle throughout the entire biomechanical range, the spatial threshold of that muscle should be moved beyond the upper limit of the biomechanical range of the joint ($TSRT_-$ and $TSRT_+$ in Fig. 3a; Levin and Feldman 1994). The postulated mechanism by which this is done is through hyperpolarization of motoneurons. In order to balance a particular external load at a shorter muscle length within the biomechanical range, the muscle should be appropriately activated. This is achieved by shifting the threshold to the left, into this range. In order to activate the muscles even at the shortest muscle length, the threshold should be set below the lower limit of the biomechanical range. By regulating the threshold position, the nervous system can specify any level of joint torque at any muscle length.

In neurological patients, damage to transmission in descending pathways has been shown to result in limitations in the ability of the nervous system to regulate the spatial thresholds in the normal range. After neurological injury, at rest, the tonic stretch reflex (spatial) threshold (TSRT) abnormally lies inside of the biomechanical range when measured in the elbow flexors in patients with stroke (Levin et al. 2000; Musampa et al. 2007; Calota et al. 2008), Parkinson's disease (Mullick et al. 2013) and Cerebral palsy (Jobin and Levin 2000) as well as in the ankle joint of patients with stroke (Blanchette et al. 2016). The implication of the reduced or absent modulation of the angular limits of TSRTs is a limitation in the range in which a patient can produce active movement of that joint. Thus, for a patient with spatial thresholds of flexors and extensors fixed between 100 and 150° elbow extension, voluntary movement (i.e., elbow flexion and extension) can only be controlled within that limited joint range (Fig. 3b). Attempts to extend the joint beyond the TSRT of the flexors by contraction of the elbow extensors, result in abnormal flexor muscle coactivation. However, small amplitude movements that do not require extension of the elbow beyond the 150° upper limit are characterized by normal muscle reciprocal activation patterns. According to the principle of

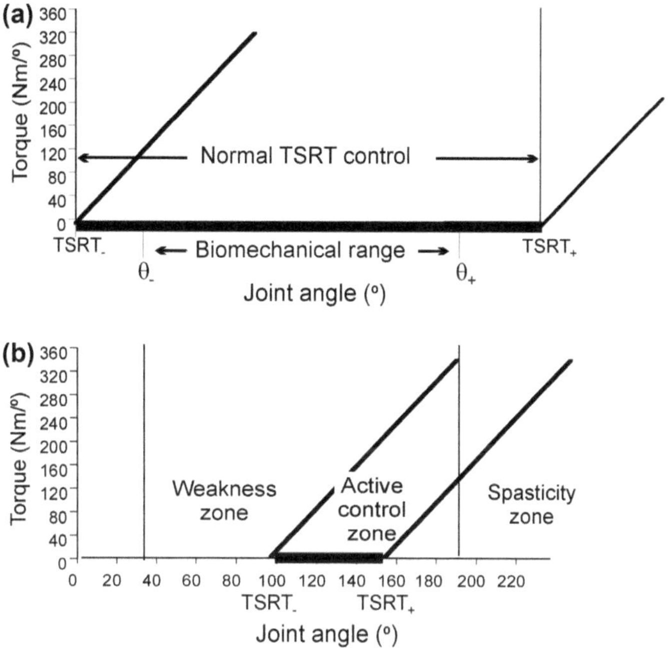


**Fig. 3** Principle of Threshold Control. **a** In the healthy nervous system, the tonic stretch reflex threshold (TSRT) can be regulated in the range $TSRT_-$ to $TSRT_+$ that exceeds the biomechanical range of the joint ($\Theta_-$ to $\Theta_+$). Torques can be balanced at any muscle length by shifting the TSRT along with the joint torque/angle characteristic (*diagonal black line*) into the biomechanical range and beyond. **b** After neurological injury, one or both limits of the TSRT abnormally lie inside of the biomechanical range. In the example shown, the range of joint angles between $TSRT_-$ and $TSRT_+$ is the angular zone in which normal muscle activation occurs in both flexors and extensors (Active control zone). In the angular zone, to the *right* of the $TSRT_+$, muscles cannot be relaxed (Spasticity zone) and in the area to the *left* of the $TSRT_-$, muscle weakness occurs

Threshold Control, the angular zone in which normal agonist/antagonist muscle activation occurs is termed the "active control zone," while that in which abnormal agonist/antagonist coactivation occurs is called the "spasticity zone." This terminology should not be confused with the clinical term "spastic restraint," which has traditionally been used to describe abnormal resistance occurring in antagonist muscles without consideration of the spatial zone in which it occurs. For the flexors, the angular range below 100°, would be characterized by weakness because of spastic restraint of the extensors.

Spatial thresholds in stretched elbow flexors were increased by approximately 40° during slow active elbow extension compared to those measured at rest in 10 patients with chronic stroke. This increase in threshold position during voluntary antagonist (extensor muscle) contraction is likely due to reciprocal inhibition from the extensors onto the flexors (Hultborn 2006). This suggests that reciprocal inhibitory mechanisms may be preserved in the forearm of patients with stroke that can reduce the angular range in which spasticity occurs during voluntary movement.

Deficits in threshold regulation may also partially explain movement slowness in people who have had a stroke. Aside from decreases in the speed of cognitive processes, movements may be slowed when attempts to extend the arm beyond the threshold for activation of the elbow flexors result in coactivation of antagonists, due to movements into the spasticity zone.

Deficits in spatial threshold regulation may also account for some of the observed deficits in interjoint and/or intersegment coordination. Thus, it is likely that active reaching movements may be affected by the limitation in the range of regulation of the spatial thresholds in some or all of the muscles moving the limb, resulting in loss of normal interjoint coordination and/or the appearance of new interjoint coupling relationships (e.g., abnormal synergies).

## Activity Level

Limitations at the ICF *Activity level* include difficulties in executing functional motor tasks such as opening a jar, turning a key, or holding a pen. Despite intensive rehabilitation, motor problems persist into the chronic stage after stroke and patients still underutilize the more-affected arm in everyday activities (Brooks et al. 1999; Platz and Denzler 2002; Rand and Eng 2012, 2015). It has been repeatedly shown that even though patients may score high on clinical tests measuring impairment or activity limitations; they do fully use their upper limb, as monitored with accelerometers, in daily activities in the home setting (e.g., Rand and Eng 2015). Furthermore, daily use of the less-affected arm has been reported to be up to three times greater than that of the more-affected upper limb.

Loss of coordination is a key barrier to regaining functional arm use, yet motor control aspects of coordination, such as the spatial and temporal structure of coordinated movement, have not been sufficiently characterized and targeted in stroke rehabilitation. Indeed, little attention has been paid to the recovery of higher order motor control skills, which include those that have higher attentional and or decisional components (Cisek 2007; Cos et al. 2012).

Specifically, although daily activities often involve reaching around obstacles, upper limb obstacle avoidance behavior has received relatively little attention. Reaching trajectories are controlled so that the mid-trajectory curvature maximum is located just as the obstacle is cleared (Abed et al. 1982), with minimal clearance points organized within a preferred axis regardless of target and obstacle orientations (Dean and Brüwer 1994; Sables et al. 1998). Reaches are thought to be planned using anisotropic "near points" to the obstacle in consideration of limb kinematic and dynamic properties (Sables et al. 1998). When grasping is involved, apertures are decreased and limb transport is adjusted so that aperture closure is maintained at a constant distance from the target regardless of trajectory adaptations made for obstacle avoidance (Alberts et al. 2002; Saling et al. 1998) while movement time is prolonged (Biegstraaten et al. 2003). Thus, the CNS uses hand and object spatial information in hand transport planning to produce closely coordinated grasp and transport components (for review, see Sangole and Levin 2007).

Effects of distracting objects flanking a target have been studied during reaching in healthy subjects. Changes in transport trajectories occur even when flanking objects do not block the movement path (Chapman and Goodale 2008; Tresilian 1998), but their orientation and location influence reaching behavior differently (Chapman and Goodale 2008; Mon-Williams and McIntosh 2000; Mon-Williams et al. 2001). Grasp apertures, however, are only affected when the object is directly obstructed. Changes are speed dependent and occur even in children as young as 8-years old (Tresilian et al. 2005). However, very few studies have characterized obstructed reaching behavior in patients with stroke. The few studies in patients with neglect (McIntosh et al. 2004), optic ataxia (Schindler et al. 2004) or visual agnosia (Rice et al. 2006) suggest that hand guidance to reach around an obstacle relies on dorsal stream visual processing while reaching between obstacles (bisecting) requires more visual attention involving ventral stream processing. In healthy subjects, even complex movements made in real environments are relatively automatic, requiring limited attention (Doyon and Benali 2005). Movement automaticity has been studied using dual-task paradigms, in which a motor task competes with a secondary (often cognitive) task for the same attentional resources (Huang and Mercer 2001). Most studies of dual-tasking involve locomotor activities combined with increased cognitive loading. When the motor task requires disproportionate attention, decreased performance in the form of slower or less accurate movement occurs due to dual-task interference (Kahneman 1973). In one of the few studies of the effects of dual-tasking on upper limb function in stroke, Houwink et al. (2013) found that increasing cognitive load via an auditory Stroop task impaired circle drawing performance in a small group of subjects with moderate upper limb paresis. More severely affected stroke patients required more attentional resources to control the limb causing a detriment in dual-task performance. They suggested that improvement in dual-task performance occurred due to improved movement automaticity, but this motor element was not objectively quantified. They also noted a discrepancy between motor capacity and actual arm use and indicated the need to identify underlying processes responsible for spontaneous arm use.

We recently identified spatiotemporal coordination deficits during an obstacle avoidance task (Baniña et al. 2016) in patients with chronic stroke resulting in mild-to-moderate hemiparesis. Patients had persistent deficits in their ability to avoid hitting a sliding door that partially obstructed a target object (juice can) on a shelf behind it when reaching from sitting. They also had smaller margins of error when reaching around the obstacle and lower success rates than healthy age-matched controls. These results suggested that diminished arm use in patients with stroke may be related to reduced arm and hand coordination to produce effective interactions with a changing environment (Baniña et al. 2016; Dancause et al. 2002; Feldman et al. 2007). We also identified that the ability to rapidly correct errors and respond to changing environmental conditions was related to the person's level of self-confidence (self-efficacy) in using the upper limb in daily activities.

Successful obstacle avoidance behavior may rely on many factors including cognitive skills such as mental flexibility to make rapid decisions and choose appropriate responses. Knowledge gaps remain about higher order motor control deficits in stroke and how deficits in multiple systems (motor, cognitive, psychosocial) relate to diminished arm use.

## *Explanation of Motor Deficits at the Activity Level Based on Equilibrium-Point Theory*

The ability to interact with one's arm in a meaningful way in real-world, complex and/or unpredictable environments, depends on the coordination of multiple degrees of the freedom (DFs) in a dexterous way (Bernstein 1967; Latash 2012). Reaching movements involve the coordination of multiple (>7) DFs of the arm to transport the hand from one position to another in 3D Cartesian space. Rather than controlling each joint rotation separately, the nervous system shifts the endpoint (usually the hand for reaching movements) from one stable position to another, while the coordination between multiple DFs emerges from the biomechanical and intermuscular neural interactions between body segments and the physical environment (Feldman 2011). The action of displacing the endpoint from one position to another is described by a shift in the referent arm configuration so that movement emerges due to the deviation of the actual arm configuration from the referent one specified by the nervous system (Asatryan and Feldman 1965; Feldman 2011; see Chap. 6).

The concept of referent control has solid empirical support and provides a solution to the classical posture-movement problem described by Von Holst (1954). Referent control is based on the predetermination of a "referent body configuration" by the brain in anticipation of actual movement. It extends the principle of Threshold Control from a single muscle or joint to a whole body segment or whole body posture. Muscle activation occurs according to the deviation of the actual body configuration (Q; i.e., the current position of the body) from the referent body configuration (R). The R is a tool that allows the nervous system to control multiple muscles of the body as a coherent unit. In the absence of muscle coactivation, R represents a common threshold configuration for a system with multiple DFs, i.e., the body position at which muscles begin to be recruited or de-recruited. This common threshold configuration can be shifted by reciprocal central influences on motoneurons (MNs) of agonist and antagonist muscles. By means of central co-facilitation of agonist and antagonist MNs, muscles can be coactivated at position R and throughout the range defined $R_-$ to $R_+$ (Asatryan and Feldman

1965). The referent body configuration is specified by the brain in a task-specific way. If the resulting action is not satisfactory, the referent position is modified to minimize the movement error.

The production of upper limb reaching or pointing movements can be considered within the framework of referent control. To produce a pointing movement, the nervous system changes the referent body configuration from the initial to a final position, resulting in a referent trajectory ($R_T$) of the endpoint (e.g., the tip of index finger, Fig. 4). Because of the inertia of body segments, external forces and neuromuscular dynamics, the actual endpoint trajectory ($Q_T$) may not necessarily coincide with the referent one, but the same system tends to minimize the deviation of $Q_T$ from $R_T$. Transmitted to motoneurons, the difference between the $Q_T$ and $R_T$ influence muscle activation thresholds (λs), thus producing motion of body segments. In dynamical terms, the referent trajectory represents an attractor to which the actual trajectory tends to converge (Won and Hogan 1995; Ghafouri and Feldman 2001). The convergence process does not imply that the target position is represented by the final $R_T$. In addition, the system may not always reach the absolute minimum (zero) at the end of the movement because of external and internal constraints (e.g., gravitational forces, additional external loads and stability requirements). Within these limitations, the nervous system adjusts the rate, direction, and extent of changes in $R_T$ such that the endpoint reaches an equilibrium position ($eQ_T$) at the target position without an error, at which muscle and external forces (torques) are balanced.

The minimization process can involve the arm as well as other body segments. For example, when reaching to a target located beyond the arm length, changes in

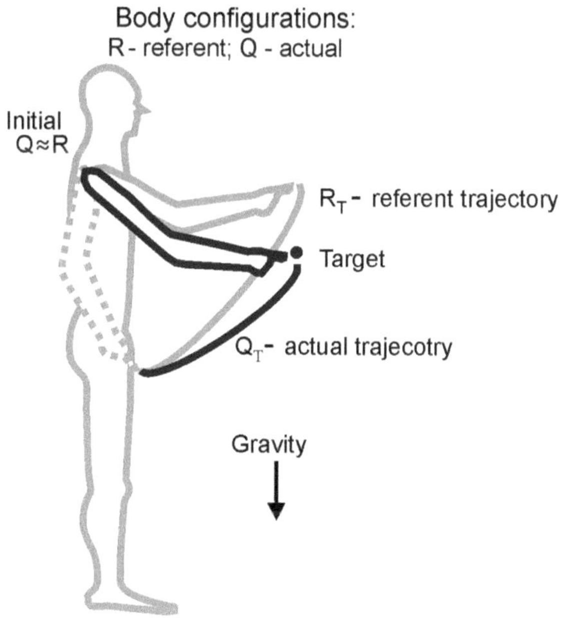


**Fig. 4** Referent body configuration for pointing. Representation of referent (R) and actual (Q) arm configurations for producing a pointing movement. From the initial arm configuration (R, *grey dashed line*), the central nervous system changes the referent arm configuration resulting in a referent trajectory ($R_T$) of the fingertip. Due to gravity and other external forces, the actual arm trajectory ($Q_T$) is deflected downward from $R_T$, bringing the fingertip to the target. The difference between $R_T$ and $Q_T$ leads to the generation of muscle activity

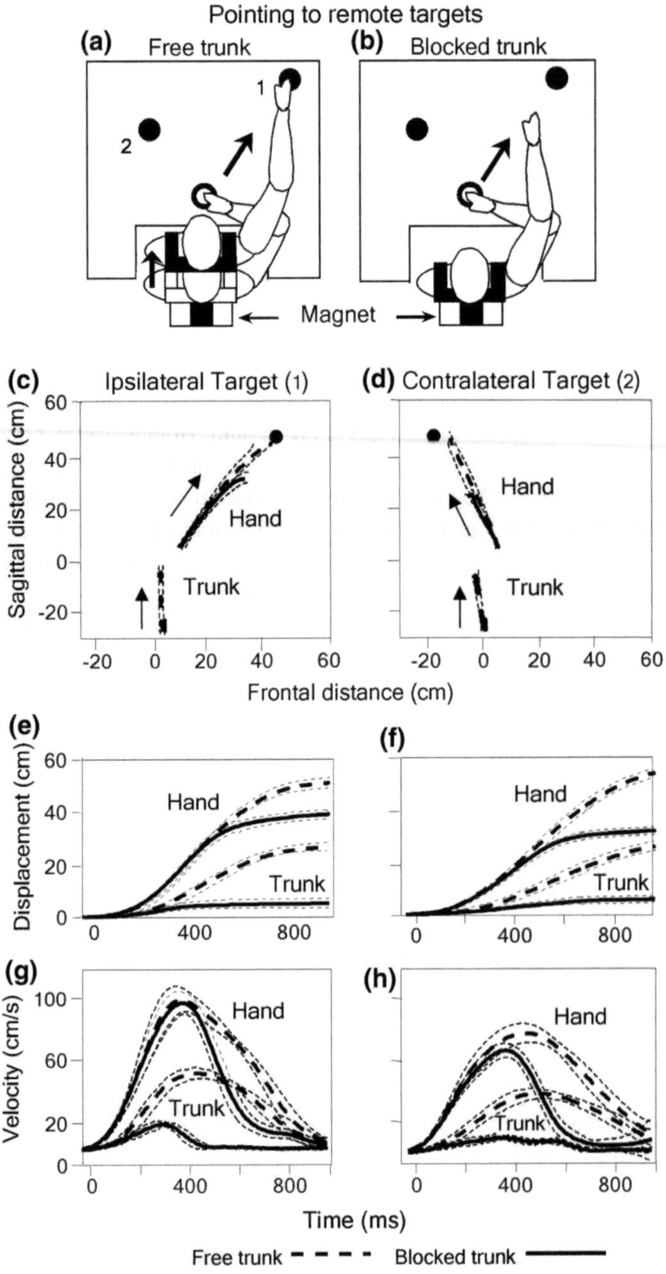


the referent trajectory are provided by appropriate alterations in R which include trunk movement to contribute to endpoint motion extent (Rossi et al. 2002). The minimization rule thus suggests that the same pointing trajectory can be produced

◀ **Fig. 5** Referent control of reaching for targets beyond the arm reach. Although they move simultaneously, the arm and trunk contribute sequentially to the hand motion extent. **a**, **b** Subject moved the hand to target placed beyond the arm reach, which necessitated forward trunk displacement. In randomly chosen trials, trunk motion was prevented by an electromagnet. Example shown is for pointing to the ipsilateral target. **c–h** The hand and trunk began to move simultaneously (**e**, **f**). Blocking the trunk had no effect on the hand trajectory or velocity profile until about the peak hand velocity. Thus, the trunk began to contribute to the hand movement extent after the offset of referent arm control but before that it was accelerated in preparation for its future contribution. Reproduced with permission from Rossi et al. (2002). Copyright 2002 John Wiley and Sons

regardless of the number of DFs involved, which provides an explanation of the concept of "motor equivalence" introduced by Lashley (1951). Accordingly, the actual degrees of freedom recruited for a reaching movement depend on the biomechanical relationships between the joints or segments, preferences of the individual and environmental constraints. Thus, the system self-organizes the recruitment of different degrees of freedom according to environmental or contextual constraints to ensure that the hand reaches the desired target. In a study of trunk-assisted reach-to-grasp movements Yang and Feldman (2010) showed that the minimization process consists of gradual, continuous, task-specific shifts in the referent positions of the arm DFs, while preparing the future involvement of other DFs (i.e., trunk, hand).

In healthy subjects making reaching movements involving trunk displacement (i.e., trunk-assisted reaching in sitting), a specific arm and trunk recruitment sequence ensures the production of accurate endpoint movement. The initial contribution of the trunk movement to hand displacement is neutralized by compensatory shoulder and elbow rotations (Fig. 5; Rossi et al. 2002). Trunk movement begins to contribute to hand displacement only after the peak hand velocity is reached. This elegant coordination is generally preserved in some patients who have had a stroke (with mild-to-moderate sensorimotor impairment) and used to produce invariant hand trajectories even when reaching involves trunk displacement or when reaching from standing. In contrast, in patients with mild-to-moderate stroke, trunk displacement makes a larger and earlier contribution to hand transport (compensatory movement) for both within and beyond arm length reaches (Cirstea et al. 2003a; Michaelsen et al. 2006; Levin et al. 2000, 2002; Shaikh et al. 2014). The incorporation of the trunk segment into the kinematic pattern of reaching movement may be a consequence of the need for the nervous system to reorganize the recruitment of the available DFs due to limitations in some of them.

The ability of the damaged nervous system to preserve invariant hand trajectories was tested by comparing hand trajectories and elbow–shoulder interjoint coordination patterns during beyond the reach pointing movements in healthy and stroke subjects when the trunk was free to move or unexpectedly arrested (Radanovich et al. 2001; Archambault et al. 1999; Shaikh et al. 2014; Fig. 6). In healthy subjects, trajectory invariance was maintained by changing the arm interjoint coordination. However, in most stroke subjects, the invariance of the hand trajectories was not completely preserved and trajectories deviated from the desired

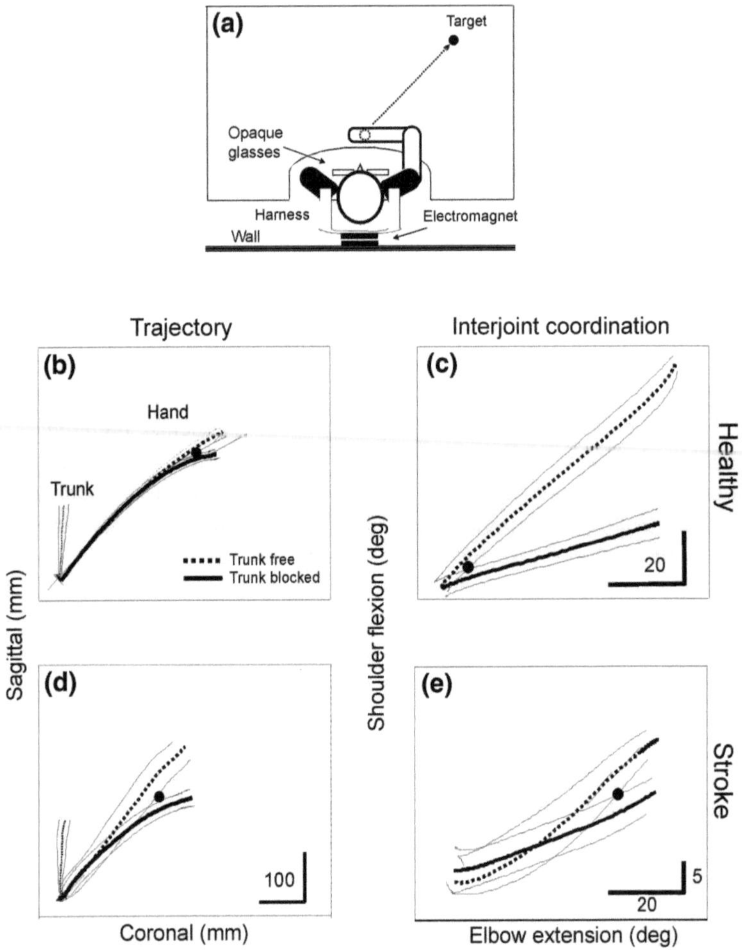


**Fig. 6** Arm-trunk incoordination during reaching in a post-stroke subject. Comparison of reaching movements made to a beyond-the-reach target (**a**) in a representative healthy subject (**b**, **d**) and in a patient with moderate post-stroke hemiparesis (**c**, **e**). Reaching was done with eyes closed to the remembered target and on unexpected trials, the trunk movement was blocked with an electromagnet (**a**). Mean (SD) hand and trunk trajectories (**b**, **c**) for trunk free (*dotted lines*) and trunk blocked (*solid lines*) trials are shown. Hand trajectories of both types of trials initially coincided and then diverged. In the patient with stroke (**d**), the trajectory divergence occurred earlier compared to the healthy subject (**b**). Healthy subject maintained the trajectory invariance by using a different shoulder elbow coordination in the two conditions (**c**). In the patient with stroke (**e**), the interjoint coordination pattern was not modified according to the task

path earlier than in healthy subjects. The earlier trajectory deviation was associated with a later divergence of interjoint coordination patterns indicating a lack of flexibility in adapting arm joint rotations to account for the added trunk movement. Using a mathematical analysis of kinematic variability (uncontrolled manifold) during whole arm reaching movements from sitting, Reisman and Schultz (2003) found that people with mild-to-moderate hemiparesis had deficits coordinating specific joint movements and were only partially able to rapidly compensate movement errors. Problems in adapting movements (error correction) to unexpected external conditions has also been shown for single-joint elbow movements by Dancause et al. (2002) who further related the error compensation deficits to impairments in executive functioning in patients with chronic stroke. Deficits have also been reported in error compensation for regaining interim phase relationships after arm perturbations during arm swinging (Ustinova et al. 2006).

## Take Home Message

This chapter provides examples of how transfer of knowledge from basic science to clinical applications can fill important gaps in our understanding of disordered upper limb movement, loss of coordinated movement, diminished use of the arm as well as mechanisms limiting motor recovery after neurological injury or disease. Impairments in single-joint control at the Body Structures and Function level of the ICF and deficits in the production of higher order coordinated movement at the Activity level of the ICF are considered in light of two interrelated principles of the Equilibrium-Point Theory: Threshold Control and Referent Control. Explanations of motor deficits based on these principles provide knowledge about the underlying mechanisms of disordered control of movement beyond simple descriptions of the consequences of these disorders (e.g., decreased force output) that can be used to develop better clinical tests to identify and quantify upper limb coordination deficits, which is currently lacking in clinical practice. At the same time, a better understanding of motor deficits in patients with neurological lesions may also advance our understanding of normal motor control. Understanding upper limb motor control deficits as diminished threshold control suggests that interventions aimed at increasing this range may result in better motor recovery. To this end, different types of interventions that can be explored include pharmacological and physical training approaches.

## References

Abend W, Bizzi E, Morasso P (1982) Human arm trajectory formation. Brain 105:331–348

Adamovich SV, Archambault PS, Ghafouri M, Levin MF, Poizner H, Feldman AG (2001) Hand trajectory invariance in reaching movements involving the trunk. Exp Brain Res 138:288–303

Alberts JL, Saling M, Stelmach GE (2002) Alterations in transport path differentially affect temporal and spatial movement parameters. Exp Brain Res 143:417–425

Archambault P, Pigeon P, Feldman AG, Levin MF (1999) Recruitment and sequencing of different degrees of freedom during pointing movements involving the trunk in healthy and hemiparetic subjects. Exp Brain Res 126:55–67

Asatryan DG, Feldman AG (1965) Functional tuning of the nervous system with control of movements or maintenance of a steady posture: I. Mechanographic analysis for the work of the joint on execution of a postural tasks. Biophysics 10:925–935

Ballard C, Stephens S, Kenny R, Kalaria R, Tovee M, O'Brien J (2003) Profile of neuropsychological deficits in older stroke survivors without dementia. Dement Geriatr Cogn Disord 16:52–56

Blanchette AK, Mullick AA, Moïn-Darbari K, Levin MF (2016) Tonic stretch reflex threshold as a measure of ankle plantar-flexor spasticity after stroke. Phys Ther 96(5):687–95

Bandura A (1986) Social foundations of thoughts and actions: a social cognitive theory. Prentice-Hall, New Jersey

Bandura A (1997) Self-efficacy: the exercise of control. Freeman, New York

Baniña MC, Mullick AA, McFadyen BJ, Levin MF (2016) Upper limb obstacle avoidance behaviour in individuals with chronic stroke. Neurorehabil Neural Repair, pii: 1545968316662527. [Epub ahead of print], PMID: 27542986

Beer RF, Dewald JP, Rymer WZ (2000) Deficits in the coordination of multijoint arm movements in patients with hemiparesis: evidence for disturbed control of limb dynamics. Exp Brain Res 131:305–319

Bernstein NA (1967) The coordination and regulation of movements. Pergamon Press, London

Biegstraaten M, Smeets JB, Brenner E (2003) The influence of obstacles on the speed of grasping. Exp Brain Res 149:530–534

Boyd LA, Quaney BM, Pohl PS, Winstein CJ (2007) Learning implicitly: effects of task and severity after stroke. Neurorehabil Neural Repair 21:444–454

Broeks JG, Lankhorst GJ, Rumping K, Prevo AJH (1999) The long-term outcome of arm function after stroke: results of a follow-up study. Disabil Rehab 21:357–364

Brunnstrom S (1966) Motor testing procedures in hemiplegia: based on sequential recovery stages. Phys Ther 46:357–375

Calota A, Feldman AG, Levin MF (2008) Spasticity measurement based on tonic stretch reflex threshold in stroke using a portable device. Clin Neurophysiol 119:2329–2337

Chapman CS, Goodale MA (2008) Missing in action: the effect of obstacle position and size on avoidance while reacing. Exp Brain Res 191:83–97

Cirstea MC, Mitnitski AB, Feldman AG, Levin MF (2003a) Interjoint coordination dynamics during reaching in stroke. Exp Brain Res 151(3):289–300

Cirstea MC, Ptito A, Levin MF (2003b) Arm reaching improvements with short-term practice depend on the severity of the motor deficit in stroke. Exp Brain Res 152:476–488

Cirstea CM, Ptito A, Levin MF (2006) Feedback and cognition in arm motor skill reacquisition after stroke. Stroke 37:1237–1242

Cisek P (2007) Cortical mechanisms of action selection: the affordance competition hypothesis. Phil Trans Roy Soc B 362:1585–1599

Clarke P, Marshall V, Black SE, Colantonio A (2002) Well-being after stroke in Canadian seniors: findings from the Canadian study of health and aging. Stroke 33:1016–1021

Cos I, Medley F, Cisek P (2012) The modulatory influence of endpoint controllability on decisions between actions. J Neurophysiol 108:1764–1780

Dancause N, Ptito A, Levin MF (2002) Error correction strategies for motor behavior after unilateral brain damage: short-term motor learning processes. Neuropsychologia 40: 1313–1323

De Ryck A, Fransen E, Brouns R, Geurden M, Peij D, Marien P, De Deyn PP, Engelborghs S (2014) Psychosocial problems associated with depression at 18 months poststroke. Int J Ger Psych 29:144–152

Dean J, Brüwer M (1994) Control of human arm movements in two dimensions: paths and joint control in avoiding simple linear obstacles. Exp Brain Res 97:497–514

Dewald JPA, Pope PS, Given JD, Buchanan TS, Rymer WZ (1995) Abnormal muscle coactivation patterns during isometric torque generation at the elbow and shoulder in hemiparetic subjects. Brain 118:495–510

Doyon J, Benali H (2005) Reorganization and plasticity in the adult brain during learning of motor skills. Curr Opin Neurobiol 15:161–167

Feldman AG (1966) Functional tuning of the nervous system with control of movement and maintenance of a steady posture: II. Controllable parameters of the muscle. Biophysics 11: 565–578

Feldman AG (2011) Space and time in the context of the equilibrium-point theory. Wiley Interdiscip Rev Cogn Sci 2:287–304

Feldman AG (2015) Referent control of action and perception. Springer, New York

Feldman AG, Goussev V, Sangole A, Levin MF (2007) Threshold position control and the principle of minimal interaction in motor actions. In: Cisek P, Drew T, Kalaska JF (ed) Progress in brain research, vol 165, pp 267–281

Fugl-Meyer AR, Jääskö L, Leyman I, Olsson S, Steglind S (1975) The post-stroke hemiplegic patient. I. A method for evaluation of physical performance. Scand J Rehab Med 7:13–31

Gage M, Valiant Cook J, Fryday-Field K (1997) Understanding the transition to community living after discharge from an acute care hospital: an exploratory study. Am J Occ Ther 51(2):96–103

Ghafouri M, Feldman AG (2001) The timing of control signals underlying fast point-to-point arm movements. Exp Brain Res 137:411–423

Gowland C, Stratford P, Ward M et al (1993) Measuring physical impairment and disability with the Chedoke-McMaster stroke assessment. Stroke 24:58–63

Hacke W (2003) European stroke initiative recommendations for stroke management—update 2003. Cerebrovasc Dis 16:311–337

Hallett M, Pascual-Leone A, Topka H (1996) Adaptation and skill learning: evidence for different neural substrates. In: Bloedel JR, Ebner TJ, Wise S (ed) The acquisition of motor behavior in vertebrates, MIT Press, Cambridge, pp 289–30

Hellstrom K, Lindmark B, Wahlberg B, Fugl-Meyer AR (2003) Self-efficacy in relation to impairments and activities of daily living disability in elderly patients with stroke: a prospective investigation. J Rehabil Med 35:202–207

Hermsdorfer J, Laimgruber K, Kerkhoff G, Mai N, Goldenberg G (1999) Effects of unilateral brain damage on coordination, and kinematics of ipsilesional prehension. Exp Brain Res 128:41–51

Heruti RJ, Lusky A, Dankner R et al (2002) Rehabilitation outcome of elderly patients after a first stroke: effect of cognitive status at admission on the functional outcome. Arch Phys Med Rehabil 83:742–749

Hochstenbach J, Mulder T, van Limbeek J, Donders R, Schoonwaldt H (1998) Cognitive decline following stroke: a comprehensive study of cognitive decline following stroke. J Clin Exp Neuropsychol 20:503–517

Houwink A, Steenbergen B, Prange GB, Buurke JH, Geurts ACH (2013) Upper-limb motor control in patients after stroke: attentional demands and the potential beneficial effects of arm support. Hum Movement Sci 32:377–387

Huang JH, Mercer VS (2001) Dual-task methodology: applications in studies of cognitive and motor performance in adults and children. Pediatric Phys Ther 13:133–140

Hultborn H (2006) Spinal reflexes, mechanisms and concepts: from Eccles to Lundberg and beyond. Prog Neurobiol 78:215–232

Jaillard A, Naegele B, Trabucco-Miguel S, LeBas J, Hommel M (2009) Hidden dysfunctioning in subacute stroke. Stroke 40:2473–2479

Jobin A, Levin MF (2000) Regulation of stretch reflex threshold in elbow flexors in children with cerebral palsy: a new measure of spasticity. Dev Med Child Neurol 42:531–540

Kahneman D (1973) Attention and effort. Prentice-Hall International, New Jersey

Kizony R, Katz N, Weiss PL (2004) Virtual reality based intervention in rehabilitation: relationship between motor and cognitive abilities and performance within virtual environments for patients with stroke. In: Proceedings of the 5th international conference on disability, virtual reality and associated technology. Oxford, UK

Kwakkel G, Kollen BJ, van der Grond J, Prevo AJ (2003) Probability of regaining dexterity in the flaccid upper limb: impact of severity of paresis and time since onset in acute stroke. Stroke 34:2181–2186

Lai SM, Studenski S, Richards L, Perera S, Reker D, Rigler S, Duncan PW (2006) Therapeutic exercise and depressive symptoms after stroke. J Am Geriatr Soc 54:240–247

Lamontagne A, Malouin F, Richards CL (2001) Locomotor-specific measure of spasticity of plantarflexor muscles after stroke. Arch Phys Med Rehabil 82:1696–1704

Lance JW (1980) Pathophysiology of spasticity and clinical experience with baclofen. In: Feldman RG, Young RR, Koella WP (eds) Spasticity: disordered motor. Year Book Medical Publisher Control, Chicago, pp 185–220

Lashley KS (1951) The problem of serial order in behavior. In: Cerebral Mechanisms in Behavior. Wiley, New York

Latash ML (2012) The bliss (not the problem) of motor abundance (not redundancy). Exp Brain Res 217(1):1–5

Levin MF (1996) Interjoint coordination during pointing movements is disrupted in spastic hemiparesis. Brain 119:281–294

Levin MF, Feldman AG (1994) The role of stretch reflex threshold regulation in normal and impaired motor control. Brain Res 657:23–30

Levin MF, Selles RW, Verheul MH, Meijer OG (2000) Deficits in the coordination of agonist and antagonist muscles in stroke patients: implications for normal motor control. Brain Res 853:352–369

Levin MF, Michaelsen S, Cirstea C, Roby-Brami A (2002) Use of the trunk for reaching targets placed within and beyond the reach in adult hemiparesis. Exp Brain Res 143:171–180

Mayo NE, Mayo NE, Fellows LK, Scott SC et al (2009) A longitudinal view of apathy and its impact after stroke. Stroke 40:3299–3307

McIntosh RD, Dijkerman HC, Mon-Williams M, Milner AD (2004) Grasping what is graspable: evidence from visual form agnosia. Cortex 40:695–702

Michaelsen SM, Luta A, Roby-Brami A, Levin MF (2001) Effect of trunk restraint on the recovery of reaching movements in hemiparetic patients. Stroke 32:1875–1883

Michaelsen SM, Jacobs S, Roby-Brami A, Levin MF (2004) Compensation for distal impairments of grasping in adults with hemiparesis. Exp Brain Res 157:162–173

Michaelsen SM, Dannenbaum R, Levin MF (2006) Task-specific training with trunk restraint on arm recovery in stroke: randomized control trial. Stroke 37:186–192

Mon-Williams M, McIntosh RD (2000) A test between two hypotheses and a possible third way for the control of prehension. Exp Brain Res 134:268–273

Mon-Williams M, Tresilian JR, Coppard VL, Carson RG (2001) The effect of obstacle position on reach-to-grasp movements. Exp Brain Res 137:497–501

Mullick AA, Musampa NK, Feldman AG, Levin MF (2013) Stretch reflex threshold measure discriminates between spasticity and rigidity. Clin Neurophysiol 124(4):740–51

Musampa NK, Mathieu PA, Levin MF (2007) Relationship between stretch reflex thresholds and voluntary arm muscle activation in patients with spasticity. Exp Brain Res 181:579–593

Platz T, Denzler P (2002) Do psychological variables modify motor recovery among patients with mild arm paresis after stroke or traumatic brain injury who receive the Arm Ability Training? Restor Neurol Neurosci 20:37–49

Rand D, Eng JJ (2012) Disparity between functional recovery and daily use of the upper and lower extremities during subacute stroke rehabilitation. Neurorehabil Neural Rep 26:76–84

Rand D, Eng JJ (2015) Predicting daily use of the affected upper extremity 1 year after stroke. Stroke Cerebrovasc Dis 24:274–283. doi:10.1016/j.jstrokecerebrovasdis.2014.07.039

Reisman DS, Scholz JP (2003) Aspects of joint coordination are preserved during pointing in persons with post-stroke hemiparesis. Brain 126:2510–2527

Rice NJ, McIntosh RD, Schindler I, Mon-Williams M, Demonet JF, Milner AD (2006) Intact automatic avoidance of obstacles in patients with visual form agnosia. Exp Brain Res 174: 176–188

Robinson-Smith G (2002) Self-efficacy and quality of life after stroke. J Neurosci Nursing 34: 91–98

Robinson-Smith G, Johnston MV, Allen J (2000) Self-care, self-efficacy, quality of life, and depression after stroke. Arch Phys Med Rehabil 81:460–464

Roby-Brami A, Feydy A, Combeaud M, Biryukova EV, Bussel B, Levin M (2003) Motor compensation and recovery of reaching in stroke patients. Acta Neurol Scand 107:369–381
Rochat L, Van der Linden M, Renaud O, Epiney JB, Michel P, Sztajzel R, Spierer L, Annoni JM (2013) Poor reward sensitivity and apathy after stroke: implication of basal ganglia. Neurology 81:1674–1680
Rodrigues MRM, Slimovitch M, Blanchette AK, Levin MF (2015) Temporal and spatial upper-limb interjoint coordination in chronic stroke subjects versus healthy individuals when reaching. Society for neuroscience abstracts, Chicago, IL
Rossi E, Mitnitski A, Feldman AG (2002) Sequential control signals determine arm and trunk contributions to hand transport during reaching. J Physiol 538:659–671
Sabes PN, Jordan MI, Wolpert DM (1998) The role of inertial sensitivity in motor planning. J Neurosci 18:5948–5957
Salbach NM, Mayo NE, Robichaud-Ekstrand S, Hanley JA, Richards CL, Wood-Dauphinee S (2005) The effect of task-oriented walking intervention on improving balance self-efficacy poststroke: a randomized, controlled trial. J Am Geriatr Soc 53:576–582
Saling M, Alberts J, Stelmach GE, Bloedel JR (1998) Reach-to-grasp movements during obstacle avoidance. Exp Brain Res 118:251–258
Sangole A, Levin MF (2007) A new perspective in the understanding of hand dysfunction following neurological injury. Top Stroke Rehab 14:80–94
Schindler I, Rice NJ, McIntosh RD, Rossetti Y, Vighetto A, Milner AD (2004) Automatic avoidance of obstacles is a dorsal stream function: evidence from optic ataxia. Nat Neurosci 7:779–784
Shaikh T, Goussev V, Feldman AG, Levin MF (2014) Arm-trunk coordination for beyond the reach movements in adults with hemiparesis. Neurorehabil Neural Repair 28:355–366
Skidmore ER, Whyte EM, Holm MB, Becker JT, Butters MA, Dew MA, Munin MC (2010) Cognitive and affective predictors of rehabilitation participation after stroke. Arch Phys Med Rehabil 91:203–207
Subramanian SK, Levin MF (2013) Influence of cognitive deficits on the ability to use feedback for arm motor recovery in chronic stroke. Stroke 44(12):E200
Teasell R, Bayona N, Salter K, Hellings C, Bitensky J (2006) Progress in clinical neurosciences: stroke recovery and rehabilitation. Cdn J Neurol Sci 33:357–364
Toglia J, Fitzgerald KA, O'Dell MW, Mastrogiovanni AR, Lin CD (2011) The mini-mental state examination and montreal cognitive assessment in persons with mild sub-acute stroke: relationship to functional outcome. Arch Phys Med Rehabil 92:792–798
Tresilian JR (1998) Attention in action or obstruction of movement? A kinematic analysis of avoidance behavior in prehension. Exp Brain Res 120:352–368
Tresilian JR, Mon-Williams M, Coppard VL, Carson RG (2005) Developmental changes in the response to obstacles during prehension. J Mot Behav 37:103–110
Urbin MA, Waddell KJ, Lang CE (2015) Acceleration metrics are responsive to change in upper extremity function of stroke survivors. Arch Phys Med Rehabil 96(5):854–861
Ustinova KI, Feldman AG, Levin MF (2006) Central resetting of neuromuscular steady states may underlie rhythmical arm movements. J Neurophysiol 96:1124–1134
Von Holst H (1954) Relations between the central nervous system and the peripheral organs. Br J Anim Behav 2:89–94
Wagle J, Farner L, Flekkoy K, Wyller T, Sandvik L et al (2011) Early post-stroke cognition in stroke rehabilitation patients predicts functional outcome at 13 months. Dementia Ger Cog Dis 31:379–387
Wenzelburger R, Kopper F, Frenzel A, Stolze H, Klebe S, Brossmann A, Kuhtz-Buschbeck J, Golge M, Illert M, Deuschl G (2005) Hand coordination following capsular stroke. Brain 128:64–74
Wissel J, Manack A, Brainin M (2013) Toward an epidemiology of poststroke spasticity. Neurology 80:S13–S19

Won J, Hogan N (1995) Stability properties of human reaching movements. Exp Brain Res 107:125–136

World Health Organization (2001) International classification of functioning, disability and health. World Health Organization, Geneva

Yang F, Feldman AG (2010) Reach-to-grasp movement as a minimization process. Exp Brain Res 201(1):75–92. doi:10.1007/s00221-009-2012-1

Zorowitz RD, Gillard PJ, Brainin M (2013) Poststroke spasticity: sequelae and burden on stroke survivors and caregivers. Neurology 80:S45–S52

# Part III
# Neurophysiological Mechanisms of Motor Control

# What Do TMS-Evoked Motor Potentials Tell Us About Motor Learning?

**Richard G. Carson, Kathy L. Ruddy and Emmet McNickle**

**Abstract** Thirty years ago, the first magnetic device capable of stimulating the human brain without discomfort through the intact skull was unveiled in Sheffield, England (Barker et al. in Lancet 1:1106–1107, 1985). Since that time, transcranial magnetic stimulation (TMS) has become the tool of choice for many scientists investigating human motor control and learning. In light of the fact that there are limits to the information that can be provided by any experimental technique, we first make the case that the necessarily restricted explanatory scope of the TMS technique—and the motor-evoked potentials to which it gives rise, is not yet reflected adequately in the research literature. We also argue that this inattention, coupled with the pervasive adoption of TMS as an investigative tool, may be restricting the elaboration of knowledge concerning the neural processes that mediate human motor learning. In order to make these points, we use as an exemplar the study of cross-education—the interlimb transfer of functional capacity.

**Keywords** Transcranial · Magnetic stimulation · Interlimb transfer · Functional capacity · Cross-education · Motor-evoked potentials · MEP amplitude · Corticospinal · Primary motor cortex

R.G. Carson (✉) · E. McNickle
Trinity College Institute of Neuroscience and School of Psychology, Trinity College Dublin, Dublin 2, Ireland
e-mail: Richard.Carson@tcd.ie

R.G. Carson
School of Psychology, Queen's University Belfast, Northern Ireland, UK

K.L. Ruddy
Neural Control of Movement Lab, ETH, Zurich, Switzerland



J. Laczko and M.L. Latash (eds.), *Progress in Motor Control*,
Advances in Experimental Medicine and Biology 957,
DOI 10.1007/978-3-319-47313-0_8

## The Thesis to Be Developed

Within the field of motor learning, there is a long tradition of employing metaphors derived from contemporary technological culture as a means to conceptualise the work of the brain. Franklin Henry adopted the term memory drum—a magnetic data storage device invented by Gustav Tauschek in 1932 and used as computer memory throughout the 1950s and 1960s—to capture the sense of a stored program moved to a temporary buffer and thereafter translated into motor commands (Henry and Rogers 1960). In this respect, Henry was echoing Sherrington's famously poetic allusion to an enchanted (Jacquard) loom in describing the complex patterns woven by the cerebral cortex as "the body is up and rises to meet its waking day" (Sherrington 1942). While such analogies undoubtedly help convey ideas effectively, there is a broader issue that concerns the ways in which technology may constrain the generation of knowledge (Harrington 1992). In some cases, the severity of the constraint may be such that hypothesis generation becomes biased. In an era in which brain imaging has become the tour de force of cognitive neuroscience, we are not alone in noting that the application of these technologies imposes limits (often inadvertently) on theory development and testing. Given tools that map activity in specific brain regions, the tendency has been to proceed with model building on the basis that each has its fundamental computation (Anderson 2014). With the contemporary development of imaging technologies and analytical methods that permit estimates of structural and functional brain connectivity, theories that encapsulate the integration (of neural information) arising from generic properties of parallel, distributed networks receive fresh impetus (e.g. Sporns 2011). As each new technology creates its own constituency, there emerges the associated risk that the technology begins to run the science (Fodor 1999).

Thirty years ago, the first magnetic device capable of stimulating the human brain without discomfort through the intact skull was unveiled in Sheffield, England (Barker et al. 1985). Since that time, transcranial magnetic stimulation (TMS) has become the tool of choice for many scientists investigating human behaviour, particularly those for whom their interest extends to the neural processes that mediate motor control and learning. Among the proliferation of empirical reports based on the application of this technique, there are many that have made truly novel contributions to our knowledge of the human motor system. Nonetheless, there are limits to the information that can be provided by any experimental technique. The thesis to be developed in the present chapter has two key elements. Our first point is that the restricted explanatory scope of the TMS technique—and the motor-evoked potentials to which it gives rise—is seldom reflected in the research literature. We furthermore contend that this inattention, coupled with the pervasive adoption of TMS as an investigative tool, is restricting the elaboration of knowledge concerning the neural processes that mediate human motor learning.

With a view to presenting our case succinctly, we will not attempt to be comprehensive with respect to the wide variety of contexts in which motor learning is expressed, or consider the multitude of paradigms that have been used in its experimental study. Rather, we will focus upon transfer, i.e. generalisation of acquired capability, which along with retention is considered demonstrative of effectual learning (e.g. Magill 2004). Furthermore, we deal primarily with the interlimb transfer of functional capacity. This expression of motor learning is also referred to commonly as cross-education.

## Motor-Evoked Potentials

As there are authoritative and comprehensive reviews dealing with the biophysics and neurophysiology of TMS (e.g. Di Lazzaro and Rothwell 2014; Lemon et al. 2002; Rothwell 1997; Ziemann and Rothwell 2000), we hereby provide only a summary pertinent to our present purpose that draws in part upon these previous works. The essence of the technology is that a brief electric current is passed through a conducting coil (positioned over the scalp), creating a transient ($\sim$200 μs) magnetic field which in turn induces an eddy current in the underlying cortical tissue (Fig. 1). The induced electrical current tends to flow parallel to the

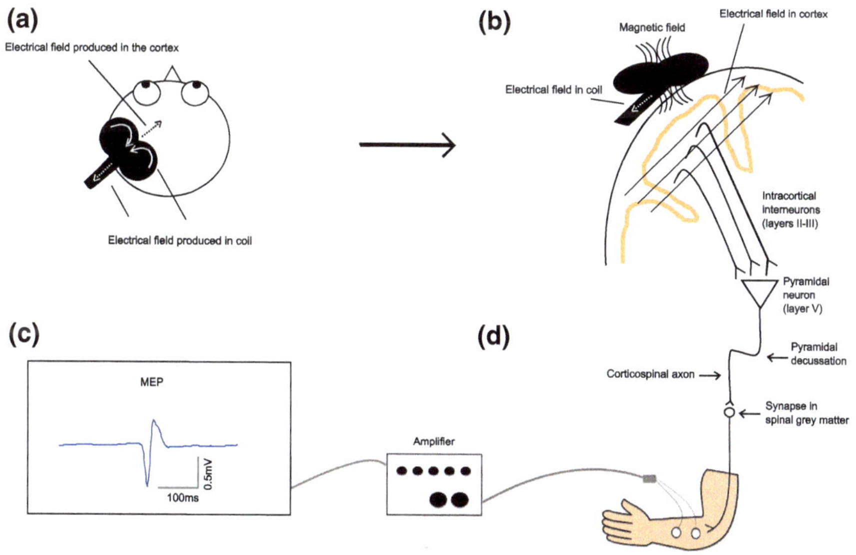


**Fig. 1** Schematic of transcranial magnetic stimulation (TMS). Panel **a** depicts the TMS coil placed over the scalp. The current direction generated in the cortex is opposite in direction to that produced by the coil. **b** The magnetic field generated by the coil is perpendicular to the current direction. It traverses skull and other layers between the scalp and cortex, producing an electrical current opposite in direction to the original coil current. This generates action potentials in horizontal interneurons synapsing on pyramidal neurons and thus activates the corticospinal tract. The descending corticospinal tract then synapses on to spinal motor neurons, carrying the signal to the muscle. **c** Shows the electromyographic (EMG) recording of the resultant muscle activity via surface electrodes as the corticospinal volley reaches the periphery. **d** Shows the amplified EMG signal as a stereotypic motor-evoked potential (MEP). (Adapted from Brown et al. 2014)

plane of the TMS coil, and to the horizontal fibres in superficial layers of cortex (Terao and Ugawa 2002). If the induced current is of sufficient magnitude, cortical neurons are depolarised and action potentials generated. When the coil is positioned over the primary motor cortex (M1), pyramidal tract neurons may be stimulated directly or indirectly, giving rise to a descending corticospinal volley. In some circumstances, this volley may result in the depolarization of spinal motoneurons, which can be registered as compound motor action potentials in electromyographic recordings of skeletal muscle. The short latency component (e.g. 12.9–21.6 ms

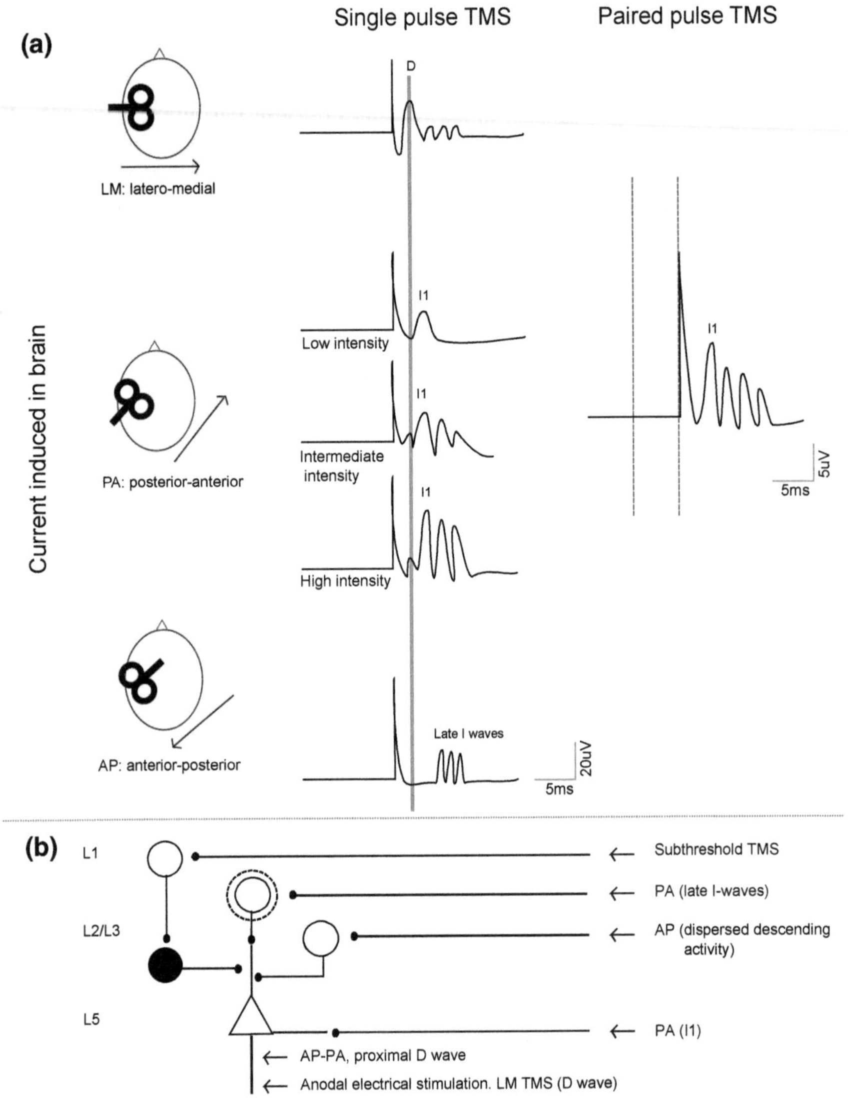

◂ **Fig. 2** Physiological underpinnings of the MEP. Panel **a** depicts a schematic of stereotypical descending volleys evoked by single and paired-pulse TMS, which can be recorded from the spinal epidural space. Low intensity magnetic stimulation with a posterior–anterior (PA) induced current in the brain evokes a single descending wave with a latency about 1 ms longer than the D wave (evoked by transcranial electrical stimulation), termed I1 wave. At intermediate intensity later I waves are evoked and at high intensity, an earlier small wave with the same latency as the D wave evoked by electrical anodal stimulation appears. Magnetic stimulation with a latero-medial (LM) induced current in the brain preferentially evokes D wave activity. On the *right*, paired-pulse stimulation—epidural volleys evoked by a test magnetic stimulus preceded by a subthreshold conditioning stimulus at 3 ms interstimulus interval (*dotted trace*). The test stimulus evokes multiple descending waves that are suppressed in amplitude by the conditioning stimulus. Panel **b** —Schematic representation of motor cortex circuits. This model proposed by Di Lazzaro and Rothwell (2014) indicates possible preferential site of activation using the different techniques of transcranial brain stimulation. *Open circles* indicate excitatory neurons while *filled circles* indicate inhibitory neurons. A superficial inhibitory circuit composed of layer 1 (L1) neurons have connections with layer 2 (L2) and 3 (L3) interneurons (*filled circle*) which inhibit the distal apical dendrites of layer 5 pyramidal neurons. L2 and L3 bursting (*open circle* inside a *dotted circle*) and non-bursting excitatory interneurons project upon the distal apical dendrites of layer 5 pyramidal neurons; cortico-cortical axons project upon basal dendrites of layer 5 pyramidal neurons. Anodal stimulation and LM magnetic stimulation directly activate the corticospinal axons of pyramidal tract neurons (PTNs) evoking a D wave. The wave evoked by low intensity magnetic stimulation that appears 1 ms later than the D wave evoked by electrical anodal stimulation is suggested to be produced by monosynaptic activation of basal dendrites of PTNs by cortico-cortical axons activated by the magnetic stimulus. The late I waves evoked at higher intensities might be produced by a circuit that involves cortico-cortical axons that activate L2 and L3 bursting neurons, and in turn activate PTN apical dendrites. Descending activity evoked by AP magnetic stimulation is produced by a different circuit that might include non-bursting L2 and L3 interneurons projecting upon PTN apical dendrites. Inhibitory effects produced by a subthreshold conditioning stimulus may originate from a selective enhancement of the excitability of the GABAergic circuit originating from L1 neurons and projecting upon PTNs apical dendrites, resulting in a selective suppression of the late I waves. (Figure and legend adapted from Di Lazzaro and Rothwell 2014)

onset in flexor carpi radialis (FCR) motor units, Olivier et al. 1995) of this response is typically referred to as a motor-evoked potential (MEP).

While reasonably taken to reflect the excitability of the corticospinal tract, the MEP is a compound measure consisting of several descending waves (Fig. 2) that can be discriminated in epidural recordings (Di Lazzaro et al. 2004). The long-standing interpretation is that TMS over M1 activates chains of interneurons with fixed temporal characteristics that produce a periodic bombardment of corticospinal neurons (Amassian et al. 1987). When a figure-of-eight shaped stimulating coil is used, and current at the junction of the coil windings flows initially from an anteromedial to latero-posterior direction, the successive components of motor-evoked potentials elicited at ("threshold") intensities required to generate a descending volley are thought to first reflect the activation of cortico-cortical axons projecting onto corticospinal neurons, or axon collaterals of corticofugal systems [e.g. motor areas such as dorsal premotor cortex (PMd), dorsal premotor cortex (PMv) and supplementary motor area (SMA)] with corresponding projections, i.e. onto corticospinal neurons (Di Lazzaro et al. 2008), and subsequently transmission via polysynaptic networks or recurrent synaptic networks (cf. Rusu et al. 2014). As the intensity of stimulation is increased, neurons in addition to those activated at threshold, which are intrinsically less excitable or are spatially removed from the peak of the magnetic field, also contribute to the descending volley. Collectively, these are referred to as the "indirect" I-wave components. If the intensity of stimulation is increased sufficiently, or the currents induced in the brain flow in a lateral to medial direction, it is also possible to activate the corticospinal axons directly to produce a shorter latency "direct" (D wave) component (Rothwell 1997).

The MEP is therefore not simply a 'read-out' of the post-synaptic state of pyramidal tract neurons. It reflects the excitability of intracortical and transcortical networks of interneurons that synapse upon corticospinal neurons. The magnitude of the compound motor action potential is also influenced by the state of the spinal segment of the pathway (Taylor 2006) and by all factors that alter the post-synaptic excitability of spinal motoneurons. At least in monkey, there are direct corticospinal projections onto spinal motoneurons from supplementary and premotor areas (Dum and Strick 1996, 2002) and from parietal regions (Murray and Coulter 1981) that exhibit this potential.

Paired-pulse protocols (Kujirai et al. 1993)—in which two magnetic stimuli are delivered in quick succession through a single coil—have been used to determine the contribution of component I waves to the MEP in varying experimental conditions. Changing the direction of the induced current flow by manipulating the orientation of the stimulating coil with respect to the underlying cortical tissue may also alter the balance between the various I-wave components (Di Lazarro and Ziemann 2013). Approaches of this type (to which we will return) permit some of the factors that contribute to an experimentally induced change in the amplitude of the MEP to be examined. Nonetheless, the dynamic properties of the MEP, and the factors that determine its composition and modification, remain incompletely understood. As Bestmann and Krakauer (2015) highlight, MEPs do not result from the stimulation of a uniform set of neural elements; they are a compound measure subject to the influence of multiple known (and probably several unknown) factors. In most experimental contexts, it is not a trivial matter to delineate the influences of these various factors on the magnitude of the MEP.

In summary, when an empirical observation is made that there has been a reliable change in the magnitude of the MEP, we can conclude that *something* has changed. That which is less clear in most circumstances, is exactly that which has changed and what, if anything, is the functional relevance of the change.

## Changes in MEP Amplitude in the Context of Motor Learning

Variations in the amplitude of the MEP have been used frequently to draw inferences concerning the contribution of the primary motor cortex to processes underlying motor learning, in the context of a wide range of behavioural paradigms (e.g. Lotze et al. 2003; Muellbacher et al. 2001; Pascual-Leone et al. 1994, 1995). Should the basis for such inferences be valid, the monitoring of MEPs would appear to offer particular promise if it can be used to discriminate differences in the level of engagement of specific circuits local to M1 as learning evolves. The most influential experimental reports have tended to suggest that in humans, corticospinal excitability (CSE) is elevated during and following the completion of training regimes that give rise to an expression of motor learning (Muellbacher et al. 2001; Pascual-Leone et al. 1994, 1995; Perez et al. 2004). Nonetheless, the opposite pattern of change, i.e. decreases in MEP amplitude during motor skill learning, has also been documented (Coxon et al. 2014; Holland et al. 2015). This contrariety alone suggests that generalisations concerning the relationship between changes in MEP amplitude and the processes of motor learning (particularly when considered across task contexts) should be made with caution. There is however a more substantial concern. While it may be the case that performance of the tasks that give rise to motor learning is accompanied by systematic and reliable changes in CSE, evidence of an instrumental relationship between the degree of change in CSE, and the learning that accrues to an individual, is extremely sparse (Ljubisavljevic 2006). Among the small proportion of the most prominent studies that have included measures of association, null findings are the norm (e.g. Liepert et al. 1999). Although positive correlations between changes in MEP amplitude and short-term improvements in performance have on occasion been noted (e.g. Muellbacher et al. 2001), we are aware of only a few instances in which retention or transfer tests have been employed to consider the potential presence of a relationship. Even in studies in which "follow-up" measurements were obtained, this step is seldom taken (e.g. Muellbacher et al. 2001). A conspicuous exception in this regard is the positive finding reported by Lundbye-Jensen et al. (2005). In a carefully crafted investigation, their participants were required to perform 16 min of visuomotor tracking (by means of elbow flexion-extension movements), in each of 13 training sessions conducted over a period of 4 weeks. Increases in the maximum amplitude of the MEP that could be produced by TMS, and decreases in the MEP threshold over this period (both measures being obtained prior to the training movements), were correlated with the retained improvements in performance.

Within the sub-domain of motor learning in which the transfer of acquired capacity has been the focus of interest, and in particular those investigations in which interlimb transfer has been examined, measures of association (i.e. between changes in MEP amplitude and accrued improvements in performance) have been generated more diligently. This corpus of work therefore provides a helpful basis upon which to examine critically the assumption that variations in MEPs provide an index of processes that are instrumentally related to motor skill acquisition.

## Changes in MEP Amplitude in the Context of Interlimb Transfer

Cross-education is the process whereby training of one limb gives rise to increases in the subsequent performance of its opposite counterpart. Motor learning thus 'transfers' from a trained to an untrained effector following a period of practice that is exclusively unilateral. In view of the fact that the execution of many unilateral tasks is associated with increased excitability of corticospinal projections to the opposite limb (Carson et al. 2004; Hortobagyi et al. 2003), it has been proposed that interlimb transfer of training-induced performance gains may be subserved by interactions between the primary motor cortices (e.g. Hinder et al. 2011). The more general conjecture is that bilateral cortical activity generated during unilateral training drives concurrent neural adaptations in both cerebral hemispheres (Hellebrandt 1951). This has been dubbed as the 'cross-activation' hypothesis (e.g. Lee et al. 2010) (Fig. 3).

As a consequence, there are a number of studies that have been conducted on the basis of the (sometimes implicit) assumption that functional adaptations within the M1 that provides the (majority of) corticospinal projections to the untrained (transfer) limb, should be evident as increases in the amplitude of MEPs following completion of the training regime. In the context of ballistic tasks in which short-term unilateral practice of finger or thumb movements brings about bilateral increases in movement velocity or acceleration, rises in the excitability of corticospinal projections to the muscles of the untrained limb have indeed been reported

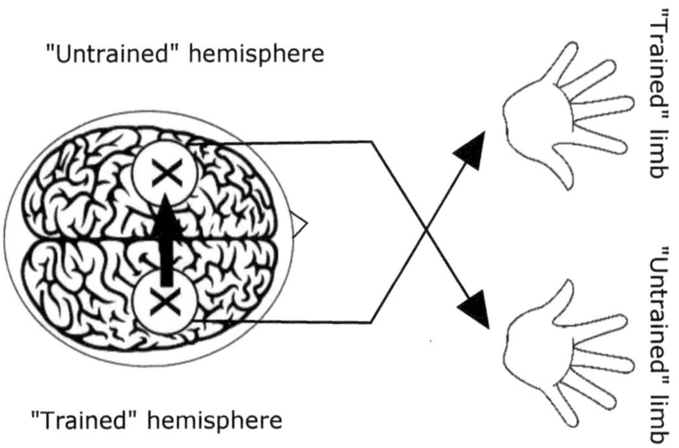


**Fig. 3** Cross-activation model of cross-education. This hypothesis postulates that during unilateral training, activation of the homologous motor network gives rise to bilateral adaptations that facilitate subsequent performance by the untrained limb. The 'X' represents the putative locus of training related adaptations. *White circles* indicate lateralised motor networks in their entirety. *Solid arrows* represent processes that occur during unilateral training. Adapted from Ruddy and Carson (2013)

(Carroll et al. 2008; Lee et al. 2010; Hinder et al. 2011; Poh et al. 2013; Dickins et al. 2015; Reissig et al. 2015). This does not however constitute evidence that any such changes are instrumentally related to the level of interlimb transfer (Ruddy and Carson 2013). It has been highlighted, for example, that increases in corticospinal excitability (CSE) can also be present for projections to homologues of muscles that do not make a *direct* mechanical contribution to the action that is trained (Carroll et al. 2008).

In many of the studies cited above, however, the investigators also took the trouble to calculate measures of association (typically correlation coefficients) to determine whether a relationship existed between variations in the degree of cross-education expressed across participants, and the extent to which elevations in the excitability of corticospinal projections to the muscles of the untrained limb were manifested. In all of the instances of which we are aware (Carroll et al. 2008; Hinder et al. 2011; Dickins et al. 2015; Reissig et al. 2015), null findings were reported. This was also the case for a recent study conducted in our laboratory (Ruddy et al. 2016). In circumstances in which the performance (peak acceleration) of the training limb increased by more than 40 %, and the median level of interlimb transfer was greater than 80 %, there was no association with individual changes in the amplitude of MEPs recorded in the transfer limb. In short, with exception of circumstances in which augmented visual feedback has been provided (Nojima et al. 2012), there is very little evidence to indicate that changes in MEP amplitude —at least when obtained at rest (cf. Hortobágyi et al. 2011)—are predictive of the level of learning exhibited by the transfer limb.

In the context of the particular ballistic training task we employed (Ruddy et al. 2016)—requiring wrist flexion movements, there was no net elevation in the amplitude of MEPs measured in the untrained limb following training. Thus, a substantial (≈35 %) improvement in the performance of a task that demanded maximum engagement of the wrist flexor muscles was entirely dissociated from variations in the excitability of corticospinal projections to one of the principal agonists (FCR), as assessed using TMS. A finding of this nature is not without precedent. Using a variant of the serial reaction-time task (SRTT), Perez et al. (2007) noted that increases in the performance of the transfer hand were not accompanied by changes in MEP amplitude (or threshold).

None of these observations precludes the instrumental involvement of other circuits local to primary motor cortex in mediating interlimb transfer of learning. Hortobágyi et al. (2011) noted that for a high intensity training task, the level of cross-education was correlated (across participants) with changes in the TMS-derived measure of interhemispheric inhibition (IHI)—which is elicited through paired stimulation of the primary motor cortices. Perez et al. (2007) reported that following unilateral SRTT training, there was a decrease in IHI from the M1 contralateral to the training limb, to the M1 contralateral to the transfer limb. The extent of this decrease was correlated with the amount of non-specific performance transfer to the untrained limb. In the SRTT, this is typically expressed as decreased reaction times in all aspects of the task, including random blocks that have no sequential component (Robertson 2007).

By and large, in the studies referred to above, the magnetic stimulating coil has been oriented such that current in the coil flows initially in an anteromedial to latero-posterior direction. The current induced in the brain thus moves in the opposite direction—from latero-posterior to anteromedial [termed posterior-anterior (PA)]. It is possible that learning-related changes in the excitability of intracortical and transcortical networks of interneurons that synapse upon corticospinal neurons are not readily detected with this configuration. Different populations of fibres are likely to be excited by anterior–posterior (AP) as opposed to PA currents (Di Lazzaro et al. 2001). Although the most likely neuronal site for activation by TMS is at the fibre terminal, it is also possible that large afferent axons from premotor and somatosensory areas may be especially sensitive to AP currents (Esser et al. 2005). These fibres constitute the main cortical input to M1 (DeFelipe et al. 1986; Sutor et al. 2000), entering in a bend approximately 0.5 mm away from layer 5 and 1.5 mm from layer 3 (Rockel et al. 1980).

A more general and critical issue is thus highlighted. In the event that associations can be shown between the variations in the magnitude of the evoked response to AP currents, and the degree of cross-education that is expressed across participants, these are unlikely to be amenable to interpretation in terms of changes in the state of circuits local to M1. A similar point can be made with respect to variations in IHI, and indeed in relation to other measures based on paired-pulse TMS. The origins of the modulatory inputs that ultimately influence the magnitude of the "test response"—whether to AP current stimulation or following a conditioning stimulus—cannot readily be determined. In most cases, it is impossible to exclude the involvement of areas other than M1.

## Reflection

We are not seeking to imply that the primary motor cortex is not intimately involved in processes underlying motor learning. Rather, our singular aim is to demonstrate that changes in MEP amplitude cannot be relied upon as an unambiguous indication of variations in the state of specific circuits within M1, and that there is very little empirical evidence to support the assumption that there exists an instrumental relationship between modifications in the state of neural elements that influence the size of the MEP, and the quality of motor learning. Perhaps most obviously, the modifications of behaviour persist long after the changes in the amplitude or threshold of the MEP have dissipated (Ljubisavljevic 2006).

The MEP is a compound measure of excitability that is influenced by multiple neural elements. The majority of the short latency responses to TMS evoked in healthy adults are mediated by large corticospinal neurons with fast-conducting axons (Lemon et al. 2002). Slower conducting axons also make monosynaptic connections with upper limb motoneurons (Porter and Lemon 1995) and it remains possible that cortical circuits projecting onto these cells exhibit distinct patterns of adaptation in the mediation of motor learning. In principle, a component of the descending motor command may also be mediated through propriospinal relays, potentially subject to the influence of reticulospinal projections (Pierrot-Deseilligny 1996; Rothwell 2002). Beyond the recognition that TMS is capable of sampling just a subset of circuits in primary motor cortex, it is also becoming apparent that not only the orientation of the induced current flow, but also the intensity of stimulation, determines the composition of this subset—in terms of the balance of inhibitory and excitatory inputs to the corticospinal output cells (Di Lazzaro and Rothwell 2014). The precise implications of changes in MEP amplitude for voluntary motor behaviour are thus far from clear (Bestmann and Krakauer 2015).

Motor learning may be associated with increases or decreases in CSE. There are also instances in which motor learning occurs in the absence of any change in CSE. Evidently, therefore, a general causal relationship cannot be inferred. Although it is beyond the scope of this chapter to critique the assumptions that are made typically in relating changes in corticospinal excitability to the potential for neuroplastic adaptation, we note in passing that interventions such as aerobic exercise (Singh et al. 2016) or associative cortical stimulation (Lopez-Alonso et al. 2015; McNickle 2016) that have the capacity to increase corticospinal excitability may do so without any concomitant positive effect on motor learning.

There is nonetheless evidence that changes in the amplitude of motor-evoked potentials may relate to retained improvements in visuomotor tracking skill over extended training periods (Hirano et al. 2015; Lundbye-Jensen et al. 2005). There appears however to be very little evidence of similar associations, in circumstances in which retention is assessed in other task domains. In seeking illustrative cases, the principal emphasis of the present piece has been upon interlimb transfer. While it is often implicitly assumed that measures of retention and transfer provide interchangeable indices of learning, it may in fact be the case that they sample at least partially independent processes. In this regard, interlimb transfer may also represent a special case.

## Take Home Message

In our view, oft reported variations in MEP amplitude, whether obtained in the context of interlimb transfer or other manifestations of motor learning, simply indicate that at least one of the many elements of the motor system that can influence corticospinal excitability has changed as an accompaniment. There is very little empirical evidence to support the assumption that there exists an instrumental relationship between modifications in the state of neural elements that influence the size of the MEP, and the quality of motor learning. In the absence of further information—whether derived from extensions of TMS or from complementary methodologies, variations in MEP amplitude are in themselves entirely ambiguous with respect to the engagement of specific circuits acting through primary motor cortex.

In spite of the fact that this much would be readily acknowledged by most practitioners of the technique—at least when the points are made in such terms—it remains the case that an observation of changes in MEP amplitude is most often the initial finding upon which all subsequent avenues of investigation are predicated. We believe that this widespread practice may account for the ambiguity and inconsistency of the research literature as it concerns the outcomes of derivative investigations in which such methods as paired-pulse TMS have been applied. The consistent absence of associations in terms of these measures further corroborates the view that the changes in MEP amplitude fail to exhibit a clear functional relationship to the processes of adaptation that mediate improvements in performance. We are forced to conclude that the motor-evoked potential generated by single-pulse TMS is the wrong tool for the purpose of discriminating differences in the level of engagement of specific circuits local to M1, as learning evolves.

Our further argument is that the fixation on TMS as a method of first choice, coupled with interpretations of TMS outcome measures that lack sufficient constraint, has biased recent theorising in the sub-domain of cross-education in particular, and motor learning in general. We believe that this bias is towards models that place unwarranted emphasis upon a limited set of the neural elements within primary motor cortex that is recruited at rest by single-pulse TMS.

## References


Amassian VE, Stewart M, Quirk GJ, Rosenthal JL (1987) Physiological basis of motor effects of a transient stimulus to cerebral cortex. Neurosurgery 20(1):74–93

Anderson ML (2014) After phrenology: neural reuse and the interactive brain. MIT Press, Cambridge, MA

Barker AT, Jalinous R, Freeston IL (1985) Non-invasive magnetic stimulation of human motor cortex. Lancet 1(8437):1106–1107

Bestmann S, Krakauer, JW (2015) The uses and interpretations of the motor-evoked potential for understanding behaviour. Exp Brain Res 233(3):679–689. http://doi.org/10.1007/s00221-014-4183-7

Brown KE, Neva JL, Ledwell NM, Boyd LA (2014) Use of transcranial magnetic stimulation in the treatment of selected movement disorders. Degenerative Neurol Neuromuscul Dis 4: 133–151. http://doi.org/10.2147/DNND.S70079

Carroll TJ, Lee M, Hsu M, Sayde J (2008) Unilateral practice of a ballistic movement causes bilateral increases in performance and corticospinal excitability. J Appl Physiol 104: 1656–1664. doi:10.1152/japplphysiol.01351.2007

Carson R, Riek S, Mackey D, Meichenbaum D, Willms K, Forner M, Byblow W (2004) Excitability changes in human forearm corticospinal projections and spinal reflex pathways during rhythmic voluntary movement of the opposite limb. J Physiol-Lond 560(3):929–940

Coxon JP, Peat NM, Byblow WD (2014) Primary motor cortex disinhibition during motor skill learning. J Neurophysiol 112(1):156–164. http://doi.org/10.1152/jn.00893.2013

DeFelipe J, Conley M, Jones EG (1986) Long-range focal collateralization of axons arising from corticocortical cells in monkey sensory-motor cortex. J Neurosci 6(12):3749–3766

Di Lazzaro V, Oliviero A, Saturno E, Pilato F, Insola A, Mazzone P et al (2001) The effect on corticospinal volleys of reversing the direction of current induced in the motor cortex by

transcranial magnetic stimulation. Exp Brain Res. Experimentelle Hirnforschung. Expérimentation Cérébrale, 138(2):268–273
Di Lazzaro V, Oliviero A, Pilato F, Saturno E, Dileone M, Mazzone P et al (2004) The physiological basis of transcranial motor cortex stimulation in conscious humans. Clin Neurophysiol 115(2):255–266
Di Lazzaro V, Rothwell JC (2014) Corticospinal activity evoked and modulated by non-invasive stimulation of the intact human motor cortex. J Physiol 592(Pt 19):4115–4128. http://doi.org/10.1113/jphysiol.2014.274316
Di Lazzaro V, Ziemann U, Lemon RN (2008) State of the art: physiology of transcranial motor cortex stimulation. Brain Stimulation 1(4):345–362. http://doi.org/10.1016/j.brs.2008.07.004
Di Lazzaro V, Ziemann U (2013) The contribution of transcranial magnetic stimulation in the functional evaluation of microcircuits in human motor cortex. Front Neural Circuits 7:18. http://doi.org/10.3389/fncir.2013.00018
Dickins DSE, Sale MV, Kamke MR (2015) Intermanual transfer and bilateral cortical plasticity is maintained in older adults after skilled motor training with simple and complex tasks. Front Aging Neurosci 7:73. doi:10.3389/fnagi.2015.00073
Dum RP, Strick PL (1996) Spinal cord terminations of the medial wall motor areas in macaque monkeys. J Neurosci 16:6513–6525
Dum RP, Strick PL (2002) Motor areas in the frontal lobe of the primate. Physiol Behav 77: 677–682
Esser SK, Hill SL, Tononi G (2005) Modeling the effects of transcranial magnetic stimulation on cortical circuits. J Neurophysiol 94(1):622–639
Fodor J (1999) Diary: why the brain? London Rev Books 21(19):68–69
Harrington A (1992) So human a brain: knowledge and values in the neurosciences. Birkhauser Verlag AG, Basel
Hellebrandt F (1951) Cross education: ipsilateral and contralateral effects of unimanual training. J Appl Physiol 4:136–144
Henry FM, Rogers DE (1960) Increased response latency for complicated movements and a "memory drum" theory of neuromotor reaction. Res Q. Am Assoc Health Phys Educ Recreation 31(3):448–458. http://doi.org/10.1080/10671188.1960.10762052
Hinder MR, Schmidt MW, Garry MI, Carroll TJ, Summers JJ (2011) Absence of cross-limb transfer of performance gains following ballistic motor practice in older adults. J Appl Physiol 110(1):166–175. http://doi.org/10.1152/japplphysiol.00958.2010
Hirano M, Kubota S, Tanabe S, Koizume Y, Funase K (2015) Interactions among learning stage, retention, and primary motor cortex excitability in motor skill learning. Brain Stimulation 8(6):1195–1204
Holland L, Murphy B, Passmore S, Yielder P (2015) Time course of corticospinal excitability changes following a novel motor training task. Neurosci Lett 591:81–85. http://doi.org/10.1016/j.neulet.2015.02.022
Hortobagyi T, Taylor J, Petersen N, Russell G, Gandevia S (2003) Changes in segmental and motor cortical output with contralateral muscle contractions and altered sensory inputs in humans. J Neurophysiol 90(4):2451–2459
Hortobágyi T, Richardson SP, Lomarev M, Shamim E, Meunier S, Russman H, et al (2011) Interhemispheric plasticity in humans. Med Sci Sports Exerc 43(7):1188–1199. http://doi.org/10.1249/MSS.0b013e31820a94b8
Jensen JL, Marstrand PCD, Nielsen JB (2005) Motor skill training and strength training are associated with different plastic changes in the central nervous system. J Appl Physiol 99(4):1558–1568
Kujirai T, Caramia MD, Rothwell JC, Day BL, Thompson PD, Ferbert A et al (1993) Corticocortical inhibition in human motor cortex. J Physiol 471:501–519
Lee M, Hinder MR, Gandevia SC, Carroll TJ (2010) The ipsilateral motor cortex contributes to cross-limb transfer of performance gains after ballistic motor practice. J Physiol 588:201–212. doi:10.1113/jphysiol.2009.183855

Lemon R, Pascual-Leone A, Davey N, Rothwell J, Wassermann E, Puri B (2002) Basic physiology of transcranial magnetic stimulation. Handbook of transcranial magnetic stimulation, pp 61–77

Liepert J, Terborg C, Weiller C (1999) Motor plasticity induced by synchronized thumb and foot movements. Exp Brain Res 125:435–439

Lopez-Alonso V, Cheeran B, Fernandez Del Olmo M (2015) Relationship between non-invasive brain stimulation-induced plasticity and capacity for motor learning. Brain Stimulation 8(6):1209–1219

Lotze M, Braun C, Birbaumer N, Anders S, Cohen LG (2003) Motor learning elicited by voluntary drive. Brain 126(Pt 4):866–872

Ljubisavljevic M (2006) Transcranial magnetic stimulation and the motor learning-associated cortical plasticity. Exp Brain Res. Experimentelle Hirnforschung. Expérimentation Cérébrale, 173(2):215–222. http://doi.org/10.1007/s00221-006-0538-z

Magill RA (2004) Motor learning and control: concepts and applications, 7th edn. McGraw-Hill, New York

McNickle E (2016) Electrophysiological and functional responses to associative brain stimulation. PhD thesis. Trinity College Dublin

Muellbacher W, Ziemann U, Boroojerdi B, Cohen L, Hallett M (2001) Role of the human motor cortex in rapid motor learning. Exp Brain Res. Experimentelle Hirnforschung. Expérimentation Cérébrale, 136(4):431–438

Murray EA, Coulter JD (1981) Organization of corticospinal neurons in the monkey. J Comp Neurol 195:339–365. doi:10.1002/cne.901950212

Nojima I, Mima T, Koganemaru S, Thabit MN, Fukuyama H, Kawamata T (2012) Human motor plasticity induced by mirror visual feedback. J Neurosci 32:1293–1300. doi:10.1523/JNEUROSCI.5364-11.2012

Olivier E, Bawa P, Lemon RN (1995) Excitability of human upper limb motoneurones during rhythmic discharge tested with transcranial magnetic stimulation. J Physiol 485(Pt 1):257–269

Pascual-Leone A, Grafman J, Hallett M (1994) Modulation of cortical motor output maps during development of implicit and explicit knowledge. Science 263(5151):1287–1289

Pascual-Leone A, Nguyet D, Cohen LG, Brasil-Neto JP, Cammarota A, Hallett M (1995) Modulation of muscle responses evoked by transcranial magnetic stimulation during the acquisition of new fine motor skills. J Neurophysiol 74(3):1037–1045

Pierrot-Deseilligny E (1996) Transmission of the cortical command for human voluntary movement through cervical propriospinal premotoneurons. Prog Neurobiol 48:489–517

Perez MA, Lungholt BKS, Nyborg K, Nielsen JB (2004) Motor skill training induces changes in the excitability of the leg cortical area in healthy humans. Exp Brain Res. Experimentelle Hirnforschung. Expérimentation Cérébrale, 159(2):197–205. http://doi.org/10.1007/s00221-004-1947-5

Perez MA, Wise SP, Willingham DT, Cohen LG (2007) Neurophysiological mechanisms involved in transfer of procedural knowledge. J Neurosci 27:1045–1053. doi:10.1523/JNEUROSCI.4128-06.2007

Poh E, Riek S, Carroll TJ (2013) Ipsilateral corticospinal responses to ballistic training are similar for various intensities and timings of TMS. Acta Physiol (Oxf) 207:385–396. doi:10.1111/apha.12032

Porter R, Lemon R (1995) Corticospinal function and voluntary movement. Oxford University Press, Oxford

Reissig P, Stöckel T, Garry MI, Summers JJ, Hinder MR (2015) Age-specific effects of mirror-muscle activity on cross-limb adaptations under mirror and non-mirror visual feedback conditions. Front Aging Neurosci 7:222. doi:10.3389/fnagi.2015.00222

Robertson EM (2007) The serial reaction time task: implicit motor skill learning? J Neurosci 27(38):10073–10075. http://doi.org/10.1523/JNEUROSCI.2747-07.2007

Rockel AJ, Hiorns RW, Powell TP (1980) The basic uniformity in structure of the neocortex. Brain 103(2):221–244

Rothwell JC (1997) Techniques and mechanisms of action of transcranial stimulation of the human motor cortex. J Neurosci Methods 74(2):113–122

Rothwell JC (2002) Spinal interneurones: re-evaluation and controversy. Adv Exp Med Biol 508:259–263

Ruddy KL, Carson RG (2013) Neural pathways mediating cross education of motor function. Front Human Neurosci 7:397. http://doi.org/10.3389/fnhum.2013.00397

Ruddy KL, Rudolf AK, Kalkman B, King M, Daffertshofer A, Carroll TJ, Carson RG (2016) Neural adaptations associated with interlimb transfer in a ballistic wrist flexion task. Front Human Neurosci. http://doi.org/10.3389/fnhum.2016.00204

Rusu CV, Murakami M, Ziemann U, Triesch J (2014) A model of TMS-induced I-waves in motor cortex. Brain Stimulation 7(3):401–414. http://doi.org/10.1016/j.brs.2014.02.009

Sherrington CS (1942) Man on his nature. Cambridge University Press, Cambridge, p 178

Sporns O (2011) Networks of the brain. MIT Press, Cambridge, MA

Singh AM, Neva JL, Staines WR (2016) Aerobic exercise enhances neural correlates of motor skill learning. Behav Brain Res 301:19–26. http://doi.org/10.1016/j.bbr.2015.12.020

Sutor B, Schmolke C, Teubner B, Schirmer C, Willecke K (2000) Myelination defects and neuronal hyperexcitability in the neocortex of connexin 32-deficient mice. Cereb Cortex 10(7):684–697

Taylor JL (2006) Stimulation at the cervicomedullary junction in human subjects. J Electromyogr Kinesiol: Official J Int Soc Electrophysiological Kinesiol 16(3):215–223. http://doi.org/10.1016/j.jelekin.2005.07.001

Terao Y, Ugawa Y (2002) Basic mechanisms of TMS. J Clin Neurophysiol 19(4):322

Ziemann U, Rothwell JC (2000) I-waves in motor cortex. J Clin Neurophysiol 17(4):397–405

# Motor Control of Human Spinal Cord Disconnected from the Brain and Under External Movement

**Winfried Mayr, Matthias Krenn and Milan R. Dimitrijevic**

**Abstract** Motor control after spinal cord injury is strongly depending on residual ascending and descending pathways across the lesion. The individually altered neurophysiology is in general based on still intact sublesional control loops with afferent sensory inputs linked via interneuron networks to efferent motor outputs. Partial or total loss of translesional control inputs reduces and alters the ability to perform voluntary movements and results in motor incomplete (residual voluntary control of movement functions) or motor complete (no residual voluntary control) spinal cord injury classification. Of particular importance are intact functionally silent neural structures with residual brain influence but reduced state of excitability that inhibits execution of voluntary movements. The condition is described by the term discomplete spinal cord injury. There are strong evidences that artificial afferent input, e.g., by epidural or noninvasive electrical stimulation of the lumbar posterior roots, can elevate the state of excitability and thus re-enable or augment voluntary movement functions. This modality can serve as a powerful assessment technique for monitoring details of the residual function profile after spinal cord injury, as a therapeutic tool for support of restoration of movement programs and as a neuroprosthesis component augmenting and restoring movement functions, per se or in synergy with classical neuromuscular or muscular electrical stimulation.

W. Mayr (✉) · M. Krenn
Center for Medical Physics and Biomedical Engineering, General Hospital Vienna, 4L, Medical University of Vienna, Waehringer Guertel 18–20, 1090 Vienna, Austria
e-mail: winfried.mayr@meduniwien.ac.at

M. Krenn
e-mail: matthias.krenn@outlook.com

M.R. Dimitrijevic
Baylor College of Medicine Houston, Houston, TX, USA

M. Krenn
Institute of Electrodynamics, Microwave and Circuit Engineering, Vienna University of Technology, Gusshausstrasse 25–27, 1040 Vienna, Austria



J. Laczko and M.L. Latash (eds.), *Progress in Motor Control*,
Advances in Experimental Medicine and Biology 957,
DOI 10.1007/978-3-319-47313-0_9

**Keywords** Injury · Sublesional control · Spinal reflex · Motor unit · EMG · Surface polielectromyography · Discomplete · Supraspinal control · Neurorehabilitation · Functional electrical stimulation (FES)

## Introduction

Features of the human spinal cord motor control can be described using surface electrodes and the two spinal injury model: (i) the spinal cord completely separated from brain motor structures by accidental injury models; (ii) the spinal cord receiving reduced and altered supraspinal input due to an incomplete lesion. Systematic studies using surface electrode polielectromyography, i.e., simultaneous EMG recording from multiple muscle groups, were carried out to assess skeletal reflex responses to a single and repetitive stimulation in a large number of subjects. In complete spinal cord, the functional integrity of three different neuronal circuits below the lesion level is demonstrated: first, simple monosynaptic and oligosynaptic reflex arcs and polysynaptic pathways; second, propriospinal interneuron system with their cells in the gray, the axons in the white matter of the spinal cord, conducting activity to different spinal cord segments; and third, internuncial gray matter neurons with short axons and dense neurons in contact within spinal gray matter. All of these three systems participate continuously in the generation of spinal cord reflex output activating muscles. The integration of these systems and the relative degree of excitation and setup produces characteristic functions of motor control and depends on relative degree of excitation and setup produced characteristic function of motor control. In incomplete spinal cord injury patients, the implementation of brain motor control depends on the profile of residual brain descending input and its integration with functional neuronal circuits below the lesion. Locomotors patterns of this new structural relationship are expressed as an alternative, but characteristic and consistent neurocontrol. The more we know about how brain governs spinal cord networks, the better we can describe human motor control. On the other hand, such knowledge for restoration of residual functions and for construction of new cord circuitry is crucial to expend the function of injured spinal cord (Dimitrijevic et al. 2005).

These observations highlight another long-standing issue as to the nature of a motor complete injury (SCI). Some evidence suggests that most of the individuals with clinically motor complete paralysis do not have a complete anatomical separation of the proximal and distal ends of the spinal cord at the site of the lesion (Kakulas 1987). The absence of a clear functional dichotomy of a spinally complete and incomplete injury has been demonstrated to be dependent on the thoroughness and procedures of the clinical examination. For example, in individuals that are considered to be motor complete based on a standard clinical assessment, a more detailed assessment can reveal some residual motor function (Kakulas and Bedbrook 1976). Thus, from a functional perspective, there are probably gradations of completeness of the injury that can easily remain undetected. Could remaining descending axons that cross the lesion be 'reawakened', and if so, what is the mechanism? Such a scenario would emphasize the importance of there being some remaining anatomical continuity of axons across the lesion, even though they may have no remaining functionally detectable synaptic function before beginning of epidural stimulation. On the other hand, they could have some remaining functional influence recognizable in clinical assessment. Perhaps, when voluntary control was recovered in the presence of epidural stimulation, the stimulation modulated the circuitry to a higher level of excitability, which allowed for some residual descending motor input to exceed the motor threshold of some motor neurons. However, even if this was the case there had to be some fundamental reorganization of the supraspinal and/or spinal circuitry that occurred after months of stimulation and training. Or was the underlying mechanism recovery of the supraspinal and/or spinal circuitry that occurred after months of stimulation and training? Or was the recovery of voluntary control attributable to a combination of stimulation and repeated voluntary efforts generated simultaneously that induced axons to grow through or around the lesion and make functional connections in the proximal stump? If this did occur, then a remarkable ability of the descending axons to form functional connections to interneurons (probably propriospinal neurons) that can then generate coordinated movements would be rather impressive and certainly encouraging.

## Transected Spinal Cord Complete and Discomplete Spinal Cord Injury Model

The term "discomplete spinal cord injury" was proposed on the basis of apparently isolated spinal cord in which clinical signs indicated complete motor paralysis, whereas the neurophysiological evaluation documented the presence of spinal reflex activity, or even volitionally induced facilitation of motor unit activity, restoring augmentation of motor output (Dimitrijevic 1987; Sherwood et al. 1992). Human neuroscientists, after epidural or transcutaneous posterior root stimulation are speculating on posterior lumbosacral stimulation to result in "electro enabling" motor control, when observing regain of voluntary control over single paralyzed motor functions in selected cases during administration of continuing trains of electrical stimuli (Edgerton and Harkema 2011).

By recording EMG activity from leg muscles, changes in segmental excitability of the plantar reflex elicited by cutaneous stimulation of the plantar surface have been studied. Using surface polielectromyography, 50 SCI patients were examined for their ability to volitionally suppress the plantar reflexes by recording motor units activity from their paralyzed muscles. The patients, who had no voluntary EMG activity in the monitored muscles, were able to volitionally suppress the plantar reflex responses by 45 % in the tibialis anterior, hamstring, and triceps surae muscles and to suppress the quadriceps response by 72 %. In this patient group, 73 of 100 tibialis anterior muscle groups showed suppression of more than 20 % compared with the control response. On reexamination, these findings were consistent during a period of 2 years in six patients (Cioni et al. 1986). Thus, it is possible to volitionally suppress excitability in paralyzed muscles below a complete spinal cord lesion.

The above observation supported the hypothesis that the absence of volitional motor activity does not preclude the existence of CNS axons traversing the injury site and influencing the activities below the level of the lesion.

We are proposing to use for this model the term "discomplete" which combines the notion that in absence of motor control brain excitatory and inhibitory influence on the nervous system below the level of the lesion is preserved.

## Model of Human Spinal Cord Injury with Reduced and Altered Motor Control

Systematic studies of motor units activity from the lower thoracic, lumber and upper sacral segments of the spinal cord during reflex, automatic and volitional motor tasks in the chronic SCI subjects, led us to describe in this chapter a clinical neurophysiological classification of spinal cord injury. Furthermore, we also reviewed supportive anatomical evidence for the proposed clinical neurophysiological classification and we especially addressed newly reported clinical-pathomorphological evidence for the "discomplete" model of the human spinal cord.

In this study of the model of the human spinal cord with "reduced and altered motor control," we applied the identical approach used in the "discomplete" model. This means that we carried out a study of the qualitative features of motor control recorded by BMCA in subjects with residual volitional control of motor units in paretic and paralyzed muscles below the spinal cord injury (Dimitrijevic et al. 1997). This "reduced functional anatomy of spinal cord injury" is just the opposite of the "discomplete" SCI model in which residual motor control is absent and is replaced by the brain's facilitatory and suppressive influence on segmental motor output.

Results were obtained from 91 chronic SCI patients and 15 neurologically healthy individuals. Of these 91, 23 were women and 68 were men, with ages ranging from 11 to 77 years, with a mean of 33.6 ± 1.7 (mean ± SE). All suffered from post-traumatic closed SCI between C2 and T10. The time between injury and the recording of motor control ranged from 3 months to 27 years with a mean duration of 78.5 ± 10.9 months. 21 of these 91 patients showed no signs of volitional movement or volitionally induced motor unit activity below the level of injury, 22 could move voluntarily but could not ambulate, and 48 were ambulatory with a variety of assistive devices.

We analyzed the EMG patterns during gait and two discrete volitional motor tasks in the supine position: the first consisted of volitional efforts to perform full flexion and extension of the lower limbs (multi-joint movement) and the second consisted of dorsal and plantar flexion of single ankle (single-joint) movements.

The main characteristic of integrated EMG activity during gait is the repetitive pattern of burst discharges followed by silent periods. Another general feature of EMG activity is that EMG bursts are lower amplitude and shorter duration in proximal and trunk muscles than in the leg muscles. The third general feature is the reciprocal relationship of bursts of EMG and activity between antagonistic leg and thigh muscle groups during the stance and swing phases of gait. Bearing in mind these general EMG features for the adult patient with an intact nervous system, it is easy to recognize alterations in motor output (Dimitrijevic et al. 1997).

Unlike the semi-automatic and stereotyped motor control during steady gait, if we record EMG activity in a SCI patient during a volitional motor task of single- or multi-joint movements in the supine position, it is possible to expand observations on motor control in paralyzed patients (Dimitrijevic et al. 2012).

We have found in EMG studies of motor control during volitional motor tasks that there is a consistency in the different features of motor unit activity. This was the case in 11 wheelchair-bound and 23 ambulatory patients. In this group of 11 and 23 SCI patients, we analyzed the percentage of the presence of volitional multi-joint flexion, isolation of single- from multi-joint movement, and isolation of agonist from antagonist EMG activity. All 3 features indicate discrete motor control and are observed more frequently in ambulatory patients than in the wheelchair-bound population. On the other hand, diffuse patterns of motor output and delayed onset are much more frequent in the wheelchair-bound than in the ambulatory group.

The fact that we can recognize different features of motor control during single-joint and multi-joint volitional motor tasks and that these features are consistent, repetitive, and related to the severity of the lesion led us to hypothesize that a new organization of spinal circuits is established and that this new structural and functional relationship can generate definite patterns of residual motor functions.

In incomplete SCI, the organization of brain motor control depends on the profile of the residual brain descending system and its integration with the spinal network. In the past, our understanding was that motor activity after incomplete SCI was inconsistent and considered abnormal because it differed from normal motor activity. However, we have shown that locomotor patterns in SCI people are consistent. Thus, the functional properties of these newly established sensory motor integration mechanisms are expressed in the repeatable neurophysiological characteristics of motor unit activity during single- and multi-joint volitional tasks. A thorough understanding of this new motor control will play an important role in establishing the scientific basis for neurobiological interventions to restore impaired SCI functions.

Human SCI models based on studies with surface polielectromyography (BMCA) play an important role in assessing and understanding the underlying mechanisms of newly developing protocols for the modifications of the already established clinical output in chronic SCI subjects. An example of this is found in locomotor training with or without FES for the improvement of walking speed and capacity (Cardenas 2010).

Furthermore, the new spinal cord organization established as a result of spinal cord injury can develop different potentials for neuroplasticity and, as a result of this activity-dependent capacity, the CNS can respond by increasing motor performance (Muir and Steeves 1997). Moreover, another factor, which will influence the capacity of neuroplasticity will be the profile of residual supraspinal control. Such residual profile can have components of postural, locomotor and volitional activity, or only one of these components can prevail. It has been shown, for instance, that interactive locomotor training using body weight support can improve walking capabilities in patients and not change the capacity for volitional muscle contraction (Wernig 2015).

Thus, the assessment of motor control in SCI subjects can reveal subclinical brain influence below the level of the lesion in otherwise clinically complete SCI, or it can describe and measure the relation between different components of residual suprasegmental control in incomplete SCI patients. This new possibility will help in defining clinical and laboratory inclusion criteria of SCI patients for a variety of available restorative procedures and neurorehabilitation protocols (Dimitrijevic et al. 2012).

## Repetitively Elicited Spinal Reflexes

The emphases in human reflex studies have been on reflex pathways, their latencies and their reciprocal relation. Thus, repetitively elicited spinal reflex responses in humans lacking brain motor control can reveal behavioral motor units modification depending from rate of repetitive afferent stimulation. Epidural stimulation was carried out with 4-polar linear or 16-polar array electrode configurations (Sayenko et al. 2014).

## Spinal Cord Stimulation and Elicited Motor Outputs

While stimulating effectively posterior lumbar cord structures, it is possible to generate rhythmical activity with synchronized EMG discharges and repetitive withdrawal flexor movement in the lower limbs. Tonic EMG activity was elicited by stimulating same spinal cord structure using different stimulating parameters.

Jilge et al. (2004) provided evidence that the human spinal cord is able to respond to external afferent input and to generate a sustained extension of the lower extremities when isolated from brain control. The presented study demonstrated that sustained, non patterned electrical stimulation of the lumbosacral cord—applied at a frequency in the range of 5–15 Hz and a strength above the thresholds for twitches in the thigh and leg muscles—can initiate and retain lower limb extension in paraplegic subjects with a long history of complete spinal cord injury. The authors hypothesized that the induced extension is due to tonic input applied by the epidural stimulation to primary sensory afferents. The induced volleys elicit muscle twitches (posterior root reflex responses) at short and constant latency times and coactivate the configuration of the lumbosacral interneuronal network, presumably via collaterals of the primary sensory neurons and their connectivity with this network. They further speculate that the volleys induced externally to the lumbosacral network at a frequency of 5–15 Hz initiate and retain an "extension pattern generator" organization. Once established, this organization would recruit a larger population of motor units in the hip and ankle extensor muscles as compared to the flexors, resulting in an extension movement of the lower limbs. In the electromyograms of the lower limb muscle groups, such activity is reflected as a characteristic spatiotemporal pattern of compound motor unit potentials.

Another report by Fisher et al. (2008) describes the performance of a 16-channel implanted neuroprosthesis for standing and transfers after spinal cord injury, including four-contact nerve-cuff electrodes stimulating the femoral nerves for knee extension. Responses to the stimulation via nerve-cuffs were stable and standing times increased by 600 % over time-matched values with a similar 8-channel neuroprosthesis utilizing muscle-based electrodes on vastus lateralis for knee extension.

Meanwhile, sensory and motor complete spinal cord injury (SCI) has been considered functionally complete resulting in permanent paralysis with no recovery of voluntary movement, standing, or walking. Rejc et al. (2015) refer on a previous demonstration that lumbosacral spinal cord epidural stimulation can activate the spinal neural networks in one individual with motor complete, but sensory incomplete SCI, who achieved full body weight-bearing standing with independent knee extension, minimal self-assistance for balance, and minimal external assistance for facilitating hip extension. The group now presented a more recent study showing that two clinically sensory and motor complete participants were able to stand over-ground bearing full body weight without any external assistance, using their hands to assist balance. Two clinically motor complete, but sensory incomplete participants also used minimal external assistance for hip extension. Standing with the least amount of assistance was achieved with individual-specific stimulation parameters, which promoted overall continuous EMG patterns in the lower limbs' muscles. Stimulation parameters optimized for one individual resulted in poor standing and additional need of external assistance for hip and knee extension in the other participants. During sitting, little or negligible EMG activity of lower limb muscles was induced by epidural stimulation, showing that the weight-bearing related sensory information was needed to generate sufficient EMG patterns to effectively support full weight-bearing standing. In general, electrode configurations with cathodes selected in the caudal region of the array at relatively higher frequencies (25–50 Hz) resulted in the more effective EMG patterns for standing. The authors conclude that human spinal circuitry can generate motor patterns effective for standing in the absence of functional supraspinal connections; however, the appropriate selection of stimulation parameters is critical.

As well, in an earlier study Minassian et al. (2004) showed that epidural stimulation of the posterior lumbosacral cord recruited lower limb muscles in a segmental-selective way, which was characteristic for posterior root stimulation. Continuous stimulation with 2.2 Hz elicited stimulus-coupled compound muscle action potentials (CMAPs) of short latency, which were approximately half that of phasic stretch reflex latencies for the respective muscle groups. EMG amplitudes were stimulus-strength dependent. Stimulation at 5–15 and 25–50 Hz elicited sustained tonic and rhythmic activity, respectively, and initiated lower limb extension or stepping-like movements representing different levels of muscle synergies. All EMG responses, even during burst-style phases were composed of separate stimulus-triggered CMAPs with characteristic amplitude modulations. During burst-style phases, a significant increase of CMAP latencies by about 10 ms was observed.

Angeli et al. (2014) point to a previous report on one individual who had a motor complete, but sensory incomplete spinal cord injury regained voluntary movement after 7 months of epidural stimulation and stand training, where they had presumed that the residual sensory pathways were critical in this recovery. In the same article, they now reported on three more individuals, where voluntary movement occurred with epidural stimulation immediately after implant even in two who were diagnosed with a motor and sensory complete lesion. They conclude that neuromodulating the spinal circuitry with epidural stimulation, enables completely paralyzed individuals to process conceptual, auditory, and visual input to regain relatively fine voluntary control of paralyzed muscles. Further they assume that neuromodulation of the sub-threshold motor *state of excitability* of the lumbosacral spinal networks was the key to recovery of intentional movement in four of four individuals diagnosed as having complete paralysis of the legs. Finally, they claim to have uncovered a fundamentally new intervention strategy that can dramatically affect recovery of voluntary movement in individuals with complete paralysis even years after injury.

In a recent review article, Grecco et al. (2015) analyze the state of the art of transcutaneous spinal cord stimulation as therapeutic option after spinal cord injury. Though they still see limited effects in treatments for spinal cord injury, they found evidences that electrical stimulation might facilitate plastic changes in affected spinal circuitries that may be beneficial in improving motor function and spasticity, and reducing SCI-related neuropathic pain. They critically reviewed the physiological basis and therapeutic action of transcutaneous spinal cord stimulation in SCI, based on available animal and clinical studies. They analyzed the literature published on PubMed to date, looking for the role of three main noninvasive stimulation techniques in the recovery process of SCI and focusing mainly on transcutaneous spinal stimulation. This review discusses the main clinical applications, latest advances, and limitations of noninvasive electrical stimulation of the spinal cord. Although most recent research in this topic has focused on transcutaneous spinal direct current stimulation (tsDCS), they also reviewed the technique of potential methods to modulate spinal cord plasticity. They also developed a finite element method (FEM) model to predict current flow in the spinal cord when using different electrode configurations. Identifying gaps in the knowledge on noninvasive electrical stimulation for modulation of spinal neuronal networks in patients with SCI they nevertheless conclude that their is encouraging preliminary evidence that tsDCS, TENS, and NMES have a positive influence on the promotion of plasticity in SCI. They see FEMs useful to predict how in tsDCS electrode size and placement can be used to shape, which spinal segments are modulated and what detailed areas of spinal anatomy can concentrate current density (e.g., spinal roots). In conclusion, they see tsDCS as a technique that can influence conduction along ascending tracts in the spinal cord, so could modulate supraspinal activity.

In their review on technical aspects and therapeutic applications Nardone et al. (2015) discuss instrumentation for noninvasive spinal cord stimulation. Electrical and magnetic trans-spinal stimulation can be used to increase the motor output of multiple spinal segments and modulate cortico-spinal excitability. The application of direct current through the scalp as well as repetitive transcranial magnetic stimulation are known to influence brain excitability, and hence can also modulate other central nervous system structures including spinal cord. They conclude that noninvasive electrical and magnetic spinal stimulation may provide reliable means to characterize important neurophysiologic and pathophysiologic aspects of spinal cord function. Moreover, transcutaneous direct current stimulation and repetitive magnetic stimulation may hold therapeutic promise in patients with spinal cord disorders, although future well-controlled studies are needed to corroborate and extend the preliminary findings.

Even though there is growing evidence from clinical studies that tsDCS could serve as an effective therapeutic tool for influencing *central state of excitability* in the central nervous system, the method raises safety issues that need more attention. There are non-negligible risks for electrochemical tissue damage at electrode tissue interfaces, in particular if DC is administered, which needs systematic investigations and, even more concerning, axon damage under application of similarly low direct current, as applied in clinical tsDCS, has been reported in a recent animal study (Ravid and Prochazka 2012).

## Functional Electrical Stimulation Application for Functional Improvement of Upper Motor Neuron Function

Clinical studies of FES in subjects with upper motor neuron dysfunction revealed that in addition to the stimulation of motor fibers of peripheral nerves for "correction" of central paresis (Mayr 2015). Functional Electrical Stimulation (FES) can be a versatile tool for support of movement rehabilitation. On basis of noninvasive (electrodes at the skin surface) or invasive (implanted electrodes) techniques there are three options for reactivation of lost muscle functions:

(1) Stimulation of efferent nerves—neuromuscular stimulation—as established method,
(2) Direct activation of muscle fibers—muscle stimulation—in absence of the motor nerve, which can lead to similar contraction responses, and
(3) Stimulation of afferent nerve structures, which can lead to improvement in impaired movement patterns by artificially substitution of missing neural

inputs to central neuronal networks in the spinal cord—movement pattern generators and adaptive control functions.

Applications of electrical stimulation for restoration of movement reach back in the eighteenth century. At this time stimulators based on electrostatic discharge became available and were applied for treatment of post-stroke movement disorders. Though over the years we see a constant growth in the knowledge of physiology and pathophysiology of movement, technically supported rehabilitation methods and associated equipment progress was limited till the invention of the bipolar transistor, mid twentieth century that suddenly enabled the construction of compact battery-powered stimulators. Numerous noninvasive and implantable devices for a variety of applications were developed since, for experimental and clinical research, only few got transferred to market and permanent availability for clinical treatment and application by patients in need.

Whereas initial treatments with electrical discharge had obviously targeted cutaneous afferences, earlier concepts with mobile miniaturized equipment focused on stimulation of efferent motor nerves for direct reactivation of paralyzed muscles or muscle groups to make FES induced muscle contractions act as a kind of biological braces. Extension of these concepts to multiple channels and inclusion of biosignal and sensor feedback led to closed-loop approaches with the intention to come to more physiological execution of complex "more natural" movement patterns, reduce fatigue phenomena, and optimize functional benefit. Many difficulties were experienced regarding handling of equipment, costs and achievable stability of control, persisting problems remain on stable and selective recording of biosignals, as well as stable and selective control of motor units and muscle fibers, which mostly limits the achievable quality of artificial motor control.

The recent new developments started to rely more in taking advantage from remaining intact neural structures in and below the lesion zone in the spinal cord. There are promising developments in using noninvasive stimulation of posterior roots at the lumbar spinal cord level to assess, in combination with multichannel EMG and sensor recordings, functional injury profiles. These can be of value for primary assessment and intervention planning as well as for monitoring of applied protocols and biological developments with time. Similar setups can act as neuroprosthesis for enabling, augmenting, or modifying movements by influencing altered central control mechanisms via noninvasively or epidurally placed electrode arrangements. Contemporary instrumentation for invasive or noninvasive spinal cord stimulation can provide only very limited selectivity in activating neural structures. Applied stimulus patterns are mostly continuous trains with unvaried impulse amplitude, pulse width, and frequency kept constant in an application session after initial calibration. Nevertheless a variety of responses can be elicited by just adjusting amplitude and frequency of stimuli, ranging from tonic to phasic or mixed tonic and phasic muscle activation as well as inhibitory influence on spontaneous muscle activity. Based on careful assessment of functional lesion profiles these limited options can already be applied beneficially in movement rehabilitation. It can be expected that refinement of stimulation selectivity, more complex variation of stimulus patterns with phasic parameter variation, and extension to multiple channel solutions with coordinated but independently adjustable parameter variation will open many more options for refined control of movement functions via centrally administered afferent inputs.

Still, in certain applications, "classical" direct neuromuscular or muscular stimulation methods remain important in sole or complementary use, e.g., for phrenic pacing (Mayr et al. 1993, 2001) or strengthening and maintenance of muscles. The latter is relevant for paralysis after upper motorneuronlesion and in particular in case of lower motor neuron lesion the only option for avoidance of severe degeneration associated long-term complications (Mayr et al. 2002).

## Human Motor Control and External Control of Movement Stability, Skilled Movement Without Attention

How movements are controlled is just one of many questions and influencing factors. Of equal importance is how we maintain stability of the motor system. In abnormal conditions, acquired skills might be disrupted by in part missing feedback and actuation components as disturbing factors. Learning of compensatory skills takes time and is strongly depending on autonomous and heteronomous circumstances and limited influence on their manifestation.

Besides individually altered physiology, time-dependent due to regenerative and degenerative processes, and personal live situation and motivation, multiple factors of rehabilitation environment and conditions are acting and decisive for development and outcome. It begins with level of analysis, which is carried by computational level, proceedings level, and implementation.

Research on human motor control relies on various fields contributing as neurophysiology, medicine, biology, engineering, physics, statistics, behavioral science, and cognitive and human factors. Detailed physiological foundations are essential, e.g., on muscle spindle, Golgi tendon organ, joint receptors, cutaneous-receptors, spinal reflexes, servo theory, Alfa-Gama coactivator, recurrent inhibition, reciprocal inhibition, smart spinal cord, tuning of spinal cord, and many more to be understood and holistically considered.

By decreasing the excitability of spinal reflexes with application of electrical and magnetic trains, spinal stimulation could be helpful in normalizing reflexes and treating hypertonia in subject with lesion in the upper motor neuron. Transcutaneous spinal cord stimulation can influence ascending and descending pathways, spinal reflex excitability, and functional neurotrophic changes. In addition, peripheral transcutaneous stimulation can be used to target the spinal cord, and provide specific afferent inputs; many have potential for neuromodulation in cord-injured subjects although future well-controlled studies are needed to corroborate and extend the preliminary findings.

Let us finally mention that the review showed that only few reported, eight studied subjects, on safety or complications of the spinal cord stimulation (SCS) (Bendersky et al. 2014; Benfield et al. 2016; Walsh et al. 2015). There is an urgent need for such studies even in the widely applied application of SCS for pain treatment (Cameron 2004).

## Summary and Conclusion

We see a number of preliminary findings on epidural or transcutaneous spinal cord stimulation of paralyzed persons with chronic spinal cord injury, but based on limited observation and neurophysiological studies. Furthermore, overall theoretical and practical knowledge on human motor control as well as neurophysiology of upper motor neuron dysfunction and external intervention with sustained electrical stimulation are waiting for appropriate studies with science design. In short, we are at present at the beginning of learning how to perform external modification of upper motor neuron modification by external control of afferents to spinal cord.

## References

Angeli CA, Edgerton VR, Gerasimenko YP, Harkema SJ (2014) Altering spinal cord excitability enables voluntary movements after chronic complete paralysis in humans. Brain 137(Pt 5):1394–1409. And: Angeli C, Edgerton VR, Gerasimenko Y, Harkema SJ (2015) Reply: no dawn yet of a new age in spinal cord rehabilitation. Brain 138(Pt 7):e363

Bendersky D, Claudio Yampolsky C (2014) Is spinal cord stimulation safe? A review of its complications. World Neurosurg 82:1359–1368

Benfield J, Maknojia A, Epstein F (2016) Progressive paraplegia from spinal cord stimulator Lead fibrotic encapsulation. Am J Phys Med Rehabil 95:e30–e33

Cameron T (2004) Safety and efficacy of spinal cord stimulation for the treatment of chronic pain: a 20-year literature review. J Neurosurg: Spine 100(3):254–267

Cardenas D (2010) Spinal cord medicine, second edition: principles & practice, 2nd edn. Demos Medical Publishing. ISBN-10: 1933864192

Cioni B, Dimitrijevic MR, McKay WB, Sherwood AM (1986) Voluntary supraspinal suppression of spinal reflex activity in paralyzed muscles of spinal cord injury patients. Exp Neurol 93 (3):574–583

Dimitrijevic MR (1987) Neurophysiology in spinal cord injury. Paraplegia 25(3):205–208

Dimitrijevic MR, Gerasimenko Y, Pollo FE (1997) Sustained epidural electrical stimulation of the spinalized lumbar cord in humans can facilitate or suppress spinal reflex and central pattern generator activities. In: Gurfinkel VS, Levik Yu S (eds) Proceedings of the international symposium on brain and movement, vol 60. St. Petersburg, Moscow

Dimitrijevic MR, Persy I, Forstner C, Kern H, Dimitrijevic MM (2005) Motor control in the human spinal cord. Artif Organs 29(3):216–219

Dimitrijevic, MR, Kakulas BA, McKay WB, Vrbová G (eds) (2012) Restorative neurology of spinal cord injury. Oxford University Press, NY, 336 pp. ISBN: 978-0-19-974650-7

Edgerton VR, Harkema S (2011) Epidural stimulation of the spinal cord in spinal cord injury: current status and future challenges. Expert Rev Neurother 11(10):1351–1353

Fisher LE, Miller ME, Bailey SN, Davis JA Jr, Anderson JS, Rhode L, Tyler DJ, Triolo RJ (2008) Standing after spinal cord injury with four-contact nerve-cuff electrodes for quadriceps stimulation. IEEE Trans Neural Syst Rehabil Eng 16(5):473–478

Grecco LH, Li S, Michel S, Castillo-Saavedra L, Mourdoukoutas A, Bikson M, Fregni F (2015) Transcutaneous spinal stimulation as a therapeutic strategy for spinal cord injury: state of the art. J Neurorestoratology 3:73–82

Jilge B, Minassian K, Rattay F, Pinter MM, Gerstenbrand F, Binder H, Dimitrijevic MR (2004) Initiating extension of the lower limbs in subjects with complete spinal cord injury by epidural lumbar cord stimulation. Exp Brain Res 154:308–326

Kakulas BA, Bedbrook GM (1976) Injuries of the spine and spinal cord. In: Vinken PJ, Bruyn GW (eds) Handbook of clinical neurology, Chap. 3. North Holland Publishing Company, Amsterdam, pp 27–42

Kakulas BA (1987) The clinical neuropathology of spinal cord injury. A guide to the future. Paraplegia 25(3):212–216

Sherwood AM, Dimitrijevic MR, McKay WB (1992) Evidence of subclinical brain influence in clinically complete spinal cord injury: discomplete SCI. J Neurol Sci 110(1–2):90–98

Mayr W, Bijak M, Girsch W, Holle J, Lanmüller H, Thoma H, Zrunek M (1993) Multichannel stimulation of phrenic nerves by epineural electrodes. Clinical experience and future developments. ASAIO J 39:M729–M735

Mayr W, Bijak M, Rafolt D, Sauermann S, Unger E, Lanmüller H (2001) Basic design and construction of the Vienna FES implants: existing solutions and prospects for new generations of implants. Med Eng Phys 23:53–60

Mayr W, Hofer C, Bijak M, Rafolt D, Unger E, Sauermann S, Lanmueller H, Kern H (2002) Functional electrical stimulation (FES) of denervated muscles: existing and prospective technological solutions. Basic Appl Myol 12:287

Mayr W (2015) Early and contemporary approaches for application of FES in movement rehabilitation after spinal cord injury. In: József Laczkó (ed) Progress in motor control X, Budapest July 22–25, 2015; International Society of Motor Control; Hungarian Sports Science Booklets—XII; p 43. ISBN: 978-615-5187-07-0

Minassian K, Jilge B, Rattay F, Pinter MM, Binder H, Gerstenbrand F, MR Dimitrijevic MR (2004) Stepping-like movements in humans with complete spinal cord injury induced by epidural stimulation of the lumbar cord: electromyographic study of compound muscle action potentials. Spinal Cord 42(7):401–416

Muir GD, Steeves JD (1997) Sensorimotor stimulation to improve locomotor recovery after spinal cord injury. Trends Neurosci 20(2):72–77

Nardone R, Höller Y, Taylor A, Thomschewski A, Orioli A, Frey V, Trinka E, Brigo F (2015) Noninvasive spinal cord stimulation: technical aspects and therapeutic applications. Neuromodulation 18(7):580–591 (discussion 590-1)

Ravid N, Prochazka A (2012) Nerve lesioning with direct current. In: Proceedings of 17th annual IFESS conference, Banff, Canada; pp 344–347, 9–12 Sept 2012

Rejc E, Angeli C, Harkema S (2015) Effects of lumbosacral spinal cord epidural stimulation for standing after chronic complete paralysis in humans. PLoS ONE 10(7):e0133998

Sayenko DG, Angeli C, Harkema SJ, Edgerton VR, Gerasimenko YP (2014) Neuromodulation of evoked muscle potentials induced by epidural spinal-cord stimulation in paralyzed individuals. J Neurophysiol 111(5):1088–1099

Walsh KM, Machado AG, Krishnaney AA (2015) Spinal cord stimulation: a review of the safety literature and proposal for perioperative evaluation and management. Spine J 15:1864–1869

Wernig A (2015) Response to 'Randomised controlled trials do not always give the results we want but that doesn't mean we should abandon randomised controlled trials'. Spinal Cord 53(12):897–898

# Anticipation in Object Manipulation: Behavioral and Neural Correlates

**Thomas Schneider and Joachim Hermsdörfer**

**Abstract** One way to foster the understanding of the impressively skilled fine motor control of human object manipulation is to investigate the rules of the underlying fundamental mechanisms. Examining the action of grasping and lifting objects of various characteristics reveals that one vital step to handle objects in a skilled and fluent way is to predict their properties and plan motor actions accordingly. A large number of behavioral studies have extracted the factors on which we rely when we interact with objects in an anticipatory mode of action. At the same time, considerable effort was taken to elucidate the neuroanatomical areas and networks involved and crucial for anticipatory behavior by conducting functional imaging and stimulation studies and examining the deficits of patients with localized brain damage. This chapter gives an overview of these studies and tries to reconcile their findings, in order to provide an insight into the basic principles of anticipatory motor control and their underlying neural substrates.

**Keywords** Torques · Lift-off · Tilts · Economic grip · Anticipatory control · Load force · Object dynamics · Sensorimotor memory

## Introduction

The ability to dexterously grasp and manipulate a vast variety of objects and tools is one of the most evident and remarkable skills of humans. One vital step of skilled interaction with objects is to incorporate the physical properties of the objects into the planning of the manipulation. Anticipatory planning of upcoming grasps in a feedforward manner allows us to fluently and economically lift objects according to task requests and our intentions.

T. Schneider · J. Hermsdörfer (✉)
Institute of Human Movement Science, Department of Sport and Health Sciences, Technical University of Munich, Georg-Brauchle Ring 60-62, 80992 Munich, Germany
e-mail: Joachim.Hermsdoerfer@tum.de



J. Laczko and M.L. Latash (eds.), *Progress in Motor Control*,
Advances in Experimental Medicine and Biology 957,
DOI 10.1007/978-3-319-47313-0_10

The importance of anticipatory grasp planning (see Table 1) becomes especially clear when one considers manual actions during shopping in a supermarket. Being confronted with an immense selection of goods of different forms, sizes, weights, and surface-properties, it is essential to take object properties into account when planning an upcoming grasp. Then you produce finger forces with a much higher rate when grasping and lifting a milk carton than when grasping a much lighter box of tissues. In doing so, the time needed until lift-off remains similar, despite big weight differences and the action can be performed in a smooth and controlled manner. Moreover, an anticipatory grasp planning is also frequently vital to handle objects securely. For instance, exerting excessive grip forces (GF) on a fragile glass of wine and thus breaking it, or not compensating the arising torques when grasping the handle of a coffee cup and spilling its hot content may lead to adverse consequences and must be avoided.

One important and feasible way established to research anticipatory motor planning during object manipulation is the analysis of grasping kinetics. Assessment of the forces and torques acting on a grasped object, as well as the positioning of the fingers on it, has given valuable insights into the mechanisms involved in predictively adapting motor output to the demands posed by the object characteristics. Moreover, functional imaging and stimulation studies and studies conducted with brain-lesioned patients have complemented the findings of behavioral studies in healthy subjects by suggesting brain regions and networks vital for anticipatory motor planning.

The objective of this chapter is to give a comprehensive overview on the behavioral, neurophysiological, and clinical findings of anticipatory grasp control.

Therefore, we will first introduce the rationale of evaluating anticipatory grasp planning and depict the most common indicators used to assess it. The main part of the chapter will thereafter summarize and synthesize the observations of behavioral,

**Table 1** Central aspects of anticipatory control during object manipulation

| Motor goals | Stable and economic grip | Smooth and well-controlled lift-off | Prevention of tilts |
|---|---|---|---|
| Physical object properties to be predicted | Surface-friction of the handle and interacting finger pads | Weight | Mass distribution |
| Effective object cues | Surface appearance and sense of finger moisture | Size, material and identity of the object | Geometric shape |
| Parameters to be anticipated | GF/LF-Ratio | GFR and LFR during the load phase | Generation of compensatory torques at object lift-off |

Note that motor goals and associated object properties as well as strategies of force generation overlap. Abbreviations (see also text): *GF* grip force, *LF* load force, *GFR* & *LFR* grip and load force rate neurophysiological, and clinical studies. We will sort these findings by assigning them to the sensorimotor and cognitive systems, which govern anticipatory motor planning and the interactions between these systems.

## Anticipation of Physical Object Properties

Anticipatory object manipulation encompasses predicting object dynamics, as well as the dynamics of the own body, in order to adjust the programming of motor actions accordingly, before somatosensory feedback is available. Reactive force corrections, on the other hand, rely on sensory feedback, which is not available for many properties (e.g., weight, weight distribution) until object lift-off. Additionally, reactive corrections always take effect with time latencies, needed to conduct sensory signals, detect errors, compute adequate corrections, and conduct motor commands to the effectors. While reactive mechanisms are important for motor refinement after lift-off, the initial success of object manipulation until lift-off and a certain time beyond mainly relies on neural anticipatory motor planning.

As a starting point to review anticipatory object manipulation, it is worthwhile to consider how the action of picking up an object is structured and which object properties have to be anticipated.

The elementary manipulation of grasping an object, lifting it from a supporting surface and replacing it subsequently is structured in serial action phases that are determined by distinct contact events between the object and the finger pads. The transitions between phases are crucial time points when a comparison of predicted and actual sensory inputs triggers preprogrammed motor commands (Johansson and Westling 1984; Johansson and Flanagan 2009).

Of special interest for the investigation of anticipatory planning is the load phase. This is the period after both finger pads have made contact with the grasp-surfaces and before the object lifts off. As many object properties cannot be detected prior to lift-off, motor control relies on anticipatory control during this phase. The load phase is characterized by a parallel increase of the GF, acting orthogonal toward the contact surfaces, and the vertical load force, which has to exceed the object mass load, in order to lift it. The GF must be high enough to generate a frictional force, which suffices to produce the required load force and prevent slippage. At the same time, the GF must not be excessively high as this might lead to damaging fragile objects, early exhaustion, as well as hampered fine manipulation. Therefore, the ratio between grip and load force characterizes the efficiency of force coordination employed to secure the grasp. The GF/LF-ratio is usually only slightly higher than necessary to prevent slippage and depends on the frictional properties of the grasp surface and finger pads. The parallel modulation of grip and load force is not due to a biomechanical coupling of the forces (Hager-Ross and Johansson 1996). Rather, we adjust our GF to predict the load force in a feedforward way (Flanagan et al. 2009).

In successfully programmed lifting trials the grip and load force rates form a smooth, one-peaked, bell-shaped trajectory during the load phase (Johansson and Westling 1988). Their peaks occur well before lift-off (thus in the absence of sensory feedback about weight) and have proven to be valid and highly sensitive indicators of the anticipation of object weight. When lifting an unexpectedly light object, the load force exceeds gravity too early during still increasing force rates, which leads to overshooting object loading with inappropriately high and late force peaks. On the other hand, lifting an unexpectedly heavy object leads to an initially insufficient force increase necessitating further force increments, which are triggered by the absence of the expected lift-off (Johansson and Westling 1984, 1987).

Besides the above-described task aspects of smoothly lifting objects in a stable, yet force economic grip, it is also vital to control object orientation after lift-off by preventing tilts. To that aim subjects must predictively exert torques to compensate for external torques emerging at lift-off due to an asymmetric object weight distribution. The exerted torque is hereby the net product of the selected finger positioning, the exerted GFs and the sharing patterns of load forces between opposing fingers (Fu et al. 2010).

## Factors Determining Anticipation and Their Interplay

Several distinctive sensorimotor and cognitive processes govern the adjustment of grasp planning to object demands. There are certain situations in which they act exclusively, whereas they are concurrently active and interfere with each other most of the time. This is reflected by the structure of the following section.

## Sensorimotor Memory

During the repeated grasping and lifting of objects, the central nervous system encodes and stores information about a variety of experienced object characteristics (e.g., friction, weight, mass distribution), as well as the motor effort employed to manipulate them. This so-called sensorimotor memory has a strong impact on the anticipatory force planning of following trials.

The best way to experimentally address questions concerning sensorimotor memory processes distinctly is to have subjects lift neutral objects that do not provide cues allowing to predict physical properties. It has repeatedly been shown that one trial is often sufficient to establish a sensorimotor memory, which reflects the experienced friction, grip shape, weight, and mass distribution of the preceding lift of the object. This allows for rapid trial-by-trial adaptation to changes of object properties. When participants repeatedly lift an unaltered object, they usually refine these memory traces within two to three lifts and do not improve significantly thereafter (Johansson and Westling 1984, 1988; Jenmalm and Johansson 1997; Salimi et al. 2000; Fu et al. 2010).

In order to detect the brain regions involved in encoding, storing and recalling these sensorimotor memory traces, neuroimaging studies employed fMRI and TMS setups, which allowed testing the grasping and lifting of unseen objects. Functional MRI studies revealed a large network of sensorimotor areas involved in this basic task (Schmitz et al. 2005; Jenmalm et al. 2006; Ehrsson et al. 2007). During grasping and lifting with the right hand the left contralateral primary somatosensory and motor cortices (S1/M1), the left dorsal premotor cortex (PMd), the left parietal operculum, the supplementary motor area (SMA), areas close to the left lateral fissure, and the right and left cerebellum were activated (Schmitz et al. 2005). Lifting a sequence of objects with varying weights induced additional activations in the right inferior frontal gyrus (IFG), the left parietal operculum, and the right supramarginal gyrus (SMG). Lifts of unexpectedly heavy or light objects in a lifting sequence triggered activity in right SMG, in the contralateral primary sensorimotor cortex and the ipsilateral cerebellum (Jenmalm et al. 2006). Recent evidence suggests that also object perception areas in the ventral stream code information about object weight during lifts of constant objects. In particular, multivoxel pattern analysis of fMRI data differentiated BOLD activity during planning lifts of light and heavy objects in object-sensitive regions within the occipitotemporal cortex, in addition to the expected M1 and PMd responses (Gallivan et al. 2014).

TMS studies affirmed that primary motor cortex and cortico-spinal system are not only crucial for finger force generation but also play an important role in the storage and recall of object-related sensorimotor memory modulating force planning. Motor evoked potential in hand muscles following a single pulse of transcranial magnetic stimulation (TMS) applied over M1 just before a lifting movement reflects the weight of the previous lift if there are no other weight cues (Loh et al. 2010). Additionally, stimulation of the primary motor cortex with perturbation protocols seems to deteriorate the memory of object characteristics obtained in the preceding lift (Chouinard et al. 2005; Nowak et al. 2005; Berner et al. 2007). Moreover, stimulation of numerous brain regions of the so-called grasping network apart from M1 effects force scaling. Inhibitory stimulation of the ventral premotor cortex (PMv) also changes the influence of the last object's weight on force scaling (Dafotakis et al. 2008). In addition, stimulation of the left anterior intraparietal area (AIP), which is strongly involved in object manipulation (Davare et al. 2011) as well as the stimulation of the left SMA affects the force scaling of both hands (Davare et al. 2007; White et al. 2013).

Intactness of peripheral sensibility and central sensorimotor integration of tactile inputs are essential to grasp and lift objects in an energy efficient, economic way by exerting GFs just high enough to prevent object slippage, and to establish and utilize sensorimotor representations of previous lifts. Healthy adults initially exert GFs only about 20 % above the minimum necessary to prevent slippage and further decrease this safety margin across subsequent trials (Johansson and Westling 1987).

In contrast, patients suffering from middle cerebral artery (MCI) stroke generally exhibit excessive GF levels and fail to adapt their force programming to given surface-friction- and weight-properties when grasping objects with their slightly to moderately paretic hand. These deficits correlate best with tests evaluating patient's surface discrimination and reactions to sensory perturbations (Hermsdörfer et al. 2003; Blennerhassett et al. 2007).

However, sensory loss does not seem to be the sole reason for uneconomic force programming. Studies investigating anticipatory force control of the ipsilesional hand of stroke patients suggested that patients suffer from a general force control deficit, which also extends to the presumably unimpaired, ipsilesional hand. Despite the absence of clinical sensory or motor deficits, GF levels were significantly elevated when subjects lifted objects with the ipsilesional hand in a two-finger precision grip (Quaney et al. 2005; Nowak et al. 2007a). The demonstration of disrupted anticipatory control in the absence of sensory deficits and the possibility of long lasting improvements after sensory augmentation interventions (Quaney et al. 2010; Hsu et al. 2012) are highly suggestive of the existence of a central sensorimotor integration deficit, causing clumsy motor control of stroke patients even when using the ipsilesional hand (Sunderland et al. 1999; Schaefer et al. 2012). However, it has to be noted that studies employing a three-finger grip did not detect elevated GF levels of the ipsilesional hand (Li et al. 2011; Eidenmüller et al. 2014; Buckingham et al. 2015).

Importantly, despite having difficulties in anticipatory force scaling, stroke patients still generate grip and load forces in parallel to each other. Although some studies revealed an slightly impaired temporal coupling between grip- and load forces (Wenzelburger et al. 2005; Blennerhassett et al. 2006; McDonnell et al. 2006), the majority of stroke patients still programs GFs in a feedforward manner anticipating load forces instead of reacting to them (Hermsdörfer et al. 2003; Nowak et al. 2003). This shows that fundamental aspects of anticipatory control are usually preserved after stroke. Different from MCI stroke, lesion of the cerebellum frequently impair the parallel production of grip and load force (Rost et al. 2005; Brandauer et al. 2008, 2010). This observation led to the conclusion that the cerebellum may contribute importantly to the anticipatory control of GF in parallel to self-produced loads (Nowak et al. 2007c). Neuroimaging paradigms supported this notion (Kawato et al. 2003; Boecker et al. 2005). Evidence was however also found in support of a contribution of parietal cortical areas and of the basal ganglia (Ehrsson et al. 2003; Imbach et al. 2015).

Sensorimotor memories can be updated. Stacking previously lifted objects of equal appearance but different weight upon each other, participants were able to predict the weight of the novel, combined object and scale their force accordingly. This suggests that internal models of object properties may be additive. Similarly, (Chang et al. 2008) reported that subjects were capable of predictive force scaling when lifting an object from the palm of the other hand. However, participants were unable to predict the weight of a water-filled, previously lifted object after having reduced its weight by drinking from it with a straw (Nowak and Hermsdörfer 2003).

Due to these findings, it was proposed that direct haptic contact between the hand and the object is indispensable to update sensorimotor memories (Nowak et al. 2013).

## *Sensorimotor Representations Can Be Higher Order Motor Plans*

As summarized above, studies investigating the lifting of objects with a symmetrical mass distribution had shown that sensorimotor memory serves to precisely program the GF/LF-ratio to surface-friction (Johansson and Westling 1984, 1987) and handle-shape (Jenmalm and Johansson 1997), as well as the grip and load force rate to object weight (Johansson and Westling 1988). More recent studies dealing with the grasping and lifting of objects with an eccentric center of mass demonstrate that sensorimotor representations are not limited to the scaling of effector-forces but contain higher order motor plans instead. For this type of task, a hidden weight is typically positioned eccentrically in the basis of an object resembling a symmetric-inverted-T (Fig. 1a). To avoid object tilt, subjects have to anticipate the external torque and compensate it at the moment of lift-off. The external torque can be compensated by employing different load forces at each side of the grasp surface and by positioning fingers at different vertical positions (and exerting a regular GF), if the geometry of the handle allows for this way of counteracting torque production (Fig. 1a). (Fu et al. 2010) showed that subjects, who can freely position their fingers on grasp-surfaces, quickly learn to anticipatorily compensate torques within two to three lifts similar to a group of subjects whose finger positioning was restricted. Although compensatory torques at lift-off remained stable in both groups in the following lifting trials, subjects in the unconstrained finger positioning group continued to modulate and refine their finger positioning across the following trials. They compensated for the increased variability of finger positioning by coordinating the asymmetric load force distribution patterns and their finger positions with an inverse relationship. This indicates that not effector-based parameters like force output and finger positioning are represented in the CNS, but a higher order motor plan concerned with the goal of the task (e.g., torques). Such a representation allows for coordination of different interacting aspects of motor execution to accomplish a certain motor task (e.g., compensatory torque production).

Another phenomenon supporting this claim is motor equivalence. Motor equivalence describes the fact that learned motor tasks can still be mastered when other effector elements are used (Lashley 1930). An instance of motor equivalence in anticipatory grasp control has been demonstrated by (Fu et al. 2011): Subjects who had learned to generate matching compensatory torques at lift-off with a certain multi-finger grip, continued to do so after changing the number of fingers employed. It is important to mention, that they did not try to adjust a previous motor plan, but applied a different combination of finger positioning and force sharing

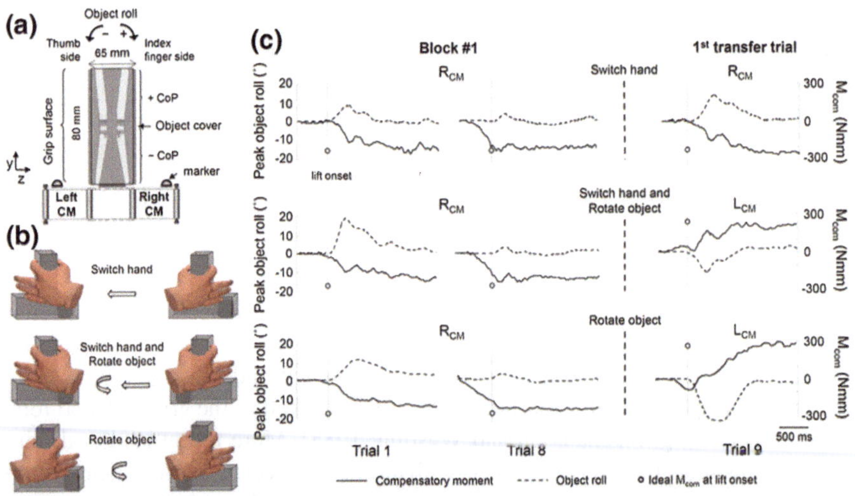


**Fig. 1** Learning and generalization of anticipatory torque production. **a** Custom-built inverted-T grip device used to measure forces and centers of pressure of the thumb and index finger. Note that 'thumb side' and 'index finger side' denote a grasp performed with the right hand. Two light-emitting diode markers were mounted on the base of the device to track object kinematics (vertical position and roll in the *x–y* plane). A 400 g mass was inserted in either the left or right compartment to change the center of mass (CM) of the object to create an external moment. A cover on the front and back of the grip device blocked view of force/torque sensors. **b** Three learning transfer conditions with "switch hand" and/or "object rotation". **c** Time course of compensatory moment ($M_{com}$; *solid line*) and object roll (*dashed line*) on pre-transfer trials (Block 1, Trials 1 and 8) and the first transfer trial (Block 2, Trial 1) from three representative subjects. The *solid vertical line* in each panel denotes object lift onset. *Circles* denote the subjects' "ideal" $M_{com}$ at object lift onset, which subjects should generate at object lift onset to neutralize the external moment generated by the added mass. The *left* and *right vertical* axes refer to object roll and $M_{com}$, respectively. Negative and positive values of object roll denote counterclockwise and clockwise roll relative to the vertical, respectively. Adapted from Fu et al. (2014)

resulting in the same torque. These results point out that task-related sensorimotor memories are not represented at the level of the fingers involved.

## *Universal Influence of the Previous Task Execution on the Current Task Performance*

The strong role of sensorimotor memory in anticipatory object manipulation of the upcoming trial is a ubiquitous trait of motor control, which affects motor planning also in unpredictable as well as inappropriate situations.

When lifting a neutral object, which weight randomly changes across consecutive trials hidden from view, healthy participants produce forces appropriate for the preceding trial (Johansson and Westling 1988). As a consequence, force profiles are only adequate if the weight remains the same. Switching to a lighter object leads to overshooting force increases and an abrupt lifting movement, whereas after switching to a heavier object several stepwise force increases and a longer load phase are observed until lift-off occurs. Concordantly, the direction of initial compensatory torques is determined by the previously encountered mass distribution when the center of mass of the object changes randomly across subsequent trials (Lukos et al. 2013).

Quaney et al. (2003) demonstrated that not only the weight of the previously lifted object impacts force scaling, but also previous task-unrelated actions. Particularly, subjects who were instructed to squeeze a pinch force transducer with a defined GF in between a series of lifting trials subsequently exerted a similar GF when lifting an object, irrespective of the actual weight of the object and the hand performing the lift. Although the load force was not influenced by previous pinching, the findings suggested that not only object properties but also a sense of effort might be memorized and can influence GF scaling.

## *Generalization of Learned Object Manipulation to New Situations and Across Hands*

Even when subjects are explicitly aware that a 180° rotation of an object with an asymmetric mass distribution has occurred or even rotate the object themselves, they fail to change the direction of the previously learned compensatory torque and thus initially even add to the external torque (Fig. 1b, c bottom row). As a consequence, large initial object tilts occur (Salimi et al. 2000, 2003; Bursztyn and Flanagan 2008; Fu et al. 2010; Zhang et al. 2010; Marneweck et al. 2015). This failure of compensatory torque anticipation is both due to not being able to modify previously learned load force partitioning as well as vertical finger positioning to the new situation. Such a lack of adaption to a new object orientation also appears when the subject is rotated around the object instead (Marneweck et al. 2015). The authors proposed that sensorimotor memories of a learned manipulation of an object with asymmetrical weight distribution is strictly specific to the frame of reference in which it is acquired and can basically not be transformed to new orientations or situations (Marneweck et al. 2015). This implies that object dynamics are not represented in a single veridical object representation but multiple representations for distinct task situations of object handling must concurrently exist from which the sensorimotor system selects the most appropriate one.

Nevertheless, being exposed to several object rotations, subjects eventually learn to change the torque-direction according to alterations of the object orientation and are able to at least partially compensate the external torque. They hereby succeed to alter finger positioning after object rotations, while persistently failing to change the direction of load force distribution (Fu et al. 2010; Fu and Santello 2015; Marneweck et al. 2015). This dissociation in learning rates of finger positioning and force sharing across context changes led to the assumption that these two aspects might be governed by independent neural processes.

Studies on intermanual transfer addressed the question if sensorimotor memories established in lifts with one hand can be accessed with the other hand. Interestingly, they yielded opposing results depending on the object property under investigation. When subjects repeatedly lift an object with a symmetric weight distribution with one hand and subsequently change to the other hand, they employ precise anticipatory control already during the first lift after the hand switch (Gordon et al. 1994).

Comparable studies conducted with stroke patients and children with cerebral palsy raised the hope that such transfer mechanisms might aid rehabilitation (Gordon et al. 1999; Raghavan et al. 2006). While employing inappropriately high GFs when lifting an object of varying weight with the impaired hand, patients are able to scale forces correctly according to object weight with their impaired hand after having learned successful manipulation with the unimpaired hand before.

However, studies examining the lifting of an object with asymmetric weight distribution contradict these findings. Participants who have learned to compensate torques with one hand failed to exert any compensatory moment when switching to the other hand (Fig. 1b, c top row) and therefore behave as if they were completely naïve to the task (Salimi et al. 2000; Bursztyn and Flanagan 2008; Fu et al. 2014). Nevertheless, one subject group in the study of (Bursztyn and Flanagan 2008) constitutes an exception. When subjects simultaneously switched hands and inverted their hand posture, considerable compensatory torques preventing initial object tilt to a large degree could be observed. Indeed, this kind of hand and posture alternations resemble the way in which we pass objects from one hand to another in daily life more. Therefore, practice might be a decisive factor in transferring information across hands.

## Associative Memory

In order to be able to learn and refine object manipulation skills in the long term, it is essential to be able to link recently acquired short-living sensorimotor memory of lifting trials to characteristic object traits. Later, these traits can be used to identify the object and retrieve the corresponding internal representation of object dynamics enabling anticipatory object manipulation (Nowak et al. 2007b; Ameli et al. 2008). The pace at which such a link is established seems to depend upon the predictability of presentation. For example, children learned to scale force rates according to different pre-cued object weights at a slower rate when exposed to a random lifting schedule compared to a more predictable blocked schedule (Gordon and Duff 1999). However, scaling precision after a one-day retention was similar for both schedule groups. When arbitrary visual or auditory cues signal the weight of an object to be lifted, healthy participants can easily use these cues to predictively scale GFs with both hands by building and accessing associations between the cues and sensorimotor memories of object weight (Ameli et al. 2008).

Importantly, both patients with right-sided and left-sided MCI stroke are clearly hampered in establishing and/or retrieving such associations when using their affected contralesional hand. While this deficiency is also prominent when grasping with the ipsilesional, non-paretic hand in the left-brain damage group, right brain-lesioned patients' force prediction resembles that of healthy controls when grasping with the ipsilesional hand (Bensmail et al. 2012). This adds further evidence to the presumed motor dominant role of the left hemisphere.

Employing a similar protocol, a recent fMRI study measured BOLD responses when subjects prepared to lift a lighter or a heavier weight indicated by a color cue using the right hand (van Nuenen et al. 2012). Bilateral activity was found in the PMd, in mesial premotor areas, in the SMA, as well as in the left parietal cortex when the lifting of the heavier weight was signaled. In this study, the GF deviation when subjects were first cued for an incorrect weight and then received a valid second cue for the correct weight was taken as the measure for the strength of the color association. Interestingly, this measure correlated with left PMd activity supporting the clinical findings of a more important role of left-brain areas in the task.

In accordance with the fMRI findings, the authors as well as other groups found impaired force scaling with arbitrary cues after TMS stimulation of the PMd (Chouinard et al. 2005; Nowak et al. 2009; van Nuenen et al. 2012). In addition to the premotor cortex, stimulation of the AIP deteriorated cue-based anticipatory activity during reaching and grasping movements (Taubert et al. 2010).

The by far most objects we encounter daily are well known and we usually have gathered extensive lifetime experience in manipulating them. Hence, one might think it is easy to guess the properties of familiar objects and adapt force development accordingly. Indeed, studies investigating the anticipatory control of grasps of common everyday objects (Fig. 2a) concordantly showed that healthy subject succeed in increasing their GF (Hermsdörfer et al. 2011) and load force (Gordon et al. 1993; Hermsdörfer et al. 2011) more rapidly when grasping heavy objects (like a book) than when lifting a lighter object (like a box of tissues) (Fig. 2b CL). Thus, the peaks of lift and GF rates before lift-off that are highly indicative of preplanning of object manipulation were almost linearly associated with the actual object weights (Fig. 2c CL). In the design of one experiment, objects were chosen so that object weights could not be derived from their size using a uniform density estimate. Consequently, a control group who lifted the same objects covered in white paper showed a by far weaker association between force-parameters and actual weight (Hermsdörfer et al. 2011). This supported the authors' thesis that successful estimation of object properties was due to the identification of the objects and the retrieval of learned object-specific internal representations of object weight and not visual size analysis.

Cole (2008) showed that subjects remain sensitive to size when lifting everyday objects. Subjects lifted two filled bottles, which slightly differed in size and weight, accordingly. Despite not realizing that they actually lifted two different bottles, they employed force rates that matched the actually different weights. This showed that despite consciously identifying the object as being the same, the sensorimotor

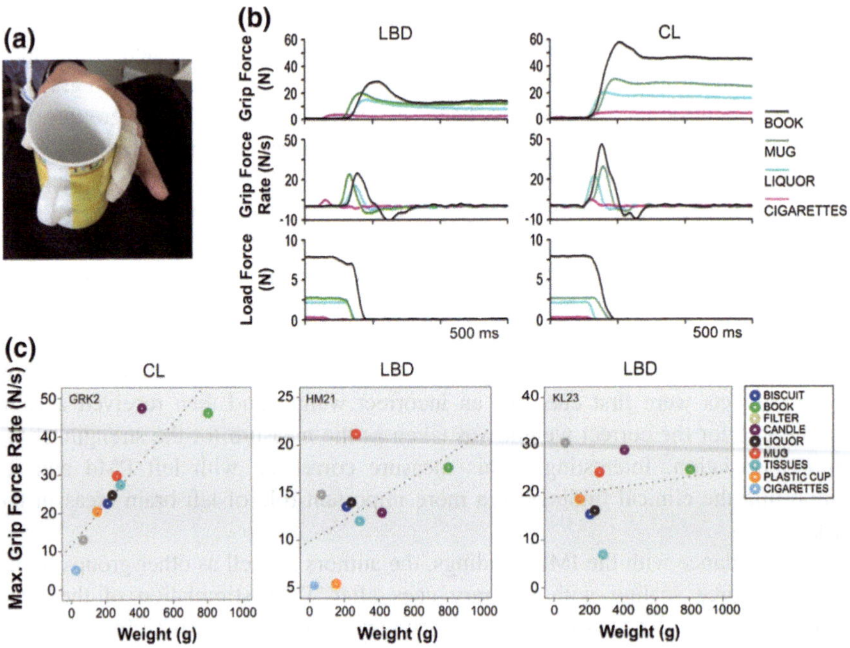


**Fig. 2** Anticipatory grip force scaling during grasping and lifting everyday objects. **a** Grip force measurement using flexible force-sensor arrays mounted above the finger pads and secured by rubber caps. **b** Examples of time courses of grip force, maximum grip force rate (1st derivative of grip force) and load force during the lifting of four everyday objects in one patient (LBD) and one control subject (CL). The load force was measured with a scale; decrease of the scale signal reflects loading of the fingers. Note that the patient scales grip force rate similarly for the book and the mug, although the book is much heavier. The patient also employed a lower grip force level, which was not typical for the group. **c** Relationship between maximum grip force rate and the weight of everyday objects lifted by a control subject (CL) and two selected patients with left-brain damage (LBD). While the control subject produced grip force rates with a nearly linear relationship with the object weight, the relationship was very scattered in the patients indicating an inability to anticipate object weight when it had to be concluded from object identity. Modified from Eidenmüller et al. (2014)

system still processes size information to modify force output. Cole proposed that not object weight is stored in internal memory traces. Instead, the sensorimotor system integrates a remembered density representation to predict weight based on visual size analysis.

Patients suffering from MCI stroke who were instructed to lift a set of everyday objects with their ipsilesional, less impaired hand showed a significantly smaller and flattened correlation compared to healthy controls (Dawson et al. 2010; Eidenmüller et al. 2014). Several patients applied clearly inappropriate force patterns, which seemed at random (Fig. 2b, c LBD). This disruption of natural force control was limited to the group of left-brain-lesioned patients. Since all patients were able to name or point to the objects, this failure cannot be attributed to general loss to identify objects, but might rather stem from an inability to retrieve or process object properties from object identity and appearance.

While (Dawson et al. 2010) suggested lesions in the left parietal and adjacent temporal areas to be related to the deficits, the voxel based lesion symptom analysis performed by (Eidenmüller et al. 2014) found symptoms-related regions in the left IFG and premotor cortex which are also frequently impaired in patients with tool using deficits.

## Visual Cues

Even when lifting novel objects for the first time, our brain tries to tune the motor program according to the expected object properties. We can process various visual cues like the size, assumed material and shape or form of the object and handle. On the basis of this information, we compute predictions in order to optimize the first lift in terms of adequate force control (Gordon et al. 1991; Jenmalm and Johansson 1997; Buckingham et al. 2009; Fu and Santello 2012). These computations rely on associations between object traits and physical properties that have developed over a lifetime of experience, e.g., the relation between size and weight based on the density of an object. We universally compute such visuomotor transformations when we encounter novel as well as familiar objects and usually obtain reliable predictions to guide motor control.

### *The Interaction of Visual Cues and Sensorimotor Memory*

Important paradigms to gain insights into the interaction between weight predictions relying on size cues and the sensorimotor system are size–weight illusion (SWI) tasks. We assume that objects with the same appearance consist of the same material and thus have the same density. Therefore, we expect the weight of such objects to increase with size. SWI paradigms investigate the effect of violations of this size–weight relationship. Asked to alternately lift two cubes of differing size but identical weight, participants initially base their weight predictions on visual size cues and increase forces faster when lifting the big cube which is unusually light than when lifting the small one which is unusually dense (Flanagan and Beltzner 2000). After lift-off the motor error is signaled to the subjects and it takes them about 5–10 lifts to gradually adapt force scaling to the true weight and program their motor output adequately, thus in the same way for both cubes. Interestingly, despite the adequate force output after a few trials, subjects perceive a strong illusion that the smaller object is heavier, which does not ameliorate across trials. Different from the perceptual system, the system controlling GF apparently learned to block predictive mechanisms based on visual size cues and to only rely on sensorimotor influences (Flanagan and Beltzner 2000). Importantly, the dominance of sensorimotor memory over visual cues in this specific situation, as inferred by adequate motor planning, could still be observed when participants returned to lift the cubes after a one-day delay. Apparently, a stable associative representation between the weight of the cubes and their visual appearance of the cubes was built and retained (Flanagan et al. 2001).

Flanagan et al. (2008) showed that the normal size–weight relationship we rely on for visuomotor transformations can even be inversed. In their study, subjects were confronted with uniformly colored objects of various forms whereby the smallest objects were the heaviest and the biggest ones the lightest. After extensive lifting practice, participants were capable to correctly anticipate the real object weight after they initially had made faulty predictions based on the regular size–weight relation. Moreover, when subsequently lifting new objects of the same color but a novel shape subjects persisted with employing a more rapid load force increase pattern for the smallest of the new objects and increasing forces more slowly when lifting the biggest one. The authors concluded that not only an associative representation of the weight of each faced object was acquired, but a new size–weight map or prior was established for a specific family of objects which could be utilized for visuomotor transformations when handling new objects of the same family. This new prior proved to be long lasting and drove motor action when participants returned to lifting the same objects several days later.

However, the adjustment of anticipatory control to misleading size cues, or rather alterations of normal size–weight priors, requires extended practice. While sensorimotor adaptation to distinct object weights in a neutral cue situation is a one-trial learning process (Johansson and Westling 1988), about five to ten trials are needed to learn to handle a pair of differently sized cubes of the same weight (Flanagan and Beltzner 2000; Flanagan et al. 2001), whereas inverting the natural size–weight relationship consumes up to 240 lifts (Flanagan et al. 2008).

Apart from the size of objects, the apparent material is another powerful cue driving anticipatory force control. In a material-weight illusion task (Buckingham et al. 2009), subjects had to lift two objects appearing to be made out of different materials with very different densities, like Styrofoam and metal, while the interior of the objects was modified so that they actually had the same weight. Similar to the findings of SWI studies, participants scaled their finger forces initially according to the expected material, applying higher force rates for the metal covered object than for the Styrofoam covered object. Yet in contrast to the size–weight studies, subjects were capable to adjust force profiles to the actual weight after only one lift. This indicates that the perceived material is, although utilized to initially drive force scaling, a less fundamental predictor of object weight in motor planning than size cues and can consequently be suppressed faster.

Studies conducted in unilateral MCI stroke patients and age-matched controls employing the size–weight (Li et al. 2011; Buckingham et al. 2015) and material-weight illusion paradigms (Buckingham et al. 2015) did not detect differences concerning initial force scaling on group level, whereas two individual cases completely lacked initial force-anticipation (Li et al. 2007, 2011). These two patients shared rarely affected lesion-sites at the junction of left posterior parietal and the superior occipital lobe. Since they also presented with a rare deviation in the perception of the SWI, the authors suggested the area at the left parieto-occipital junction to play a vital role in processing object properties for motor control and perception. While in patients with cerebellar diseases the anticipatory scaling of load forces according to object size was impaired, the scaling of GFs was preserved similarly to MCI stroke patients (Rabe et al. 2009). Therefore, the processing of visual cues to drive force scaling seems to be a highly persistent mechanism which is resistant to even severe brain damage or can be effectively reorganized in case of a brain lesion (Li et al. 2011).

A fMRI study used repetition suppression methodology to measure brain activity during lifting of seen objects in a SWI paradigm as well as other combinations of different or equal object sizes and weights (Chouinard et al. 2009). From different contrasts between the conditions the authors concluded that size information was coded in the contralateral left primary sensory areas (S1) and in the left AIP, weight information was coded in the left primary motor cortex (M1). Constant density was reflected by activity changes in the left PMv suggesting that this area integrates sensory information about the size of objects and the weight of objects. When information about object texture informed about the weight, brain activity that discriminates between both objects during lift planning was obvious in the primary somatosensory cortex (S1) and in the PMd, but also in texture-sensitive brain regions in the occipitaltemporal cortex (Gallivan et al. 2014).

## Interaction Between Object Knowledge and Sensorimotor Memory in Torque Compensation

As summed up above, healthy adults can take advantage of arbitrary- and identification cues, as well as of visual density and size cues, to establish anticipatory force scaling when lifting an object with a symmetric mass distribution. However, disparate results were found in studies focusing on the anticipatory compensation of external torques.

Although participants were able to precisely localize the center of mass of an object that consisted of two halves of differently dense materials (leading to an asymmetric weight distribution) prior to lifting it, they exerted only negligible torques at the moment of lift-off and considerable tilts occurred (Craje et al. 2013). Even hefting the object in the one hand and lifting it with the other did not allow for improvement of anticipative torque planning. A similar dissociation between explicit knowledge of mass distribution and failure to program compensatory torques was observed in studies in which a weight block positioned at one side of an object was either visible or indicated by a color cue (Salimi et al. 2003; Lukos et al. 2007). These findings stand in line with the above-mentioned studies on object rotations (Salimi et al. 2000; Fu et al. 2010; Marneweck et al. 2015).

The only cue found to be of assistance in torque anticipation is the geometric shape of an object. Initially lifting an object whose form indicates its weight distribution, subjects applied correctly directed compensatory torques and thus avoided major object tilts (Salimi et al. 2003; Fu and Santello 2012). It was concluded that object shape is the dominant cue having access to the sensorimotor system. The dominance of shape on torque anticipation might partially explain why density cues and contextual cues are ineffective for torque scaling, as they were presumably overshadowed by an overall symmetrical object shape which may have visually signaled a balanced weight distribution (Fu and Santello 2012). However, switching to an equivalent but contrarily orientated, indicatively shaped object after repeated lifts, participants were able to correctly change the torque-direction but considerably under-scaled torque height. Generating even lower torques than when initially lifting the object, they were not only unable to transfer the previously acquired torque planning but were also impaired in utilizing the provided geometrical cues to the same extent. Thus, previous lifting experience had an inhibitory effect on the CNS' ability to use visual cues for planning manipulation in an opposite context. This inhibitory interference was even observed when switching context after only a single lift and continued to persist, albeit weaker, after repeated context switches (Fu and Santello 2012).

By introducing time delays of varying length, (Fu and Santello 2015) found the inhibitory effect of previous experience on the utilization of visual cues and acquired appearance torque associations to be time dependent. They could show that the initially strong interference is fading over time and is merely detectable after 60 min. On this basis, they postulated that two distinct memory processes underlie the anticipation of object manipulation. On the one hand, a short living, use-dependent adaptation mechanism, consistent with the concept of sensorimotor memories, is influencing subsequent motor planning without being linked to explicit cues. This mechanism's effect cannot be adapted to new situation, is increasing in strength with the number of repeated lifts and vanishes after a short time. On the other hand, there is a context-sensitive, error-based learning process, which relies on explicit cues and is long lasting (Fu and Santello 2015). These conclusions are congruent with force scaling studies showing that sensorimotor memories are gradually deteriorating in determining force control over time, whereas established associations between sensorimotor memory and explicit cues can still be retrieved weeks later to precisely program force output (Gordon et al. 1993; Nowak et al. 2007b).

## Take Home Message

In conclusion, kinetic grasp-to-lift studies revealed fundamental aspects of the ways in which we interact with our environment. They could show that on the one hand our sensorimotor system is laid out to quickly adapt to new situations so that object manipulation is quickly learned by practice. On the other hand, it is capable to use a variety of cues and retrieves information of previous experiences with similar objects to optimize motor performance already at object contact. Nevertheless, diverging findings between distinct tasks give an insight into the complexity of interactions between the influences of recent actions, long-term experience and cues inferring object information (see Fig. 3 for summary). Especially, the failure to use explicit knowledge and certain cues about object properties to enhance manipulation in some tasks impressively demonstrates that our sensorimotor system sometimes follows different rules than supposed by conscious thinking.

Studies conducted with stroke patients show that although stroke has a detrimental effect on execution, fundamental aspects of anticipation are robust to brain-lesions and only partial features of predictive motor planning are regularly disrupted, especially if the left brain is affected. Studies employing neuroimaging methods as well as stimulation paradigms have identified key areas being particularly involved in anticipatory control during object manipulation, but their interaction, specialization and possible redundancies are far from clear.

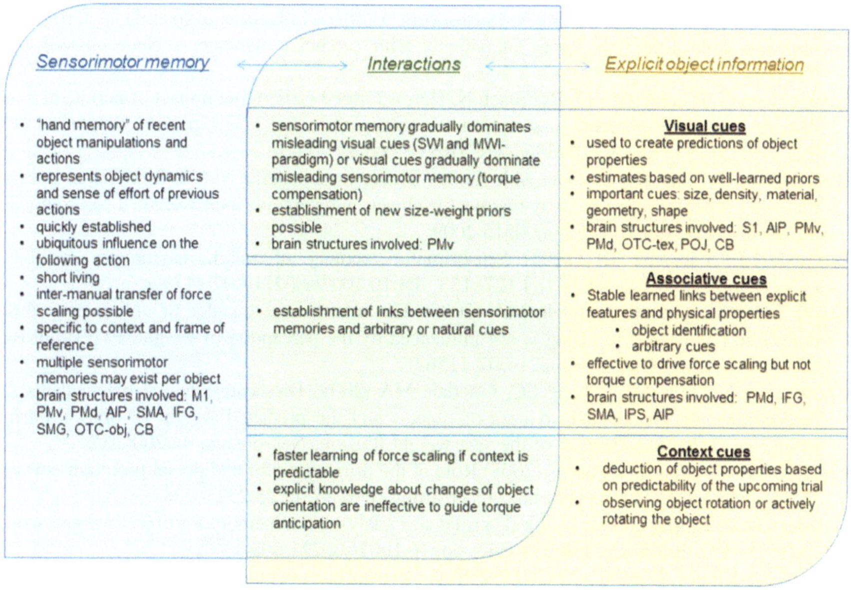


**Fig. 3** Factors influencing anticipatory object control and their interaction. Associated brain areas as revealed by neuroimaging and clinical patient studies are indicated (see text). Abbrevations: *M1* primary motor cortex, *S1* primary somatosensory cortex, *PMv* & *PMd* ventral and dorsal premotor cortex, *AIP* anterior intraparietal area, *SMA* supplementary motor area, *IFG* inferior frontal gyrus, *SMG* supramarginal gyrus, *IPS* intraparietal sulcus, *POJ* parietal-occipital junction, object - sensitive occipitotemporal cortex (OTC-obj), texture-sensitive occipitotemporal cortex (OTC-tex), *CB* cerebellum

# References

Ameli M, Dafotakis M, Fink GR, Nowak DA (2008) Predictive force programming in the grip-lift task: the role of memory links between arbitrary cues and object weight. Neuropsychologia 46:2383–2388

Bensmail D, Sarfeld AS, Ameli M, Fink GR, Nowak DA (2012) Arbitrary visuomotor mapping in the grip-lift task: dissociation of performance deficits in right and left middle cerebral artery stroke. Neuroscience 210:128–136. doi:10.1016/j.neuroscience.2012.03.015

Berner J, Schonfeldt-Lecuona C, Nowak DA (2007) Sensorimotor memory for fingertip forces during object lifting: the role of the primary motor cortex. Neuropsychologia 45:1931–1938

Blennerhassett JM, Carey LM, Matyas TA (2006) Grip force regulation during pinch grip lifts under somatosensory guidance: comparison between people with stroke and healthy controls. Arch Phys Med Rehabil 87:418–429. doi:10.1016/j.apmr.2005.11.018

Blennerhassett JM, Matyas TA, Carey LM (2007) Impaired discrimination of surface friction contributes to pinch grip deficit after stroke. Neurorehabil Neural Repair 21:263–272. doi:10.1177/1545968306295560

Boecker H, Lee A, Mühlau M et al (2005) Force level independent representations of predictive grip force—load force coupling: A PET activation study. NeuroImage 25:243–252. doi:10.1016/j.neuroimage.2004.10.027

Brandauer B, Hermsdörfer J, Beck A, Aurich V, Gizewski ER, Marquardt C, Timmann D (2008) Impairments of prehension kinematics and grasping forces in patients with cerebellar degeneration and the relationship to cerebellar atrophy. Clin Neurophysiol 119:2528–2537

Brandauer B, Timmann D, Häußler A, Hermsdörfer J (2010) Influences of load characteristics on impaired control of grip forces in patients with cerebellar damage. J Neurophysiol 103: 698–708

Buckingham G, Bienkiewicz M, Rohrbach N, Hermsdörfer J (2015) The impact of unilateral brain damage on weight perception, sensorimotor anticipation, and fingertip force adaptation. Vis Res 115:231–237. doi:10.1016/j.visres.2015.02.005

Buckingham G, Cant JS, Goodale MA (2009) Living in a material world: how visual cues to material properties affect the way that we lift objects and perceive their weight. J Neurophysiol 102:3111–3118. doi:10.1152/jn.00515.2009

Bursztyn LL, Flanagan JR (2008) Sensorimotor memory of weight asymmetry in object manipulation. Exp Brain Res 184:127–133. doi:10.1007/s00221-007-1173-z

Chang EC, Flanagan JR, Goodale MA (2008) The intermanual transfer of anticipatory force control in precision grip lifting is not influenced by the perception of weight. Exp Brain Res 185:319–329. doi:10.1007/s00221-007-1156-0

Chouinard PA, Large ME, Chang EC, Goodale MA (2009) Dissociable neural mechanisms for determining the perceived heaviness of objects and the predicted weight of objects during lifting: an fMRI investigation of the size-weight illusion. Neuroimage 44:200–212

Chouinard PA, Leonard G, Paus T (2005) Role of the primary motor and dorsal premotor cortices in the anticipation of forces during object lifting. J Neurosci 25:2277–2284

Cole KJ (2008) Lifting a familiar object: visual size analysis, not memory for object weight, scales lift force. Exp Brain Res 188:551–557. doi:10.1007/s00221-008-1392-y

Craje C, Santello M, Gordon AM (2013) Effects of visual cues of object density on perception and anticipatory control of dexterous manipulation. PLoS ONE 8:e76855. doi:10.1371/journal.pone.0076855

Dafotakis M, Sparing R, Eickhoff SB, Fink GR, Nowak DA (2008) On the role of the ventral premotor cortex and anterior intraparietal area for predictive and reactive scaling of grip force. Brain Res in press. doi:10.1016/j.brainres.2008.06.027

Davare M, Andres M, Clerget E, Thonnard JL, Olivier E (2007) Temporal dissociation between hand shaping and grip force scaling in the anterior intraparietal area. J Neurosci 27:3974–3980. doi:10.1523/jneurosci.0426-07.2007

Davare M, Kraskov A, Rothwell J, Lemon R (2011) Interactions between areas of the cortical grasping network. Curr Opin Neurobiol 21:565–570. doi:10.1016/j.conb.2011.05.021

Dawson AM, Buxbaum LJ, Duff SV (2010) The impact of left hemisphere stroke on force control with familiar and novel objects: neuroanatomic substrates and relationship to apraxia. Brain Res 1317:124–136. doi:10.1016/j.brainres.2009.11.034

Ehrsson HH, Fagergren A, Ehrsson GO, Forssberg H (2007) Holding an object: neural activity associated with fingertip force adjustments to external perturbations. J Neurophysiol 97: 1342–1352

Ehrsson HH, Fagergren A, Johansson RS, Forssberg H (2003) Evidence for the involvement of the posterior parietal cortex in coordination of fingertip forces for grasp stability in manipulation. J Neurophysiol 90:2978–2986

Eidenmüller S, Randerath J, Goldenberg G, Li Y, Hermsdörfer J (2014) The impact of unilateral brain damage on anticipatory grip force scaling when lifting everyday objects. Neuropsychologia 61:222–234. doi:10.1016/j.neuropsychologia.2014.06.026

Flanagan JR, Beltzner MA (2000) Independence of perceptual and sensorimotor predictions in the size-weight illusion. Nat Neurosci 3:737–741. doi:10.1038/76701

Flanagan JR, Bittner JP, Johansson RS (2008) Experience can change distinct size-weight priors engaged in lifting objects and judging their weights. Curr Biol 18:1742–1747. doi:10.1016/j.cub.2008.09.042

Flanagan JR, King S, Wolpert DM, Johansson RS (2001) Sensorimotor prediction and memory in object manipulation. Can J Exp Psychol 55:87–95

Flanagan JR, Merritt K, Johansson RS (2009) Predictive mechanisms and object representations used in object manipulation. In: Nowak DA, Hermsdörfer J (eds) Sensorimotor control of grasping: physiology and pathophysiology. Cambridge University Press, Cambrige, pp 161–203

Fu Q, Choi JY, Gordon AM, Jesunathadas M, Santello M (2014) Learned manipulation at unconstrained contacts does not transfer across hands. PLoS ONE 9:e108222. doi:10.1371/journal.pone.0108222

Fu Q, Hasan Z, Santello M (2011) Transfer of learned manipulation following changes in degrees of freedom. J Neurosci 31:13576–13584. doi:10.1523/JNEUROSCI.1143-11.2011

Fu Q, Santello M (2012) Context-dependent learning interferes with visuomotor transformations for manipulation planning. J Neurosci 32:15086–15092. doi:10.1523/JNEUROSCI.2468-12.2012

Fu Q, Santello M (2015) Retention and interference of learned dexterous manipulation: interaction between multiple sensorimotor processes. J Neurophysiol 113:144–155. doi:10.1152/jn.00348.2014

Fu Q, Zhang W, Santello M (2010) Anticipatory planning and control of grasp positions and forces for dexterous two-digit manipulation. J Neurosci 30:9117–9126. doi:10.1523/JNEUROSCI.4159-09.2010

Gallivan JP, Cant JS, Goodale MA, Flanagan JR (2014) Representation of object weight in human ventral visual cortex. Curr Biol 24:1866–1873. doi:10.1016/j.cub.2014.06.046

Gordon AM, Charles J, Duff SV (1999) Fingertip forces during object manipulation in children with hemiplegic cerebral palsy. II: bilateral coordination. Dev Med Child Neurol 41:176–185

Gordon AM, Duff SV (1999) Fingertip forces during object manipulation in children with hemiplegic cerebral palsy. I: anticipatory scaling. Dev Med Child Neurol 41:166–175

Gordon AM, Forssberg H, Iwasaki N (1994) Formation and lateralization of internal representations underlying motor commands during precision grip. Neuropsychologia 32:555–568

Gordon AM, Forssberg H, Johansson RS, Westling G (1991) Visual size cues in the programming of manipulative forces during precision grip. Exp Brain Res 83:477–482

Gordon AM, Westling G, Cole KJ, Johansson RS (1993) Memory representations underlying motor commands used during manipulation of common and novel objects. J Neurophysiol 69:1789–1796

Hager-Ross C, Johansson RS (1996) Nondigital afferent input in reactive control of fingertip forces during precision grip. Exp Brain Res 110:131–141

Hermsdörfer J, Hagl E, Nowak DA, Marquardt C (2003) Grip force control during object manipulation in cerebral stroke. Clin Neurophysiol 114:915–929

Hermsdörfer J, Li Y, Randerath J, Goldenberg G, Eidenmuller S (2011) Anticipatory scaling of grip forces when lifting objects of everyday life. Exp Brain Res 212:19–31. doi:10.1007/s00221-011-2695-y

Hsu HY, Lin CF, Su FC, Kuo HT, Chiu HY, Kuo LC (2012) Clinical application of computerized evaluation and re-education biofeedback prototype for sensorimotor control of the hand in stroke patients. J Neuroeng Rehabil 9:26. doi:10.1186/1743-0003-9-26

Imbach LL, Baumann-Vogel H, Baumann CR, Sürücü O, Hermsdörfer J, Sarnthein J (2015) Adaptive grip force is modulated by subthalamic beta activity in Parkinson's disease patients. NeuroImage: Clin 9:450–457. doi:http://dx.doi.org/10.1016/j.nicl.2015.09.010

Jenmalm P, Johansson RS (1997) Visual and somatosensory information about object shape control manipulative fingertip forces. J Neurosci 17:4486–4499

Jenmalm P, Schmitz C, Forssberg H, Ehrsson HH (2006) Lighter or heavier than predicted: neural correlates of corrective mechanisms during erroneously programmed lifts. J Neurosci 26: 9015–9021. doi:10.1523/JNEUROSCI.5045-05.2006

Johansson RS, Flanagan JR (2009) Coding and use of tactile signals from the fingertips in object manipulation tasks. Nat Rev Neurosci 10:345–359. doi:10.1038/nrn2621

Johansson RS, Westling G (1984) Roles of glabrous skin receptors and sensorimotor memory in automatic control of precision grip when lifting rougher or more slippery objects. Exp Brain Res 56:550–564

Johansson RS, Westling G (1987) Signals in tactile afferents from the fingers eliciting adaptive motor responses during precision grip. Exp Brain Res 66:141–154

Johansson RS, Westling G (1988) Coordinated isometric muscle commands adequately and erroneously programmed for the weight during lifting task with precision grip. Exp Brain Res 71:59–71

Kawato M, Kuroda T, Imamizu H, Nakano E, Miyauchi S, Yoshioka T (2003) Internal forward models in the cerebellum: fMRI study on grip force and load force coupling. Prog Brain Res 142:171–188. doi:10.1016/S0079-6123(03)42013-X

Lashley KS (1930) Basic neural mechanisms in behaviour psychological review. Psychol Rev 37:1–24

Li Y, Randerath J, Goldenberg G, Hermsdörfer J (2007) Grip forces isolated from knowledge about object properties following a left parietal lesion. Neurosci Lett:187–191. doi:10.1016/j.neulet.2007.09.008

Li Y, Randerath J, Goldenberg G, Hermsdörfer J (2011) Size-weight illusion and anticipatory grip force scaling following unilateral cortical brain lesion. Neuropsychologia 49:914–923. doi:10.1016/j.neuropsychologia.2011.02.018

Loh MN, Kirsch L, Rothwell JC, Lemon RN, Davare M (2010) Information about the weight of grasped objects from vision and internal models interacts within the primary motor cortex. J Neurosci 30:6984–6990. doi:10.1523/JNEUROSCI.6207-09.2010

Lukos J, Ansuini C, Santello M (2007) Choice of contact points during multidigit grasping: effect of predictability of object center of mass location. J Neurosci 27:3894–3903. doi:10.1523/JNEUROSCI.4693-06.2007

Lukos JR, Choi JY, Santello M (2013) Grasping uncertainty: effects of sensorimotor memories on high-level planning of dexterous manipulation. J Neurophysiol 109:2937–2946. doi:10.1152/jn.00060.2013

Marneweck M, Knelange E, Lee-Miller T, Santello M, Gordon AM (2015) Generalization of dexterous manipulation is sensitive to the frame of reference in which it is learned. PLoS ONE 10:e0138258. doi:10.1371/journal.pone.0138258

McDonnell MN, Hillier SL, Ridding MC, Miles TS (2006) Impairments in precision grip correlate with functional measures in adult hemiplegia. Clin Neurophysiol 117:1474–1480. doi:10.1016/j.clinph.2006.02.027

Nowak DA, Berner J, Herrnberger B, Kammer T, Gron G, Schonfeldt-Lecuona C (2009) Continuous theta-burst stimulation over the dorsal premotor cortex interferes with associative learning during object lifting. Cortex 45:473–482

Nowak DA, Glasauer S, Hermsdörfer J (2013) Force control in object manipulation–a model for the study of sensorimotor control strategies. Neurosci Biobehav Rev 37:1578–1586. doi:10.1016/j.neubiorev.2013.06.003

Nowak DA, Grefkes C, Dafotakis M, Kust J, Karbe H, Fink GR (2007a) Dexterity is impaired at both hands following unilateral subcortical middle cerebral artery stroke. Eur J Neurosci 25:3173–3184. doi:10.1111/j.1460-9568.2007.05551.x

Nowak DA, Hermsdörfer J (2003) Sensorimotor memory and grip force control: does grip force anticipate a self-produced weight change when drinking with a straw from a cup? Eur J Neurosci 18:2883–2892

Nowak DA, Hermsdörfer J, Topka H (2003) Deficits of predictive grip force control during object manipulation in acute stroke. J Neurol 250:850–860. doi:10.1007/s00415-003-1095-z

Nowak DA, Koupan C, Hermsdörfer J (2007b) Formation and decay of sensorimotor and associative memory in object lifting. Eur J Appl Physiol 100:719–726. doi:10.1007/s00421-007-0467-y

Nowak DA, Topka H, Timmann D, Boecker H, Hermsdörfer J (2007c) The role of the cerebellum for predictive control of grasping. Cerebellum 6:7–17

Nowak DA, Voss M, Huang YZ, Wolpert DM, Rothwell JC (2005) High-frequency repetitive transcranial magnetic stimulation over the hand area of the primary motor cortex disturbs predictive grip force scaling. Eur J Neurosci 22:2392–2396. doi:10.1111/j.1460-9568.2005.04425.x

Quaney BM, He J, Timberlake G, Dodd K, Carr C (2010) Visuomotor training improves stroke-related ipsilesional upper extremity impairments. Neurorehabil Neural Repair 24:52–61. doi:10.1177/1545968309341646

Quaney BM, Perera S, Maletsky R, Luchies CW, Nudo RJ (2005) Impaired grip force modulation in the ipsilesional hand after unilateral middle cerebral artery stroke. Neurorehabil Neural Repair 19:338–349. doi:10.1177/1545968305282269

Quaney BM, Rotella DL, Peterson C, Cole KJ (2003) Sensorimotor memory for fingertip forces: evidence for a task-independent motor memory. J Neurosci 23:1981–1986

Rabe K, Brandauer B, Li Y, Gizewski ER, Timmann D, Hermsdörfer J (2009) Size-weight illusion, anticipation and adaptation of fingertip forces in patients with cerebellar degeneration. J Neurophysiol 101:569–579. doi:10.1152/jn.91068.2008

Raghavan P, Krakauer JW, Gordon AM (2006) Impaired anticipatory control of fingertip forces in patients with a pure motor or sensorimotor lacunar syndrome. Brain 129:1415–1425. doi:10.1093/brain/awl070

Rost KR, Nowak DA, Timmann D, Hermsdörfer J (2005) Preserved and impaired aspects of predictive grip force control in cerebellar patients. Clin Neurophysiol 116:1405–1414

Salimi I, Frazier W, Reilmann R, Gordon AM (2003) Selective use of visual information signaling objects' center of mass for anticipatory control of manipulative fingertip forces. Exp Brain Res 150:9–18. doi:10.1007/s00221-003-1394-8

Salimi I, Hollender I, Frazier W, Gordon AM (2000) Specificity of internal representations underlying grasping. J Neurophysiol 84:2390–2397

Schaefer S, Mutha P, Haaland K, Sainburg R (2012) Hemispheric specialization for movement control produces dissociable differences in online corrections after stroke. Cereb Cortex 22:1407–1419. doi:10.1093/cercor/bhr237

Schmitz C, Jenmalm P, Ehrsson HH, Forssberg H (2005) Brain activity during predictable and unpredictable weight changes when lifting objects. J Neurophysiol 93:1498–1509

Sunderland A, Bowers MP, Sluman SM, Wilcock DJ, Ardron ME (1999) Impaired dexterity of the ipsilateral hand after stroke and the relationship to cognitive deficit. Stroke 30:949–955

Taubert M, Dafotakis M, Sparing R, Eickhoff S, Leuchte S, Fink GR, Nowak DA (2010) Inhibition of the anterior intraparietal area and the dorsal premotor cortex interfere with arbitrary visuo-motor mapping. Clin Neurophysiol 121:408–413

van Nuenen BFL, Kuhtz-Buschbeck J, Schulz C, Bloem BR, Siebner HR (2012) Weight-specific anticipatory coding of grip force in human dorsal premotor cortex. J Neurosci 32:5272–5283
Wenzelburger R, Kopper F, Frenzel A et al (2005) Hand coordination following capsular stroke. Brain 128:64–74. doi:10.1093/brain/awh317
White O, Davare M, Andres M, Olivier E (2013) The role of left supplementary motor area in grip force scaling. PLoS ONE 8:e83812. doi:10.1371/journal.pone.0083812
Zhang W, Gordon AM, Fu Q, Santello M (2010) Manipulation after object rotation reveals independent sensorimotor memory representations of digit positions and forces. J Neurophysiol 103:2953–2964. doi:10.1152/jn.00140.2010

# Part IV
# Learning of Skilled Behavior

# Brain Plasticity and the Concept of Metaplasticity in Skilled Musicians

**Eckart Altenmüller and Shinichi Furuya**

**Abstract** Early and extensive musical training provides plastic adaptations of the nervous system and enhanced sensory, motor, and cognitive functions. Over decades, neuronal mechanism underlying the plastic adaptation through musical training has been investigated using neuroimaging and transcranial stimulation techniques. Recently, plastic changes in neuroplastic functions through musical training have gradually gained some interest, so-called metaplasticity. Metaplasticity enables faster and more stable skill acquisition for individuals with a history of prior musical training. This mechanism may also serve for prevention of developing maladaptive changes in the nervous system, being pathophysiology of focal dystonia in musicians. The present chapter introduces neurophysiological mechanisms and functional significances of brain plasticity and metaplasticity of the sensory and motor systems of musicians.

**Keywords** Neuroimaging · MRI_imaging · Voxel-based morphometry (VBM) · Diffusion tensor imaging (DTI) · Magnetoencephalography (MEG) · Diffusion tensor imaging (DTI) · Dystonia · Piano · Dexterity

E. Altenmüller (✉) · S. Furuya
Institute for Music Physiology and Musicians' Medicine (IMMM),
Hannover University of Music, Drama and Media,
Emmichplatz 1, 30175 Hannover, Germany
e-mail: eckart.altenmueller@hmtm-hannover.de

S. Furuya
e-mail: auditory.motor@gmail.com

S. Furuya
Department of Information and Communication Sciences,
Musical Skill and Injury Center (MuSIC), Sophia University,
4-4 Chiyoda-Ku, Tokyo 1020081, Japan



J. Laczko and M.L. Latash (eds.), *Progress in Motor Control*,
Advances in Experimental Medicine and Biology 957,
DOI 10.1007/978-3-319-47313-0_11

## Music as a Complex Multisensory and Rewarding Stimulus

Performing music at a professional level is one of the most complex human accomplishments. Playing an instrument requires the integration of multimodal sensory and motor information, the generation of appropriate action plans, the selection and retrieval of highly refined movement patterns from procedural motor memory, and the initiation of movement. In most instances, these movements are highly over-learned, and they depend both on feedforward programming of the anticipated—mostly audible—results and on real-time sensory feedback.

Auditory and kinaesthetic feedback is needed to improve, fine-tune, and perfect the performance. Music making, therefore, relies primarily on a highly developed auditory-somatosensory-motor integration capacity, which has been conceptualized theoretically in the "common–coding" model by Prinz (1984). Simply put, in this model movements are represented as sound patterns, and sound patterns as movements [see for a review (Zatorre et al. 2007)].

In order to acquire these specialized sensory-motor skills, musicians must undergo extensive training periods over many years, usually starting in early childhood and passing through stages of increasing physical and strategic complexities. This process of practicing involves assembling, storing, and constantly improving sensory-motor programs through prolonged and repeated execution of motor patterns under the controlled monitoring of auditory and kinaesthetic senses. To attain a professional level, as a rule of thumb, musicians have 10,000 h in 10 years of deliberate practice (Ericsson et al. 1993). Of course, time invested into the acquisition of motor skills alone is not sufficient to become an outstanding artist. Quality of practice as well as communication skills and expressive gesturing, rendering a performance "speaking," are equally important in the process of artistic perfection of publicly acclaimed and valued interpreters (Hallam et al. 2016). Furthermore, muscular strength is also associated with motor virtuosity (Furuya et al. 2015). Finally, genetic predisposition for attaining outstanding musicianship also seems to play a crucial role (Mosing et al. 2014).

There is a dark side to the increasing specialization and prolonged training of modern musicians, namely loss of control and degradation of skilled movements, a disorder referred to as musicians' cramp or focal dystonia. The first historical record, from 1830, appeared in the diaries of the ambitious pianist and composer Robert Schumann. As was probably the case for Schumann, prolonged practice, late onset with instrumental training prolonged practice and pain syndromes due to overuse can precipitate dystonia, which is developed by about 1 % of professional musicians and frequently ends their career. Neuroimaging studies point to dysfunctional (or maladaptive) neuroplasticity as its cause [for a review see (Altenmüller et al. 2015)].

When discussing brain adaptations to skilled motor behavior in music performance, it is first necessary to emphasize what is unique to making music and what renders it particularly challenging in terms of motor control. First, musical training usually starts very early, sometimes before age six, when the adaptability of the central nervous system is highest. This feature is not unique to music as other skilled activities, for example, classical ballet, also require an early start. Second, making music obviously is linked to sound production. As in speech, the auditory system provides a very precise feedback of the movement effects (van der Steen et al. 2014), with a temporal resolution superior to kinaesthetic and visual feedback. Furthermore, in the frame of classical music, which is notated and available as sheet music, the target parameters are strictly predefined: namely, temporal accurateness (correct tempo, accuracy of rhythm, swing, beat, pulse, etc.) and spatial accurateness (correct key, or finger position on fingerboard, correct sound quality). Therefore, a highly reliable reproduction of movements meeting these targets is required. This feature is unique to music: in classical ballet, for example, visual feedback is less critical in terms of spatiotemporal precision. Third, most musicians work at the upper limit of their sensory-motor capabilities and strive to push their limits in order to be faster, louder, and more expressive. Given the complexity of music, the demands of composers, especially in the last 100 years, and the role of outstanding peers as models, such as the Chinese pianist Lang Lang, or the "record braking" violinist David Garrett, musicians are aware that the only limit of movement accuracy and speed is the temporal and spatial resolution of the auditory system. And finally, the societal pressure and expectancies concerning the quality of musical performances have definitely grown over the last centuries. This is due, on the one hand, to the ubiquitous availability of music recorded in the media, such as on YouTube and CDs, but on the other hand, to collective learning processes leading to higher standards of music appreciation. This process augments anxiety, tension, and competition among musicians, making their lives increasingly stressful. Frequently even outstanding soloists have to cope with severe performance anxiety (Wilson 1997). Nevertheless, making music is frequently linked to highly positive emotions, to feelings of joy, satisfaction, and even to strong emotional reactions, known as "chill responses" (Altenmüller et al. 2013). These qualities are known to enhance plastic adaptations of the brain and can even lead to a sort of addictive behavior, causing younger musicians to over-practice and ignore fatigue and musculoskeletal pain which show their bodily limits.

## Brain Plasticity: Findings in Musicians

In general terms, brain plasticity means that our brain can be molded and formed. It is brain plasticity what allows us to learn throughout our lifetime. Mechanisms include not only strengthening or weakening of synaptic connections, but also structural changes such as growth of neurons, dendrites, myelin sheets, and neuroglia and reduced physiological loss (apoptosis) of neurons. Brain plasticity is best observed in complex tasks with high behavioral relevance for the individual such that they cause strong emotional and motivational activation. Plastic changes are more pronounced in situations where the task or activity is intense and the earlier in life it has been developed. Obviously, the continued activities of accomplished musicians provide in an ideal manner the prerequisites of brain plasticity. It is therefore not astonishing that the most dramatic brain plasticity effects have been demonstrated in professional musicians [for a classic review see (Münte et al. 2002)]. Nowadays, music induced plasticity can be assessed by electrophysiological and stimulation methods and by innovative MRI-imaging methods, such as "Voxel-Based Morphometry" (VBM) or "Diffusion Tensor Imaging" (DTI) [for recent reviews see (Wan and Schlaug 2010), and (Altenmüller and Schlaug 2015)].

The first imaging studies on brain plasticity in musicians where done by Schlaug et al. (1995) showing that professional musicians (pianists and string-players) had a larger middle section of the corpus callosum compared to a nonmusician control group. Furthermore, those musicians who began their training before the age of 7 showed a larger anterior part of the corpus callosum.

Similarly, other studies demonstrated the importance of the age of onset of musical training in influencing brain plasticity. For instance, Amunts et al. (1997) affirmed that early musical training could lead to pronounced anatomical changes in the hand motor area. Similarly, a seminal magnetoencephalography (MEG) study (Pantev et al. 1998) showed that the dipole strength associated with piano tones was greater in the auditory network of those musicians who had begun practicing before the age of 9 years thus favoring the idea that the age of inception of musical training is important to determine the degree of cortical adaptation (Elbert et al. 1995; Amunts et al. 1997). In a DTI study with pianists, Bengtsson et al. (2005) found that several white matter tracts correlated with the estimated amount of musical practice during childhood (e.g., posterior limb of internal capsule, the isthmus, and the body of corpus callosum, and some fiber tracts in the frontal lobe), although the total number of practicing hours was lower in this period than the estimated hours in adolescence and adulthood. This result supports the idea that the central nervous system exhibits greater plastic capacities during early stages of development and maturation periods, contrasting with its limited malleability during adulthood (Bengtsson et al. 2005). These were first hints to the metaplasticity concept we will discuss later.

However, the age of onset of musical training is generally confounded by the degree of expertise: early starters usually have been practicing for a larger amount of time than late starters and thus, on average, should have a greater level of expertise. In a recent study with selected professional pianists, Granert et al. (2011) measured the skill level of piano playing via the temporal accuracy during a scale-playing task. These authors found that the higher the skill level of piano playing, the smaller the volume of gray matter in the right putamen (Granert et al. 2011). In another recent example, Bailey et al. (2014) used deformation based morphometry to reveal differences (early > late starters) in the right ventral premotor cortex, which correlated with the performance on an auditory-motor synchronization task. A similar finding was reported in corpus callosum, displaying a larger volume of white matter in musicians who initiated musical training earlier (Steele et al. 2013).

Broadening the concept of expertise, Gaser and Schlaug (2003) compared professional musicians, amateur musicians, and nonmusicians and reported increased gray matter (GM) volume in primary motor, somatosensory, and premotor areas, among other regions in the musician groups. Modeling musical expertise with the same three-group population (professional musicians, amateur musicians and nonmusicians), James et al. (2014) reported an intricate pattern of increased/decreased GM. In particular, musicians showed GM density increases in areas related to higher order cognitive processes (such as the fusiform gyrus or the inferior frontal gyrus), whereas GM decreases were found in sensorimotor regions (as perirolandic and striatal areas). These reductions in GM were interpreted as reflecting a higher degree of automaticity of motor skills in more expert musicians (James et al. 2014).

Our understanding of the molecular and cellular mechanisms underlying these adaptations is far from complete. At the synaptic level, long-term depression (LTD) and long-term potentiation (LTP) account for changes in synaptic excitability. Generally speaking, brain plasticity may occur on different time axes. For example, the efficiency and size of synapses may be modified in a time window of seconds to minutes, whereas the growth of new synapses and dendrites may require hours to days. An increase in gray matter density, which mainly reflects an enlargement of neurons, needs at least several weeks. White matter density also increases as a consequence of musical training. This effect is primarily due to an enlargement of myelin cells: the myelin cells, wrapped around the nerve fibers (axons) are contributing essentially to the velocity of the electrical impulses traveling along the nerve fiber tracts. Under conditions requiring rapid information transfer and high temporal precision these myelin cells grow and as a consequence nerve conduction velocity increases. Finally, brain regions involved in specific tasks may also be enlarged after long-term training due to the growth of structures supporting the nervous function, for example, in the blood vessels that are necessary for the oxygen and glucose transportation to sustain nervous function.

## Different Types of Metaplasticity

Metaplasticity is a term originally coined by Abraham and Bear to refer to the "plasticity of synaptic plasticity" (Abraham and Bear 1996). It is a concept which first emerged from experiments in cellular neurosciences (on cell cultures of hippocampal neurons) demonstrating that the history of synaptic stimulation in a cell assembly determines the future ability of these synapses to adapt to later stimulation. Here, we define metaplasticity in a broader way: if we understand changes in synaptic activity as correlates of learning in humans, it matters, at what age learning takes place in order to facilitate further learning. The proverb "a tree must bent while it is young" denotes this idea: early learning experiences in skill acquisition will raise scaffolding for later expertise in this very skill. This seems to be crucial in musical skill learning, since it is a matter of common knowledge, that almost all outstanding virtuosos have had started their musical activities at very early age, frequently before 6 years. Musicians therefore can be considered an ideal model to probe metaplasticity of the human nervous system.

In musicians, metaplasticity can be investigated in three ways, namely (a) by behavioral assessment of perceptive and motor skills in "early" and "late" starters, (b) by electrophysiological methods, such as brain stimulation techniques, and (c) by brain imaging techniques.

*Behavioral studies* First hints towards metaplasticity were found in adult musicians when tested for two-point discrimination. Those, who initiated musical training at earlier age have superior two-point discrimination, as opposed to late starters, however, these benefitted to a larger extent from short-term tactile training (Ragert et al. 2004). In general, a prominent nature of musicians' metaplasticity is that earlier musical training provides greater gain of sensorimotor learning (Herholz and Zatorre 2012). Similarly, effects of long-term musical training on learning gain of short-term training are documented in the auditory and motor domains, such as quicker comprehension of regularities and abstract rules in tone sequences (Herholz et al. 2011).

Faster learning of sequential finger movements was also reported in pianists than musically untrained individuals (Hund-Georgiadis and von Cramon 1999). These findings commonly reported enhancement of learning gain of short-term training as well as neuroplasticity through long-term training, which may enable musicians to acquire a variety of sensorimotor skills quickly.

*Electrophysiology* Noninvasive brain stimulation has been applied to do research into metaplasticity (Siebner et al. 2004; Hamada et al. 2009). To this aim, two different stimulation techniques are normally used in succession. First, transcranial direct current stimulation (tDCS) induces either facilitation or suppression of motor cortical excitability. Then, the subsequent stimulation using excitability-diminishing repetitive transcranial magnetic stimulation (rTMS) impacts in a manner depending on the tDCS-induced state of cortical excitability; namely, rTMS triggers long-term depression (LTD) and potentiation (LTP) at the stimulated motor cortex following the excitability facilitation and suppression, respectively (Siebner et al. 2004). This activity-dependent plastic change is, however, aberrant in patients with movement disorders, as represented by task-specific focal dystonia (Quartarone et al. 2005). Since task-specific focal dystonia typically develops through extensive sensorimotor training for the prolonged period (Altenmüller 2003; Furuya and Altenmüller 2015), aberrant metaplasticity in the patients suggests alteration of neuroplastic mechanism through a history of long-term training. In these patients two further details hint at aberrant topography: first, the disorder is four times more frequent in musicians who start after the age 10 with their instruments (Schmidt et al. 2013), and second reduced inhibition in the sensory-motor systems leads to a corrupted topography in the somatosensory receptive field of the postcentral cortex (Elbert et al. 1998; Candia et al. 2003).

By contrast, a few recent studies provided evidence of enhancement of resistance of the nervous system against learning according to previous long-term musical training (Furuya et al. 2013, 2014; Vaquero et al. 2016). tDCS is a means effective for modulating cortical excitability (Nitsche and Paulus 2000, 2001). Anodal tDCS over the primary motor cortex elevates the cortical excitability, whereas cathodal one lowers the excitability. Simultaneous anodal and cathodal tDCS over the left and right motor cortices, respectively, therefore elevates the left motor cortical excitability (Paquette et al. 2011; Waters-Metenier et al. 2014). The bihemispheric tDCS while nonmusicians and expert pianists were taking rest resulted in enhancement and degradation of fine motor control of skilled finger movements, respectively (Furuya et al. 2014). In addition, the bihemispheric tDCS brought about changes in fine motor control to smaller extent for pianists who initiated musical training at earlier age, particularly before the age 8 (Furuya et al. 2013). These findings are evidence in favor of "early optimization" of the nervous system, a mechanism stabilizing the brain plasticity so as to retain acquired skills and minimize undesirable neuronal changes.

*Brain Imaging* Gaser and Schlaug (2003) could demonstrate enhancement of gray matter density in cortical sensory-motor regions, auditory regions, the left dorso-lateral prefrontal cortex, and in the cerebellum in professional instrumentalists as compared to nonmusicians and amateurs. Interestingly, these plastic adaptations depend on critical age periods: Vaquero et al. (2016) investigated a group of 36 expert pianists (among them, 21 started their musical training before age 7, and 15 started at age 7 or after that age) and another group of 17 college students without any musical background.

In line with Gaser and Schlaug, comparing with nonmusicians, pianists showed more gray matter in regions associated with learning (hippocampus), with sensory and motor control and processing (putamen and thalamus), with emotional processing and the reward system (amygdala), as well as with auditory and language processing (left superior temporal cortex); but they also showed less gray matter in regions involved in sensory and motor control (postcentral gyrus), in processing of musical stimuli (right superior temporal cortex), and structures that have been related to music-score reading (supramarginal gyrus).

Moreover, among the pianists it was observed that the right putamen correlated significantly with the age of start of music training (the later they started to play the piano, the greater was the volume of gray matter in the right putamen). The putamen is a deep subcortical structure which plays a role in motor control and automatization of movements. The volume of gray matter in this structure has been previously related with piano playing skills in the same direction with in the present study: the higher the pianistic skills (the automaticity), the smaller the volume of gray matter. One of the interpretations for this result is that pianists who started earlier in life have been able to perform an optimization process of a circuit highly trained by these musicians. This could also be the reason explaining why, in the comparison between musicians and nonmusicians, less gray matter was found in some areas related to music performance.

Regarding a behavioral scale-playing task done in these pianists, those who have started their musical training before age 7 played with higher regularity than those who started after 7 years of age, even though all the pianists participating in the study practiced the same amount of hours nowadays and have achieved the same level of proficiency. The findings are summarized in Fig. 1.

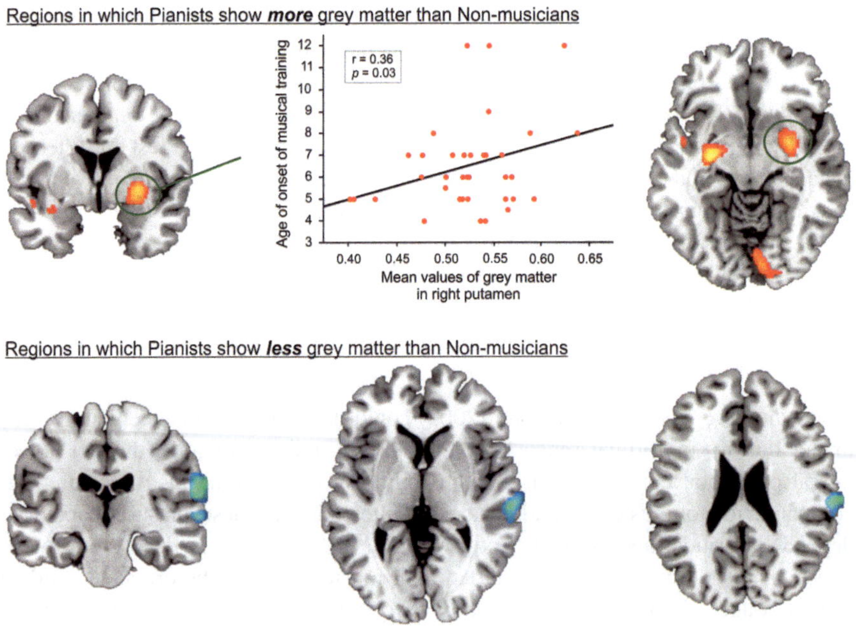


**Fig. 1** Summary oft the results of the study on pianists by Vaquero et al. (2016). Courtesy of Lucia Vaquero and Antoni Rodriguez–Fornells, with permission

As a conclusion, this new study observed that a homogeneous sample of musicians (expert pianists, without differences in the amount of practice in their lifetime or nowadays) presented increases and decreases in the volume of gray matter in different brain regions comparing with participants without music training. Furthermore, even though they exhibit the same level of proficiency, they have accumulated the same amount of hours of piano practice in their lifetime and they were currently practicing the same amount of hours at the moment of the study, those pianists who started their musical training before age 7 showed metaplasticity, that means, greater efficiency both in their brain structure and in their piano performance (observable as a smaller volume of gray matter in the right putamen and a higher temporal regularity while playing musical scales on the piano).

## Take Home Message

The findings presented here are summarized in Fig. 2. We propose that early musical training stabilizes the motor system and provides neuroprotective effect to stabilize weights in the sensorimotor networks. A plausible biological significance of this “early optimization” of the nervous system, being termed as homeostatic metaplasticity (Bienenstock et al. 1982; Muller-Dahlhaus and Ziemann 2015), is to

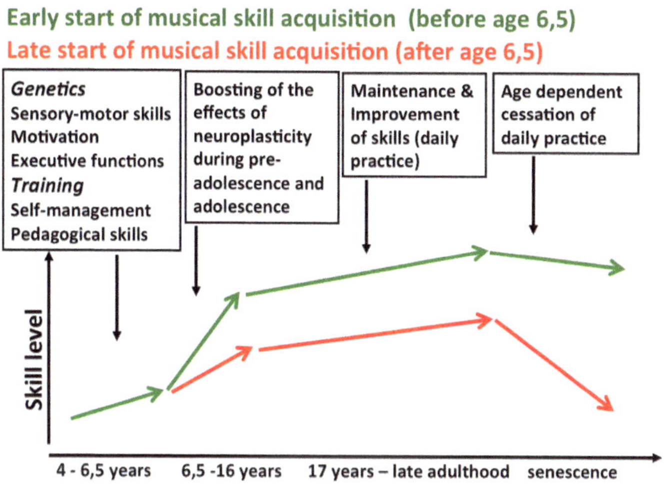


**Fig. 2** Different time courses of skill acquisition in earlier and later "starters". The neuronal networks representing a specific skill will be optimized in early childhood. This will allow a more proficient skill acquisition during preadolescence and adolescence resulting in a higher skill level. This skill level will even improve during adulthood and will be more stable with cessation of skill maintaining activities. Furthermore, these early optimized neuronal networks are more stable and less susceptible to maladaptive changes, for example in musicians dystonia

prevent maladaptive changes in the motor system via extensive sensorimotor training. This is particularly obvious in the case of Musicians' Dystonia: Musicians suffering from such a task-specific loss of motor skills ("musicians' dystonia") started playing their instruments at the later age (10.0 ± 4.7-yrs old) compared with the healthy musicians (7.7 ± 2.2 yrs old) (Schmidt et al. 2013), which suggests that early musical training lowers risks of developing MD-related maladaptation of the nervous system.

Finally, aging-induced neuroplastic changes at neuronal networks subserving sensorimotor and cognitive functions can be also ameliorated by early musical training (Krampe and Ericsson 1996; Meinz 2000). For example, a negative impact of aging on fine motor control was diminished for expert pianists as compared to nonmusicians (Krampe and Ericsson 1996). In addition, the resistance of finger dexterity against aging in the expert pianists is predictable by the amount of deliberate practice during later adulthood. These findings indicate impacts of prior extensive sensorimotor training on age-related decline of motor functions, which implicates a relation of musician's metaplasticity to aging. Similarly, adolescent neurodevelopment can be facilitated through early musical training (Tierney et al. 2015). Taken together, musical training in childhood can impact lifetime changes in both structures and functions of the sensory and motor systems, which not only enhance musical skill acquisition and prevent disorders triggered by extensive training, but also serve as ingredients better shaping fundamental neuronal development and aging.

# References

Abraham WC, Bear MF (1996) Metaplasticity: the plasticity of synaptic plasticity. Trends Neurosci 19:126–130

Altenmüller E (2003) Focal dystonia: advances in brain imaging and understanding of fine motor control in musicians. Hand Clin 19:523–538, xi

Altenmüller E, Ioannou CI, Lee A (2015) Apollo's curse: neurological causes of motor impairments in musicians. Prog Brain Res 217:89–106. doi:10.1016/bs.pbr.2014.11.022

Altenmüller E, Kopiez R, Grewe O (2013) A contribution to the evolutionary basis of music: lessons from the chill. The evolution of emotional communication: from sounds in nonhuman mammals to speech and music in man, p 313

Altenmüller E, Schlaug G (2015) Apollo's gift: new aspects of neurologic music therapy. Prog Brain Res 217:237–252. doi:10.1016/bs.pbr.2014.11.029

Amunts K, Schlaug G, Jancke L, Steinmetz H, Schleicher A, Dabringhaus A, Zilles K (1997) Motor cortex and hand motor skills: structural compliance in the human brain. Hum Brain Mapp 5:206–215. doi:10.1002/(SICI)1097-0193(1997)5:3<206:AID-HBM5>3.0.CO;2-7

Bailey JA, Zatorre RJ, Penhune VB (2014) Early musical training is linked to gray matter structure in the ventral premotor cortex and auditory-motor rhythm synchronization performance. J Cogn Neurosci 26:755–767. doi:10.1162/jocn_a_00527

Bengtsson SL, Nagy Z, Skare S, Forsman L, Forssberg H, Ullen F (2005) Extensive piano practicing has regionally specific effects on white matter development. Nat Neurosci 8: 1148–1150. doi:10.1038/nn1516

Bienenstock EL, Cooper LN, Munro PW (1982) Theory for the development of neuron selectivity: orientation specificity and binocular interaction in visual cortex. J Neurosci 2:32–48

Candia V, Wienbruch C, Elbert T, Rockstroh B, Ray W (2003) Effective behavioral treatment of focal hand dystonia in musicians alters somatosensory cortical organization. Proc Natl Acad Sci USA 100:7942–7946. doi:10.1073/pnas.1231193100

Elbert T, Candia V, Altenmüller E et al (1998) Alteration of digital representations in somatosensory cortex in focal hand dystonia. NeuroReport 9:3571–3575

Elbert T, Pantev C, Wienbruch C, Rockstroh B, Taub E (1995) Increased cortical representation of the fingers of the left hand in string players. Science 270:305–307

Ericsson KA, Krampe RT, Heizmann S (1993) Can we create gifted people? Ciba Found Symp 178: 222–231 (discussion 232–249)

Furuya S, Altenmüller E (2015) Acquisition and reacquisition of motor coordination in musicians. Ann NY Acad Sci 1337:118–124. doi:10.1111/nyas.12659

Furuya S, Klaus M, Nitsche MA, Paulus W, Altenmüller E (2014) Ceiling effects prevent further improvement of transcranial stimulation in skilled musicians. J Neurosci 34:13834–13839. doi:10.1523/JNEUROSCI.1170-14.2014

Furuya S, Nitsche MA, Paulus W, Altenmüller E (2013) Early optimization in finger dexterity of skilled pianists: implication of transcranial stimulation. BMC Neurosci 14:35. doi:10.1186/1471-2202-14-35

Furuya S, Oku T, Miyazaki F, Kinoshita H (2015) Secrets of virtuoso: neuromuscular attributes of motor virtuosity in expert musicians. Sci Rep 5:15750. doi:10.1038/srep15750

Gaser C, Schlaug G (2003) Brain structures differ between musicians and non-musicians. J Neurosci 23:9240–9245

Granert O, Peller M, Jabusch HC, Altenmüller E, Siebner HR (2011) Sensorimotor skills and focal dystonia are linked to putaminal grey-matter volume in pianists. J Neurol Neurosurg Psychiatry 82:1225–1231. doi:10.1136/jnnp.2011.245811

Hallam S, Cross I, Thaut M (2016) The Oxford handbook of music psychology. Oxford University Press, Oxford

Hamada M, Hanajima R, Terao Y et al (2009) Primary motor cortical metaplasticity induced by priming over the supplementary motor area. J Physiol 587:4845–4862. doi:10.1113/jphysiol.2009.179101

Herholz SC, Boh B, Pantev C (2011) Musical training modulates encoding of higher-order regularities in the auditory cortex. Eur J Neurosci 34:524–529. doi:10.1111/j.1460-9568.2011.07775.x

Herholz SC, Zatorre RJ (2012) Musical training as a framework for brain plasticity: behavior, function, and structure. Neuron 76:486–502. doi:10.1016/j.neuron.2012.10.011

Hund-Georgiadis M, von Cramon DY (1999) Motor-learning-related changes in piano players and non-musicians revealed by functional magnetic-resonance signals. Exp Brain Res 125:417–425

James CE, Oechslin MS, Van De Ville D, Hauert CA, Descloux C, Lazeyras F (2014) Musical training intensity yields opposite effects on grey matter density in cognitive versus sensorimotor networks. Brain Struct Funct 219:353–366. doi:10.1007/s00429-013-0504-z

Krampe RT, Ericsson KA (1996) Maintaining excellence: deliberate practice and elite performance in young and older pianists. J Exp Psychol Gen 125:331–359

Meinz EJ (2000) Experience-based attenuation of age-related differences in music cognition tasks. Psychol Aging 15:297–312

Mosing MA, Madison G, Pedersen NL, Kuja-Halkola R, Ullen F (2014) Practice does not make perfect: no causal effect of music practice on music ability. Psychol Sci 25:1795–1803. doi:10.1177/0956797614541990

Münte TF, Altenmüller E, Jancke L (2002) The musician's brain as a model of neuroplasticity. Nat Rev Neurosci 3:473–478. doi:10.1038/nrn843

Muller-Dahlhaus F, Ziemann U (2015) Metaplasticity in human cortex. Neuroscientist 21: 185–202. doi:10.1177/1073858414526645

Nitsche MA, Paulus W (2000) Excitability changes induced in the human motor cortex by weak transcranial direct current stimulation. J Physiol 527(3):633–639

Nitsche MA, Paulus W (2001) Sustained excitability elevations induced by transcranial DC motor cortex stimulation in humans. Neurology 57:1899–1901

Pantev C, Oostenveld R, Engelien A, Ross B, Roberts LE, Hoke M (1998) Increased auditory cortical representation in musicians. Nature 392:811–814. doi:10.1038/33918

Paquette C, Sidel M, Radinska BA, Soucy JP, Thiel A (2011) Bilateral transcranial direct current stimulation modulates activation-induced regional blood flow changes during voluntary movement. J Cereb Blood Flow Metab 31:2086–2095. doi:10.1038/jcbfm.2011.72

Prinz W (1984) Modes of linkage between perception and action. In: Cognition and motor processes. Springer, Berlin, pp 185–193

Quartarone A, Rizzo V, Bagnato S et al (2005) Homeostatic-like plasticity of the primary motor hand area is impaired in focal hand dystonia. Brain 128:1943–1950. doi:10.1093/brain/awh527

Ragert P, Schmidt A, Altenmüller E, Dinse HR (2004) Superior tactile performance and learning in professional pianists: evidence for meta-plasticity in musicians. Eur J Neurosci 19:473–478

Schlaug G, Jancke L, Huang Y, Staiger JF, Steinmetz H (1995) Increased corpus callosum size in musicians. Neuropsychologia 33:1047–1055

Schmidt A, Jabusch HC, Altenmüller E, Kasten M, Klein C (2013) Challenges of making music: what causes musician's dystonia? JAMA Neurol 70:1456–1459. doi:10.1001/jamaneurol.2013.3931

Siebner HR, Lang N, Rizzo V, Nitsche MA, Paulus W, Lemon RN, Rothwell JC (2004) Preconditioning of low-frequency repetitive transcranial magnetic stimulation with transcranial direct current stimulation: evidence for homeostatic plasticity in the human motor cortex. J Neurosci 24:3379–3385. doi:10.1523/JNEUROSCI.5316-03.2004

Steele CJ, Bailey JA, Zatorre RJ, Penhune VB (2013) Early musical training and white-matter plasticity in the corpus callosum: evidence for a sensitive period. J Neurosci 33:1282–1290. doi:10.1523/JNEUROSCI.3578-12.2013

Tierney AT, Krizman J, Kraus N (2015) Music training alters the course of adolescent auditory development. Proc Natl Acad Sci USA 112:10062–10067. doi:10.1073/pnas.1505114112

van der Steen MC, Molendijk EB, Altenmüller E, Furuya S (2014) Expert pianists do not listen: the expertise-dependent influence of temporal perturbation on the production of sequential movements. Neuroscience 269:290–298. doi:10.1016/j.neuroscience.2014.03.058

Vaquero L, Hartmann K, Ripolles P et al (2016) Structural neuroplasticity in expert pianists depends on the age of musical training onset. Neuroimage 126:106–119. doi:10.1016/j.neuroimage.2015.11.008

Wan CY, Schlaug G (2010) Music making as a tool for promoting brain plasticity across the life span. Neuroscientist 16:566–577. doi:10.1177/1073858410377805

Waters-Metenier S, Husain M, Wiestler T, Diedrichsen J (2014) Bihemispheric transcranial direct current stimulation enhances effector-independent representations of motor synergy and sequence learning. J Neurosci 34:1037–1050. doi:10.1523/JNEUROSCI.2282-13.2014

Wilson P (1997) The singing voice: an owner's manual: for singers, actors, dancers and musicians. MIT Press, Cambridge

Zatorre RJ, Chen JL, Penhune VB (2007) When the brain plays music: auditory-motor interactions in music perception and production. Nat Rev Neurosci 8:547–558. doi:10.1038/nrn2152

# The Coordination Dynamics of Observational Learning: Relative Motion Direction and Relative Phase as Informational Content Linking Action-Perception to Action-Production

John J. Buchanan

**Abstract** The primary goal of this chapter is to merge together the visual perception perspective of observational learning and the coordination dynamics theory of pattern formation in perception and action. Emphasis is placed on identifying movement features that constrain and inform action-perception and action-production processes. Two sources of visual information are examined, relative motion direction and relative phase. The visual perception perspective states that the topological features of relative motion between limbs and joints remains invariant across an actor's motion and therefore are available for pickup by an observer. Relative phase has been put forth as an informational variable that links perception to action within the coordination dynamics theory. A primary assumption of the coordination dynamics approach is that environmental information is meaningful only in terms of the behavior it modifies. Across a series of single limb tasks and bimanual tasks it is shown that the relative motion and relative phase between limbs and joints is picked up through visual processes and supports observational learning of motor skills. Moreover, internal estimations of motor skill proficiency and competency are linked to the informational content found in relative motion and relative phase. Thus, the chapter links action to perception and vice versa and also links cognitive evaluations to the coordination dynamics that support action-perception and action-production processes.

**Keywords** Cognitive evaluation · Coordination dynamics · Motor skills · Cognitive codes · Event coding · Limb movements · Phase transition

---

J.J. Buchanan (✉)
Human Performance Labs: Perception Action Dynamics, Department of Health and Kinesiology, Texas A&M University, College Station, TX, USA
e-mail: jbuchanan@hlkn.tamu.edu



J. Laczko and M.L. Latash (eds.), *Progress in Motor Control*,
Advances in Experimental Medicine and Biology 957,
DOI 10.1007/978-3-319-47313-0_12

## Introduction

Very young children watch their parents move and interact with objects in the world on a day-to-day basis. Parents pick things up, toss things about, move around things, and children will imitate these actions. I have a picture of myself mowing the grass and my then three-year-old son Shawn pushing a toy mower and following me around the yard. What was my son learning by imitating my behavior? Was he learning intent, general action, or specific kinematic patterns? From a social point of view, a person off the street might say my son was learning intent or socially acceptable behavior and relating it to a general outcome. A savvy bystander might say my son was extracting general and specific action information about moving and interacting with a mechanical device to develop appropriate motor commands so that the societal goal is achieved in an efficient and stable manner. When it comes to observation as a means to support motor learning, it is the viewpoint of the savvy bystander which is developed in this chapter. In other words, how does observation facilitate the development of motor commands to generate specific kinematic patterns of limb movements without (or with limited) physical practice? It is this level of analysis that observation or demonstration as a means to study action can most inform about how the central nervous system maps perception to action and action to perception with regard to motor skills.

This chapter is organized around into three sections: (1) cognitive and ecological theories of observation; (2) shared information in the form of coordination dynamics that support action understanding (perception) and action-production (control/coordination); and (3) neural areas and the link between action and action-observation. For the first section, social learning theory (Bandura 1977, 1986), the visual perception perspective (Scully and Newell 1985), and the Theory of Event coding (Hommel et al. 2001) are briefly outlined to highlight the differences between cognitive and ecological theories of observational learning. Within the second section, the attempt is made to merge the visual perception perspective of observational learning with the coordination dynamics theory of pattern formation in action-perception processes (Kelso 1995). The third section will briefly outline studies that have identified brain regions that may be functionally linked to some of the perceptual/behavioral results presented in section two.

## Cognitive and Ecological Theories of Observation

The study of observation or demonstration as a means to facilitate the learning of motor skills received a theoretical jumpstart in the late 1970s and early 1980s. In this decade, a variety of studies started to apply Bandura's social learning theory to the study of motor skill learning (Carroll and Bandura 1982, 1985). The attraction of social learning theory to some motor control individuals was seen in its emphasis on cognitive processes and the notion that motor learning in general involves the development of a conceptual representation (internal model) for actions. Thus, motor control researchers sought to use the cognitive processing features of social learning theory as a means to conceptualize how observation supports the development of symbolic codes (cognitive codes) that represent an observed action and its outcome. With these symbolic codes, the observer is then able to understand the errors made by a model and then identify successful and unsuccessful corrective strategies. Cognitive codes can be rehearsed without moving and this increases the strength of the representation and should facilitate initial attempts at action by an observer. Bandura's theory has had a major impact on the study of observation and demonstration as a means to acquire motor skill knowledge and is still playing a role today in the study of observation as a motor skill training procedure (Maslovat et al. 2010).

The visual perception perspective of observational learning (Scully and Newell 1985) was developed as an alternative to cognitive theories such as Bandura's. The visual perception perspective was developed from research that had examined the ability of people to identify human actions in minimal point light displays (Johansson 1973; Cutting et al. 1978; Cutting and Proffitt 1982). The point light work theoretically addressed the issues of information type that may be extracted when viewing biological actions. For example, when a person walks there is a common motion feature, the overall direction of motion, a relative motion feature, the directional rotation between limbs and joints, and absolute motion features, how fast and how large are the movements of the parts. The formulation of the visual perception perspective also drew heavily on the ideas of Gibson's direct perception with its emphasis on perceiving formless and timeless invariants that represent specific informational properties of objects in the world (Gibson 1979). Specifically, Scully and Newell (1985) developed their theory around the concept that the topological characteristics of the relative motion between limbs and joints may be directly picked up when watching or viewing a model. This viewpoint emphasizes the identification of what information is important for coordinating the limbs and joints, because this information is the most beneficial for an observer to identify through visual perception processes if they want to imitate an actor. The visual perception perspective shift the emphasis to identifying what information defines an action and takes it away from identifying the processing steps that observation may support.

The Theory of Event Coding is cognitively based and argues for the existence of shared representations for perceived events and the actions that are generated in response to those perceived events (Hommel et al. 2001; Hommel 2004). This theory was not specifically designed as a theory of observational learning, but was instead proposed as a general theory for perception-action planning across many contexts. The central theoretical feature that it proposes is that perception and action codes are not stored separately, but instead share a common code—the event code. On the one hand, this combination of perception and action information into an event code departs significantly from the idea of separate symbolic codes that must link action and perception to support observational learning (Carroll and Bandura 1982, 1990). On the other hand, this theory aligns with the visual perspective theory in seeking to propose that perception and action are directly linked. For the visual perspective theory, the informational link is in the form of the topological invariant features of relative motion, whereas for event coding the information link is in the form of a single cognitive representation wherein perception and action information for an event are linked. The goal of this chapter is to show that perception and action are coded in the same informational representation, wherein this representation is in the form of the self-organized coordination dynamics of relative phase and how those dynamics interact with the topological features of relative motion.

## Self-organized Coordination Dynamics

As a starting point, the theory of coordination dynamics views biological perception-action processes as akin to self-organizing phenomenon seen in many chemical and biological systems (Haken 1983; Prigogine and Stengers 1984; Schöner and Kelso 1988a; Turvey 1990; Kelso 1995). A primary emphasis of this approach is to utilize the tools of nonlinear dynamical systems and view biological systems, humans in particular, as open dissipative systems wherein an understanding of perception-action comes from an understanding of the interaction between components (nerve cells, muscles, respiration, etc.) across many levels (Kelso et al. 1980; Kugler et al. 1980; Kugler and Turvey 1987). The traditional concepts of control and coordination are to be found in the so-called order parameter (or collective variable), a variable that characterizes the temporal and spatial ordering of limbs or joints during goal directed action and changes in goal directed actions. An early study that was essential in developing certain aspects of the coordination dynamics theory was one that focused on an abrupt switch from one coordination pattern to another (Kelso 1984). The two patterns examined by Kelso required finger or wrist movements to be coordinated in either an in-phase or anti-phase pattern. In-phase was defined as moving the limbs with homologous muscles and anti-phase with nonhomologous muscles. Under the directions to (1) increase movement frequency on cue or (2) to keep pace with an external metronome, a transition from the anti-phase to in-phase pattern of coordination in both wrist and fingers emerged.

An explanation for the switching phenomenon was found in an analysis of the variability of the relative phasing ($\phi$) between the two components (fingers or wrists) when performing the two patterns. Relative phase as a measure describes the spatial–temporal relationship between two oscillating components over time. The in-phase coordination pattern is described as having a relative phase relationship of $\phi = 0°$ and the anti-phase pattern as having a relative phase relationship of $\phi = 180°$. When the variability of relative phase was examined, significant increases were found in the variability of the anti-phase pattern as movement frequency increased, whereas no significant increase was found for the in-phase pattern as frequency increased (Kelso et al. 1986). Following the transition from anti- to in-phase, variability in relative phase decreased. The transition phenomenon and variability changes were modeled at both the individual component levels (fingers or wrist as nonlinear oscillators) and the relative phase level (potential function) (Haken 1983; Haken et al. 1985; Schöner et al. 1986). In the potential function (the HKB model), 0° and 180° are stable attractors while patterns such as 90° are unstable. A primary conclusion was that relative phase was an order parameter that governed the spatiotemporal coordination of the fingers or wrists in this task and that the transition was a self-organized phenomenon that resulted from a loss of stability in the anti-phase pattern. Thus, in-phase and anti-phase coordination patterns are viewed as stable attractors of the order parameter dynamics. The demonstration of phase transitions in rhythmic human action has also been linked to single limb multi-joint tasks. With the forearm supine and the arm in the sagittal plane, an in-phase rhythmic pattern of elbow and wrist motion (wrist flex/ext. coordinated with elbow flex/ext.) is very stable, whereas an anti-phase rhythmic pattern (wrist flex/ext. coordinated with elbow ext./flex) is less stable. If movement frequency is increased, transitions from anti-phase to in-phase occur, but not vice versa (Kelso et al. 1991). With the forearm supine, the transition is from a homologous muscle action to a nonhomologous action, just as in the bimanual task. For this task and other single limb trajectory formation tasks it was concluded that relative phase acts as an order parameter for the production of rhythmic single limb multi-joint patterns (Buchanan et al. 1997; de Guzman et al. 1997).

## *Information, Order Parameter Dynamics, and Action-Perception*

How do transitions between coordination patterns relate to observational learning? The link to observational learning comes through the idea that for information to modify behavior it must impact the system at the level of the order parameter (Schöner and Kelso 1988b; Kelso 1994). In the previous examples, this would be at the level of the relative phasing between components. Within the coordination dynamics framework, learning requires the transformation of behaviorally relevant environmental information into the same space as the order parameter dynamics. This implies that the perceptual identification of actions is constrained in a manner similar to how the coordination of actions is constrained based on the order parameter dynamics. Several studies have examined this informational link with tasks that required the perceptual evaluations of animated visual displays representing the bimanual and single limb coordination patterns just described.

Kelso and Pandya (1991) examined the perception of relative phase patterns using a stick–figure animation representing the human arm (upper arm, forearm, and hand, Fig. 1a). The upper arm segment remained fixed and the elbow and wrist segments moved rhythmically. The elbow and wrist segments were programmed to produce thirteen relative phase patterns from 0° to 180° and from −180° to 0° in 30° increments (Kelso and Pandya 1991). The relative motion of the wrist to the elbow may be described as the wrist lags the elbow (Fig. 1a right side) or the wrist leads the elbow (Fig. 1a left side) through flexion and extension. The relative phase patterns from −30° to −150° are linked to the wrist lagging, while the relative phase patterns in the range 30°–150° are linked to the wrist leading. For the 0° and 180° patterns, neither the wrist nor the elbow leads. The two segments rotate in the opposite direction for 180° (as the elbow flexes the wrist extends) whereas for 0° they rotate in the same direction (both flex and extend together). The 0° and 180° animations are consistent with the coordination patterns studied by Kelso et al. (1991).

Participants were asked to categorize the different relative phase patterns as either 0° in-phase or 180° anti-phase. The categorization procedure revealed that relative phase patterns in the ranges of $0° \le \phi < \pm 90°$ were most often categorized as in-phase whereas relative phase patterns in the range of $\pm 90° < \phi < \pm 150°$ were more often categorized as 180° anti-phase. The largest variability across participants in classifying the patterns occurred for the ±90° patterns, while the variability in classification for the other patterns was not significant. The finding that the categories were defined around ±90° is consistent with the HKB model in that ±90° are symmetry partners and are unstable patterns that require extensive training to perform without very specific augmented feedback (Schöner et al. 1992; Zanone and Kelso 1992; Kovacs and Shea 2011). This experiment demonstrated that a forced choice task can produce a perceptual bias with regard to the differences in stability of relative phase patterns as defined by the HKB model. This study shows that the order parameter relative phase act as an informational variable in the perception of actions that is consistent with the coordination of actions. However, there are two important aspects of coordination that this study did not examine. First, the task did not examine the role of pattern stability in the categorization process, an essential feature of the difference between in-phase, anti-phase and other phase relationships on the action side. The role of pattern stability in visual perception of relative phase will be examined in the following paragraph. Second, the study did not determine if relative motion (wrist lag or lead) played a role in the perceptual categorization process. It could be argued that relative phase was the primary informational source in the task because of the symmetry in results. This issue will be explored later in the chapter.

This link between perception and action with regard to coordination stability (relative phase variability) was examined in a study that utilized animated dots to display relative phase patterns, e.g., 0°, 30°, 90°, etc. (Bingham et al. 1999; Zaal et al. 2000) (see Fig. 1b). Participants in the Bingham et al. (1999) study were instructed to judge how stable the motion of the dots were under manipulations of relative phase, relative phase variability, and movement frequency. Overall, the experiments demonstrated that mean relative phase influenced the judgment of the stability in the dots motions, with a 90° relative phase pattern judged as the least stable whereas 0° in-phase was judged as the most stable followed by 180° anti-phase. Different levels of variability in the visual displays were only detected for the 0° and 180° patterns. The judgments of coordination were also found to change as a function of the frequency of the moving dots. The 180° anti-phase

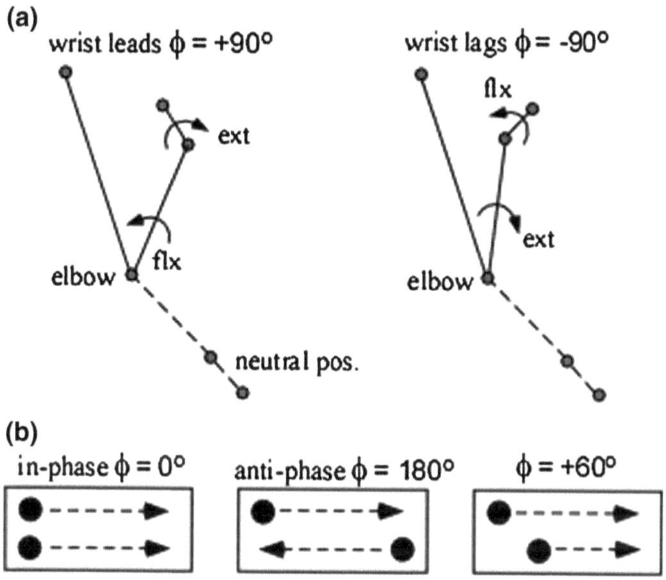


**Fig. 1** **a** Stick figures representing the animated arm display used as a stimulus in several studies. The *left side* portrays the wrist leading pattern and the *right side* portrays the wrist lagging pattern. **b** Representation of stimulus to characterize bimanual coordination and used in several perceptual evaluation studies

pattern and other relative phase patterns were judged as less coordinated as movement frequency increased, whereas movement frequency did not impact the judgments of 0° in-phase (Bingham et al. 2001). Different levels of variability in relative phase were perceived more consistently only for the in-phase pattern across different movement frequencies. These findings are consistent with the transition in bimanual coordination and with the predictions of the HKB model regarding the stability of the various coordination patterns (Bingham et al. 1999; Zaal et al. 2000). The findings from these perceptual assessments of coordination and variability suggest that action-perception is based on action capability. When a pattern cannot be produced in a stable manner, it is then perceived as less stable than patterns that can be produced. However, the issue of how perception may change with learning was not examined in these studies. Later in the chapter it is shown that observational and physical training alter perceptual processes as a function of learning and changes in coordination stability.

The finding that perceptual evaluations of actions mirror the stability of those actions when performed provides support for the application of the coordination dynamics theory to the study of observation based on the idea that order parameters are informational variables that map perceptual processes to motor processes (Kelso 1994). Application of coordination dynamics theory to the study of motor skill learning through observation or demonstration must reside in the concepts of the order parameter and its dynamics, change of stability, loss of stability, etc. To bring together the visual perspective theory of observation with the coordination dynamics theory will require the demonstration that some topological features of the relative motion direction between limbs and joints remains invariant and that such features support the informational nature of order parameters such as relative phase.

## *Information, Coordination Dynamics, and Observational Learning*

That observation or demonstration are essential to the learning of motor skills is quite evident in the often heard statements of coaches and trainers: "Watch how I do this," "Look at how so and so does it." The observer is then supposed to replicate the movement kinematics in a manner that will help insure success at the goal. The human motor system is highly redundant in that many variations in limb movement combinations can achieve the same goal, some, however, often seem more efficient than others, thus the premise for "Watch how I do this." According to Scully and Newell (1985), the thing that should be picked up is the relative motion relationships between the joints of the model performing the demonstration. In other words, as demonstrated with the work on point light displays seeing the whole body is not necessary to extracting visually relevant information. Research has investigated the concept that relative motion information can be picked up and used to support action production through observation (Scully 1986; Scully and Carnegie 1998; Al-Abood et al. 2001; Breslin et al. 2005). Studies have shown that observers can match certain kinematic features of a model's action, such as the angular displacement and velocity profiles of upper and lower arm motions in a yard-dart task (Al-Abood et al. 2001), as well as intra-limb and inter-limb segment patterns (Breslin et al. 2006). In each of these studies, the only coordination pattern (relative motion pattern) presented to the observers was that of a well-trained model, and the observers also received extensive training (physical practice ≈ observations). Thus, research shows that observers can match certain characteristic motions of a model's limbs. However, it was not specifically shown that observers selectively picked up relative motion information per se and then used that information without actual physical practice with the task. To demonstrate that relative motion information is picked up by observers and used in a learning/training context, it is necessary to show that observers will match variations in a novice model's actions and do so with minimal physical experience with the task. In other words that observers are sensitive to different relative motion patterns as viable solutions to a coordination problem that is initially unstable.

The ability of observers to distinguish variations in relative motion was found in an experiment that required observers to watch novice models learn to coordinate elbow and wrist flexion/extension motions at a 90° relative phase relationship (Buchanan et al. 2008). In the experiment models moved their arm in the sagittal motion plane with the elbow fixed to a stationary device to limit the motion of the

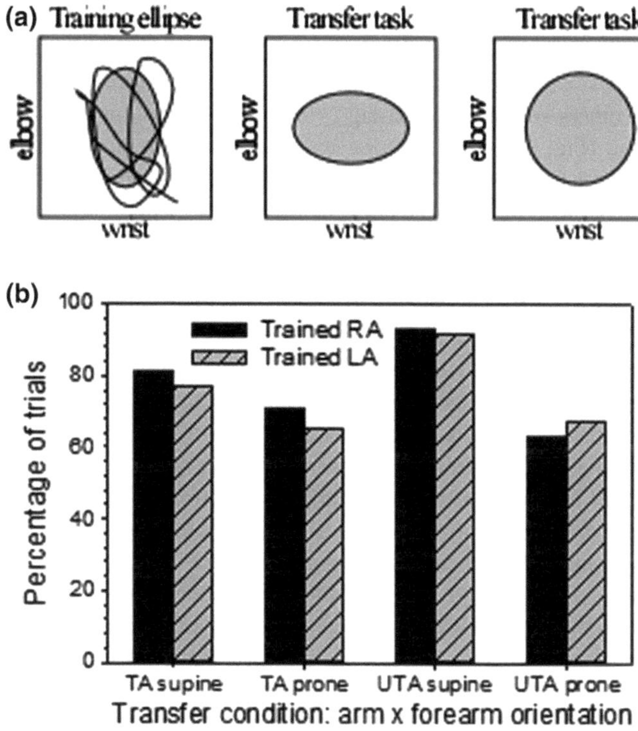


**Fig. 2** **a** The templates shown in the Lissajous plots represent both relative phase and joint amplitudes required for the initial training ellipse and the transfer tasks. **b** Percentage of trials characterized with the same relative motion direction in the observers. The percentage reported are across all transfer tasks: *TA* trained arm, *UTA* untrained arm

upper arm and allow the observers to focus on the elbow and wrist only. Each observer was yoked to an individual model. Half of the models trained with the right arm and half with the left arm and they were not instructed to have the wrist lead or lag the elbow to achieve the 90° relative phase pattern. The task required the elbow to rotate 90° as the wrist rotated 48°. The models trained for 2 days and observer's watched every training trial and were privy to concurrent and terminal feedback provided to the model. To facilitate learning, a vertically oriented ellipse template representing the difference in joint amplitudes and the required relative phase of 90° was displayed in a Lissajous plot feedback display (Fig. 2a). The feedback consisted of plotting the elbow and wrist angles along the axes of the ellipse. The models were told that their goal was to make the trace of their elbow and wrist motions create a trace around the perimeter of the ellipse. Twenty four hours after training, observers and models performed retention tests separately and only the ellipse template (without feedback) was shown in the Lissajous plot. The observers performed the retention test with the same arm that their model trained with and observers had no exposure to the elbow–wrist motion until the retention test.

A key issue examined with the above study was defining the nature of the relative motion direction selected by the models and produced by observers. Some of the models ($N = 5$) selected a wrist lead relative motion direction (+90° relative phase) while others selected a wrist-lag (−90°) relative motion direction ($N = 11$) during training (Fig. 1a). The retention tests revealed that the models performed with the same relative motion direction they trained with and that each observer selected the wrist lead or lag relative motion direction of the model they watched. The models had smaller error with respect to the target of 90° and performed with less relative phase variability compared to the observers, however, the observers had smaller error and performed with less variability compared to a control group. This experiment demonstrated that observers without any instruction pickup the relative motion direction between the joints as well as the specific relative phase pattern between the joints and use that information to help organize initial performance attempts. The results suggest that relative motion direction was important for organizing the actions, although the results did not really show that the relative motion direction is a separate source of information from relative phase.

The concept of topology in mathematics implies that the properties of geometric objects remain the same under transformations, such as stretching or bending. Within the area of human movement or biological coordination, the idea of topological features may be viewed as coordination staying the same under various transfer tasks. Scully and Newell (1985) proposed that relative motion has topological features, in other words, relative motion—both its perception and production—should remain invariant under some form of transformation. This issue was examined in a series of transfer tests that was an extension of the experiment that used the elbow–wrist task just described (Buchanan and Wright 2011). The transfer tests were designed to determine if relative motion direction and relative phase would transfer across changes in forearm orientation (supine to prone), across limbs (left-to-right and right-to-left), and changes in component amplitudes (see Table 1).[1] As with the retention test, the Lissajous plot was used to define the transfer tests which consisted of a horizontal ellipse and a circle along with the trained vertical ellipse (Fig. 2a). Performance feedback was not provided with the transfer tests. Overall there were 528 transfer trials performed by the observers, and in 400 (76 %) of the trials the observers used the same relative motion direction that their model used during training and that they used during retention. Of the various transformations, transfer across arms with the forearm supine, just as with the training, was associated with utilizing the same relative motion direction most often, in comparison to transfer across forearm positions (Fig. 2b). Thus, through observation participants picked-up relative motion direction and used that information rather consistently when performing variations of the trained task. These results show more clearly that relative motion direction is acting as an independent source of information from relative phase.

[1]The transfer results reported here are unpublished. The manuscript by Buchanan and Wright (2011) focused on relative phase performance across the transfer tasks.

**Table 1** Transfer tests and associated transformations based on arm, Forearm Orientation (FO), and Joint amplitude (JA)

| | Vertical ellipse | Horizontal ellipse | Circle |
|---|---|---|---|
| [a]RA supine | Retention | Transfer JA | Transfer JA |
| RA prone | Transfer FO | Transfer FO, JA | Transfer FO, JA |
| LA supine | Transfer arm | Transfer arm & JA | Transfer arm & JA |
| LA prone | Transfer arm & FO | Transfer arm, FO, & JA | Transfer arm, FO, & JA |

[a]This table represents the RA training condition. Substitute LA for RA for LA training condition

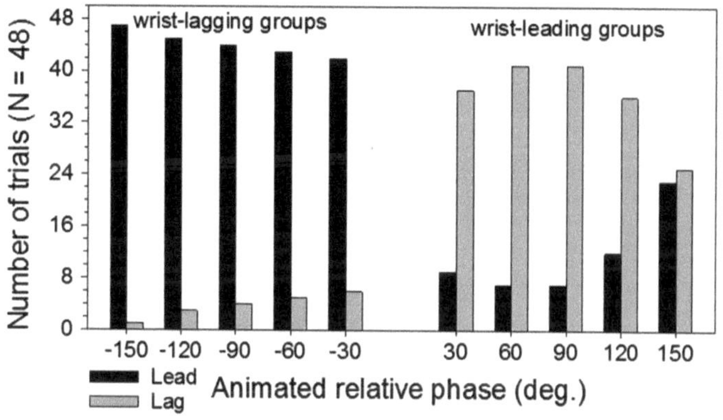


**Fig. 3** The number of trials with a given wrist lead or wrist lag classification for the wrist lagging and wrist leading test groups (labels at *top* of each side) are shown as a function of the required relative phase. Part of the data in this figure is from Fig. 2 in Buchanan (2015) *JMB* and reproduced with permission from Taylor and Francis

A more recent study using the single limb task described previously examined this relationship between various relative phase patterns (0°, ±30°, ±60°, ±90°, ±120°, ±150°, 180°) and relative motion directions using the stick figure arm animation (Fig. 1a). This task was designed to separate out the different informational aspects of relative motion direction and relative phase that may emerge at the very start of observational training (Buchanan 2015). Participants attempted to match the animated arm motions that consisted of elbow and wrist motion only in the sagittal plane. Attempts at performance matching occurred by tracking with the animation or after the animation finished 16 cycles of motion followed by a five second delay. The findings revealed that participants consistently selected the correct relative motion direction between the elbow and wrist. Those individuals exposed to wrist lagging animations (−30° to −150°) produced a negative mean relative phase (as required) for 92 % of the trials, while the individuals exposed to the wrist leading animations (+30° to +150°) produced a positive mean relative phase (as required) for 75 % of the trials (Fig. 3). Within each group, the inability to use the correct relative motion direction occurred most often in the delayed tracking groups, wrist leading group 41 of 58 incorrect trials and wrist

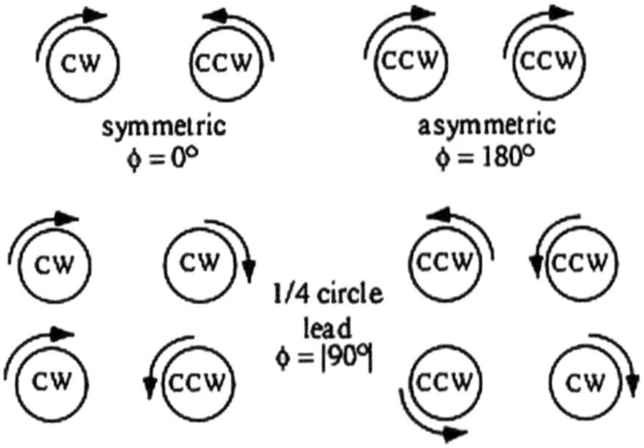


**Fig. 4** Possible relative motion directions between the hands/arms in the bimanual circle tracing task are shown. The *arrows* represent the direction around the circle from the participant's point of view

lagging group 19 of 19 incorrect trials. The results show that relative motion information is quickly identified by humans and that it can be used instantaneously to generate actions. In some of the early point light work, humans actions were identified in less than 200 ms (Johansson 1976). The results show that relative phase is perceived independently of relative motion in that the dynamics of the performance, stability and attraction, were governed by the intrinsic dynamics of in-phase and anti-phase more than either relative motion direction.

The ability of observers to distinguish various relative motion patterns has been demonstrated in a series of bimanual circle tracing tasks. As with the elbow wrist task, emphasis was placed on showing that observers' pickup relative motion direction (independent of relative phase) here defined as tracing direction around a pair of circles. Several studies have examined symmetric and asymmetric circle tracing and revealed that these two patterns of coordination have very similar dynamics to the bimanual finger patterns of in-phase and anti-phase, with asymmetric coordination (anti-phase) less stable than symmetric coordination (in-phase) (Carson et al. 1997; Byblow et al. 1999). The circle tracing task offers a means to study the pickup of relative motion direction in that a pair of circles can be traced in a variety of clockwise (CW) and counterclockwise (CCW) combinations with a left or right-hand lead, and based on the absolute value of relative phase are considered as equivalent (Fig. 4).

Symmetric and asymmetric coordination are defined in the top row for bimanual circle tracing. From a discovery learning point of view, the task of learning a 90° relative phase pattern while tracing the two circles has a variety of solutions, many more than learning a 90° relative phase using just flexion–extension motions of the two index fingers, which has only two, the left or right finger leads (Fig. 4, ¼ cycle lead). From an observation point of view, an observer may have many options to select from if they watch a novice model learning through discovery. When observers were yoked to discovery models learning a 90° relative phase in this task, around 50 % of the observers selected the rotation direction and hand lead that their yoked model practiced with the most often (Buchanan and Dean 2010). When observers watched an instructional novice learning this task that used only the CW:

CCW right-hand lead strategy or an expert using the same strategy then the observers selected the CW: CCW 100 % of the time (Buchanan and Dean 2014). Another big difference between observers that watch a discovery model, an instructional model, or an expert model is that observers that view a discovery model have a greater tendency to switch strategies within and between trials as well as switch hand leads within and between trials compared to observers that watched an instructional or expert model (Buchanan and Dean 2010).

In this section it was shown that observers detect and pickup relative motion direction between joints and limbs and use that information to help them organize outgoing motor commands in order to produce a variety of relative phase patterns. Previous work has argued that relative motion direction was the specific information that supports the coordination of rhythmic movements (Wilson et al. 2005). Here, the position is taken that relative motion information is important in defining the movement directions between joints and limbs, but that it does not carry specific information for relative phase except for the relative phase patterns of 0° (in-phase) and 180° (anti-phase) (Wilson and Bingham 2008). As the single limb and bimanual studies revealed, different relative motion directions can be associated with the same absolute relative phase value. Moreover, many relative phase patterns can be associated with the same general relative motion direction between joints and limbs (Buchanan 2015; Buchanan et al. 2015).

## *Information, Coordination Dynamics, and Self-evaluation Following Observation*

According to social learning theory, symbolic representations or codes developed through observation can be cognitively rehearsed. Such rehearsal is theoretically important because it allows for modification of the existing codes into more concise codes to reduce the load on memory and enhance retrieval (Carroll and Bandura 1982). An issue that has not been experimentally addressed from the social learning point of view is the form that this rehearsal may take. In other words, how do you test for mental rehearsal and code modification? The next two studies examined in a roundabout way the modification of cognitive processes using self-evaluation tests of motor skill proficiency and competency. The first study examined evaluations of motor skill proficiency at what might be considered the start of the observational learning process and the second study examined motor skill competency ratings at the end of an observational period that spanned two training sessions. Each study started from the assumption that relative motion and relative phase are informational variables that may be picked-up through visual perception processes. Both studies used the single limb animation portrayed in Fig. 1.

The first study used was outlined in the last section and in that study participants viewed the animated stick figure of the arm that produced eight relative phase patterns, six of which were defined as positive or negative based on the lead or lag relationship of the wrist to the elbow (Buchanan 2015). After attempting to match the coordination pattern represented in the animation participants were required to rate their level of proficiency. The proficiency rating was not an estimate of how well they thought they did at matching the animation or how coordinated they thought the animated pattern looked (Bingham et al. 1999); instead, it was a rating of how proficient (or coordinated) they felt while attempting to match the animation. A scale of 1–10 was used for the proficiency rating with 10 representing very proficient and 1 not at all proficient. Overall, proficiency ratings correlated with the relative phase variability found in the attempts at the seven different relative phase patterns (Fig. 5a) ($r^2 = 0.93$, $F = 68.2$, $p < 0.0001$). On average individuals rated themselves as more coordinated when doing 0° (in-phase), ±30° (strong attraction to 0°) and 180° (anti-phase), and less coordinated when doing ±60°, ±90°, ±120° and ±150°. The results revealed an awareness of the dynamics of relative phase in terms of intrinsic coordination stability, a finding distinct from the perceptual evaluation work that manipulated stability in the visual displays (Bingham et al. 1999, 2001; Zaal et al. 2000). The proficiency rating results are consistent with work showing that participants use proprioceptive information to judge phase variability, being most accurate for bimanual finger patterns mechanically driven at 0° and 180° compared to 90° (Wilson et al. 2003).

The second study compared both physical and observational practice and the animated arm portrayed two wrist leading to-be-learned patterns, +60° and +120° (Buchanan et al. 2015). There were two training days with 30 training trials a day. Participants in the physical group alternated every two trials in moving with the animation or moving after the animation stopped, while participants in the observation group watched the animation over the two training days. The participants were all trained separately and retention tests were performed separately. Prior to motor skill training, participants took part in a perceptual evaluation task. The task required them to view the 12 relative phase animations that were used in the Kelso and Pandya (1991) and Buchanan (2015) studies. After viewing each animated pattern (3 trials per pattern that were randomized) the participants had to rate how well they felt they could perform the pattern just animated. This task was designed to measure motor skill competency in that participants were not evaluating the level of coordination in the pattern (Bingham et al. 1999) nor were they evaluating their own performance, but were estimating their own ability to produce the just observed pattern. The rating scale was 1–10 with 10 indicating very competent and 1 not at all competent. Before practice, the perceptual ratings revealed that participants felt competent in their ability to produce the 0° and 180° patterns, but not the to-be-learned patterns with the competency ratings for all other patterns dropping off from 0° and 180° with the lowest ratings for ±60°, +90° and ±120°. After practice, the competency rating increased significantly for the two trained relative phase patterns (+60° and +120°) in both the models and observers (Fig. 5b). In fact, the observation group's competency ratings were slightly larger (not significantly) than the physical group's competency ratings. One reason the physical practice group may have had lower post-training competency ratings is because they relied more on their proprioceptive experience of the task gained through physical

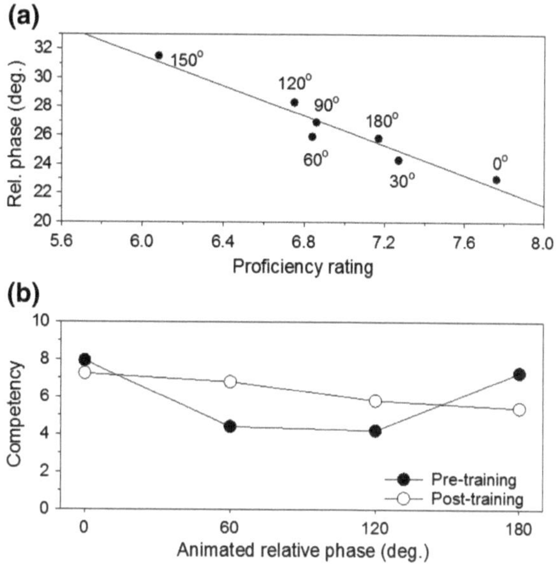


**Fig. 5** **a** Relative phase variability is plotted on the *y*-axis and proficiency rating is plotted on the *x*-axis. The *dots* are labeled based on relative phase pattern. **b** Competency rating is plotted on the *y*-axis as a function of pre- and post-training evaluation tests for four relative phases, the two trained phase of 60 and 120, and 0 and 180. Part of the data in B is from Figs. 2 and 3a Buchanan et al. (2015) *EBR* and reproduced with permission from Springer

practice, whereas the observation group were relying more on their visual knowledge when evaluating.

The above two studies when taken together show that the self-evaluation of actions is linked to information about relative motion direction and the variability in relative phase. It is proposed that any mental representation (or symbolic code or event code) formed during observation are constrained by the actual coordination dynamics that are intrinsic to the system and those dynamics that can change through learning. Thus, proposed symbolic codes or event codes formed about the kinematics of actions during observation should be entities with dynamics that provide a direct link between perception and action. If perception and action are linked through self-organized dynamic processes then there is not a need for intermediate symbolic codes. This does not mean that cognitive representations do not exist, since individuals can verbally describe the relative motion relationship between the joints. What it means is that at the level of movement kinematics there is a direct link between the production of an action and the perception of that action. The above conclusion is consistent with the idea of an event code as proposed by Hommel et al. (2001), except that coding is in the form of the dynamics of relative phase and topological features of relative motion. A more systematic way to study this idea of dynamic representations or codes and how they change through observation would be to evaluate motor skill proficiency and competency evaluations across training session and not just at the end points.

## Neural Areas Linked to Arm Movements and Competency Ratings

In this last section a brief connection to recent findings from neural imaging studies will be examined. The idea is to show that neural areas linked to arm movement kinematics have been identified and that areas linked to performance evaluation, i.e., competency, have also been identified. Within the area of neuroscience, the "mirror neurons" in the premotor cortex and the associated mirror system have been invoked as mechanisms whereby action-observation is linked to motor imitation and action-observation (Buccino et al. 2004; Rizzolatti and Craighero 2004; Calvo-Merino et al. 2005; Cattaneo and Rizzolatti 2009; Cross et al. 2009, 2012). Theoretically, the mirror neuron system and the action observation network (AON) may allow observers to map a representation of the seen action onto the motor system and gain knowledge of the action through internal execution processes.

The results summarized throughout this chapter have shown that observers can extract two specific types of action information from moving joints and limbs, relative motion and relative phase and use that information to develop stable coordination patterns with limited physical practice. Research using fMRI has demonstrated that the occipitotemporal cortex is a region selective for responding to static hand and upper limb postural positions with functional connectivity to the human middle temporal complex an area associated with motion coding (Orlov et al. 2014). Studies have shown that the human motion complex and posterior superior temporal sulcus are regions that respond to speed and distance of moving body parts (hand, arm, and head) (McAleer et al. 2014). Identification of the relative motion direction between the elbow and wrist in the animated arm studies or the relative motion directions in circle tracing and hand lead in the bimanual tasks may occur in the above regions. The rhythmic nature of the motions studied and the stability of relative phase may be processed through M1 and the cerebellum based on a MEG study that localized these areas as important for processing frequency information (Bourguignon et al. 2013). Observation as a training protocol requires the continual monitoring of a models' actions (whether human or animation) at the kinematic level and there are areas in the human nervous system that have been identified as supporting this monitoring (McAleer et al. 2014).

The proficiency and competency tasks based on the single limb motions demonstrated that what are often thought of as cognitive processes may be linked to the order parameter dynamics of relative phase and the relative motion between the joints. The competency task demonstrated that physical and observational practice can alter individuals' cognitive evaluations of their performance. Research using fMRI has revealed that expert ballet dancers are better at recognizing actions from their own motor repertoire and this ability was linked to activation differences in the greater premotor area, the parietal lobe, and the cerebellum (Calvo-Merino et al. 2006). Comparisons between experts and novices suggests that expertise is associated with configural action processing, i.e., the ability to identify transformations in actions that come from an individual's own motor repertoire (Calvo-Merino et al. 2010). The configural transformation may be viewed as similar to the single limb transfer tasks mentioned earlier. An fMRI study that examined the observation of dance moves over 5 sessions found that activity in the ventral premotor and inferior parietal lobule (areas of the AON) were linked to increases in competency ratings (Cross et al. 2006). The above studies show a link between self-evaluation and motor experience, but they did not examine whether this change in evaluation was linked to specific kinematic information as some behavioral work has demonstrated (Maslovat et al. 2010; Buchanan 2015; Buchanan et al. 2015). If the evaluation changes in the fMRI studies were linked to the kinematics, then activity in the above brain regions may support the proficiency and competency ratings that were linked to relative phase and relative motion information described in this chapter.

## Take Home Message

The chapter outlined the merging of the visual perception perspective of observational learning with the coordination dynamics theory of perception-action processes. The studies reviewed revealed that relative motion direction and relative phase both serve as informational variables that may be picked-up through visual perception processes and be used to organize and constrain outgoing motor commands with minimal physical practice. Moreover, it was shown that the processes that support motor skill proficiency evaluations and competency evaluations are influenced by the same informational variables and that these "more cognitive type processes" are modified in a manner consistent with changes in pattern stability that emerge through practice, both observational and physical. Future research needs to examine more closely the different contributions of physical and observational training to motor skill acquisition (Buchanan and Park, in press). If the phrase "Watch how I do this" has any real meaning, then future research needs to determine when this phrase is most applicable in the learning process. Several studies reviewed in this chapter suggest embedding proficiency and competency ratings into training tasks will help reveal the similar and different contributions of observation and physical practice to motor skill learning.

## References

Al-Abood SA, Davids K, Bennett SJ, Ashford D, Marin MM (2001) Effects of manipulating relative and absolute motion information during observational learning of an aiming task. J Sports Sci 19:507–520

Bandura A (1977) Social learning theory. Prentice-Hall, Englewood Cliffs, NJ

Bandura A (1986) Social foundations of thought and action: a social cognitive theory. Prentice Hall, Englewood Cliffs

Bingham GP, Schmidt RC, Zaal FTJM (1999) Visual perception of the relative phasing of human limb movements. Percept Psychophys 61:246–258

Bingham GP, Zaal FTJM, Shull JA, Collins DR (2001) The effect of frequency on the visual perception of relative phase and phase variability of two oscillating objects. Exp Brain Res 136:543–552. doi:10.1007/s002210000610

Bourguignon M, De Tiege X, de Beeck MO, Van Bogaert P, Goldman S, Jousmaki V, Hari R (2013) Primary motor cortex and cerebellum are coupled with the kinematics of observed hand movements. Neuroimage 66:500–507. doi:10.1016/j.neuroimage.2012.10.038

Breslin G, Hodges NJ, Williams AM, Curran W, Kremer J (2005) Modelling relative motion to facilitate intra-limb coordination. Hum Mov Sci 24:446–463

Breslin G, Hodges NJ, Williams AM, Kremer J, Curran W (2006) A comparison of intra- and inter-limb relative motion information in modeling a novel motor skill. Hum Mov Sci 25: 753–766

Buccino G, Vogt S, Ritzl A, Fink GR, Zilles K, Freund H-J, Rizzolatti G (2004) Neural circuits underlying imitation learning of hand actions: an event-related fMRI study

Buchanan JJ (2015) Perceptual estimates of motor skill proficiency are constrained by the stability of coordination patterns. J Mot Behav 47:453–461. doi:10.1080/00222895.2015.1008687

Buchanan JJ, Dean N (2010) Specificity in practice benefits learning in novice models and variability in demonstration benefits observational practice. Psychol Res 74:313–320

Buchanan JJ, Dean N (2014) Consistently modeling the same movement strategy is more important than model skill level in observational learning contexts. Acta Psychol 146:19–27

Buchanan JJ, Kelso JAS, de Guzman GD (1997) Self-organization of trajectory information I. Experimental evidence. Biol Cybern 76:257–273

Buchanan JJ, Park I (in press) Observation and physical practice: different practice contexts lead to similar outcomes for the acquisition of kinematic information. Psychol Res. doi:10.1007/s00426-015-0723-4

Buchanan JJ, Ramos J, Robson N (2015) The perception-action dynamics of action competency are altered by both physical and observational training. Exp Brain Res 233:1289–1305. doi:10.1007/s00221-015-4207-y

Buchanan JJ, Ryu YU, Zihlman K, Wright DA (2008) Observational practice of a relative phase pattern but not an amplitude ratio in a multijoint task. Exp Brain Res 191:157–169

Buchanan JJ, Wright DA (2011) Generalization of action knowledge following observational learning. Acta Psychol 136

Byblow WD, Summers JJ, Semjen A, Wuyts IJ, Carson RG (1999) Spontaneous and intentional pattern switching in a multisegmental bimanual coordination task. Mot Cont 3:372–393

Calvo-Merino B, Ehrenberg S, Leung D, Haggard P (2010) Experts see it all: configural effects in action observation. Psychol Res-Psychol Forsch 74:400–406. doi:10.1007/s00426-009-0262-y

Calvo-Merino B, Glaser DE, Grezes RE, Passingham RE, Haggard P (2005) Action observation and acquired motor skills: an fMRI study with expert dancers. Cereb Cortex 15:1243–1249

Calvo-Merino B, Grezes J, Glaser DE, Passingham RE, Haggard P (2006) Seeing or doing? Influence of visual and motor familiarity in action observation. Curr Biol 16:1905–1910. doi:10.1016/j.cub.2006.07.065

Carroll WR, Bandura A (1982) The role of visual monitoring in observational learning of action patterns: making the unobservable observable. J Mot Behav 14:153–167

Carroll WR, Bandura A (1985) Role of timing of visual monitoring and motor rehearsal in observational learning of action patterns. J Mot Behav 17:269–281

Carroll WR, Bandura A (1990) Representational guidance of action production in observational-learning—a causal-analysis. J Mot Behav 22:85–97

Carson RG, Thomas J, Summers JJ, Walters MR, Semjen A (1997) The dynamics of bimanual circle drawing. Q J Exp Psychol 50A:664–683

Cattaneo L, Rizzolatti G (2009) The mirror neuron system. Neurol Rev 66:557–560

Cross ES, Hamilton A, Grafton ST (2006) Building a motor simulation de novo: observation of dance by dancers. Neuroimage 31:1257–1267. doi:10.1016/j.neuroimage.2006.01.033

Cross ES, Kraemer DJM, Hamilton AFd, Kelley WM, Grafton ST (2009) Sensitivity of the action observation network to physical and observational learning. Cereb Cortex 19:315–326

Cross ES, Liepelt R, Hamilton AFD, Parkinson J, Ramsey R, Stadler W, Prinz W (2012) Robotic movement preferentially engages the action observation network. Hum Brain Mapp 33: 2238–2254. doi:10.1002/hbm.21361

Cutting JE, Proffitt DR (1982) The minimum principle and the perception of absolute, common, and relative motions. Cogn Psychol 14:211–246

Cutting JE, Proffitt DR, Kozlowski LT (1978) A biomechanical invariant for gait perception. J Exp Psychol Hum Percept Perform 4:357–372

de Guzman GC, Kelso JAS, Buchanan JJ (1997) Self-organization of trajectory formation II. Theoretical model. Biolobical Cybern 76:275–284

Gibson JJ (1979) The ecological approach to visual perception. Houghton Mifflin, Boston

Haken H (1983) Synergetics. Springer, Berlin

Haken H, Kelso JAS, Bunz H (1985) A theoretical model of phase transitions in human hand movements. Biol Cybern 51:347–356

Hommel B (2004) Event files: feature binding in and across perception and action. Trends Cogn Sci 8:494–500. doi:10.1016/j.tics.2004.08.007

Hommel B, Musseler J, Aschersleben G, Prinz W (2001) The theory of event coding (TEC): a framework for perception and action planning. Behav Brain Sci 24:849–878. doi:10.1017/s0140525x01000103

Johansson G (1973) Visual perception of biological motion and a model for its analysis. Percept Psychophys 14:201–211

Johansson G (1976) Spatio-temporal differentiation and integration in visual motion perception. Psychol Res 38:27–63

Kelso JAS (1984) Phase transitions and critical behavior in human bimanual coordination. Am J Physiol 246:1000–1004

Kelso JAS (1994) The informational character of self-organized coordination dynamics. Hum Mov Sci 13:393–413

Kelso JAS (1995) Dynamic patterns: the self-organization of brain and behavior. The MIT Press, Cambridge

Kelso JAS, Buchanan JJ, Wallace SA (1991) Order parameters for the neural organization of single limb, multijoint movement patterns. Exp Brain Res 85:432–445

Kelso JAS, Holt KG, Kugler PN, Turvey MT (1980) On the concept of coordinative structures as dissipative structures: II. Emperical lines of convergence. In: Stelmach GE, Requin EJ (eds) Tutorials in motor behavior. North-Holland Publishing Company, pp 49–70

Kelso JAS, Pandya AS (1991) Dynamic pattern generation and recognition. In: Badler NI, Barsky BA, Zeltzer D (eds) Making them move: mechanics, control, and animation of articulated figures. Morgan Kaufmann, San Mateo, pp 171–190

Kelso JAS, Scholz JP, Schoner G (1986) Nonequilibrium phase-transitions in coordinated biological motion—critical fluctuations. Phys Lett A 118:279–284. doi:10.1016/0375-9601(86)90359-2

Kovacs AJ, Shea CH (2011) The learning of 90° continuous relative phase with and without Lissajous feedback: external and internally generated bimanual coordination. Acta Psychol 136:311–320

Kugler PN, Kelso JAS, Turvey MT, Stelmach GE, Requin J (1980) On the concept of coordinative structure as dissipative structures: I. Theoretical lines of convergence. In: Tutorials in motor behavior. North-Holland, New York, pp 1–47

Kugler PN, Turvey MT (1987) Information, natural law, and the self-assembly of rhythmic movement. Erlbaum, Hillsdale

Maslovat D, Hodges NJ, Krigolson OE, Handy TC (2010) Observational practice benefits are limited to perceptual improvements in the acquisition of a novel coordination skill. Exp Brain Res 204:119–130. doi:10.1007/s00221-010-2302-7

McAleer P, Pollick FE, Love SA, Crabbe F, Zacks JM (2014) The role of kinematics in cortical regions for continuous human motion perception. Cognit Affect Behav Neurosci 14:307–318. doi:10.3758/s13415-013-0192-4

Orlov T, Porat Y, Makin TR, Zohary E (2014) Hands in motion: an upper-limb-selective area in the occipitotemporal cortex shows sensitivity to viewed hand kinematics. J Neurosci 34: 4882–4895. doi:10.1523/jneurosci.3352-13.2014

Prigogine I, Stengers I (1984) Order out of chaos: man's new dialogue with nature. Bantam Books, New York

Rizzolatti G, Craighero L (2004) The mirror-neuron system. Annu Rev Neurosci 27:169–192

Schöner G, Haken H, Kelso JAS (1986) A stochastic theory of phase transitions in human hand movement. Biol Cybern 53:247–257

Schöner G, Kelso JAS (1988a) Dynamic pattern generation in behavioral and neural systems. Science 239:1513–1520

Schöner G, Kelso JAS (1988b) A synergetic theory of environmentally-specified and learned patterns of movement coordination. 2 component oscillator dynamics. Biol Cybern 58:81–89. doi:10.1007/bf00364154

Schöner G, Zanone PG, Kelso JAS (1992) Learning as change of coordination dynamics: theory and experiment. J Mot Behav 24:29–48

Scully DM (1986) Visual perception of technical execution and aesthetic quality in biological motion. Hum Mov Sci 5:185–206

Scully DM, Carnegie E (1998) Observational learning in motor skill acquisition: a look at demonstrations. Irish J Psychol 19:472–485

Scully DM, Newell KM (1985) Observational learning and the acquisition of motor skills: toward a visual perception perspective. J Hum Mov Stud 11:169–186

Turvey MT (1990) Coordination. Amer Psychol 45:938–953

Wilson AD, Bingham GP (2008) Identifying the information for the visual perception of relative phase. Percept Psychophys 70:465–476. doi:10.3758/PP.70.3.465

Wilson AD, Bingham GP, Craig JC (2003) Proprioceptive perception of phase variability. J Exp Psychol Hum Percept Perform 29:1179–1190

Wilson AD, Collins DR, Bingham GP (2005) Human movement coordination implicates relative direction as the information for relative phase. Exp Brain Res 165:351–361

Zaal FTJM, Bingham GP, Schmidt RC (2000) Visual perception of mean relative phase and phase variability. J Exp Psychol Hum Percept Perform 26:1209–1220

Zanone PG, Kelso JAS (1992) The evolution of behavioral attractors with learning: nonequilibrium phase transitions. J Exp Psychol Hum Percept Perform 18:403–421

# Rethinking the Study of Volition for Clinical Use

**Elizabeth B. Torres**

**Abstract** Volition, the acquired voluntary control of our actions (at will), requires from birth to development and beyond a proper balance across multiple layers of the nervous systems. These levels range from the autonomic, to the automatic, to the voluntary control level, providing as well taxonomy with phylogenetic order of appearance in evolution. In the past few decades of movement research at the behavioral and systems levels, there has been a paucity of studies focusing on the possible contributions of involuntary movements to volitional control. Moreover, the work focusing on voluntary behavior has given us a valuable body of knowledge about constrained and highly over practiced activities while work involving unrestrained, naturalistic behaviors remains scarce. Perhaps in making theoretical assumptions about our data acquisition and analyses without properly empirically verifying, these assumptions have left us with a somewhat skewed notion of how we think the brain may be realizing the neural control of bodily motions; a notion that does not exactly correspond to the outcome of the extant empirical work assessing unrestrained movements as the nervous system acquires them and modifies skillful behaviors on demand. This chapter takes advantage of new technological advances and new analytics to invite rethinking some of the problems that we study in movement science by enforcing somewhat oversimplified assumptions on the data that we model, acquire, and analyze. I show that by relaxing our a priori assumptions of normality, linearity and stationarity in data from biophysical rhythms of the nervous systems, we would gain better insights into the motor phenomena. Further, we would shy away from a "*self-fulfilling prophesy*" paradigm with a tendency to a priori handcraft the outcome of our inquiry. The new lens to study natural movements and their control includes as well involuntary motions that

E.B. Torres (✉)
Department of Psychology, Cognitive Science, Neuroscience,
Biomedical Engineering, Computational Biomedical Imaging and Modeling,
Rutgers University, Building: Psychology 113/116; CBIM (Lab),
152 Frelinghuysen Road, Piscataway, NJ 08854-8020, USA
e-mail: ebtorres@rci.rutgers.edu



J. Laczko and M.L. Latash (eds.), *Progress in Motor Control*,
Advances in Experimental Medicine and Biology 957,
DOI 10.1007/978-3-319-47313-0_13

take place largely beneath deliberate awareness. I present examples of solutions amenable to the habilitation and rehabilitation of volition in patient populations and discuss a new vision for movement science in light of making a seamless transition from volitional to intentional control of actions and thoughts.

**Keywords** Afference · Voluntary · Wearable sensor · Motor noise · Autism · Obstacle avoidance · Speed · Nonlinear dynamics · Bayesian

## Introduction

There are three scientists that have inspired the work that I will refer to in this chapter: Nikolai Bernstein (Russia, 1896–1966), Erich von Holst (Germany, 1908–1962), and Esther Thelen (US, 1941–2004). The legacy that we inherited in the body of work that they created can be appreciated when we connect their theoretical frameworks and examine the concept of volition as the foundation of action ownership, agency, and cognitive autonomy over our physical actions and mental thoughts.

Starting with the developmental stages of a newborn baby, Thelen provided the framework to rethink infant neurodevelopment as a process that could be better understood in the context of complex nonlinear dynamical systems (Smith and Thelen 2003). Following up in human development, with research involving participants throughout adulthood, Bernstein posed two fundamentally interrelated problems of motor control that have a direct impact on volition from the start of life and beyond. One refers to the inherent variability in our actions and the other to the abundant degrees of freedom (DOF) in our bodies (Bernstein 1967). I posit that the "glue" connecting these ideas to enable the possibility of casting movement sensation as an integral part of cognitive processes is the principle of reafference by von Holst and Mittelstaedt (1950) stating that "*Voluntary movements show themselves to be dependent on the returning stream of afference which they themselves cause.*" This principle is used here to show examples of how to habilitate, rehabilitate, and enhance volition across different human populations.

Through the lens of a new statistical platform for the personalized analyses of physiological rhythms underlying different layers of control within natural behaviors, I will underscore the importance of the empirical estimation of variability in the bodily rhythms, the relevance of scaling up DOF in our experiments and analyses, and the overall notions that a complex, nonlinear process with multiplicative statistics shall not be treated as an oversimplified, linear phenomenon with assumed additive statistics. Careful consideration should be given to these features inherently present in our data. This is particularly so when registering physiological rhythms. Here, we aim at applying the ideas that these three giants of movement science provided us with in order to further our science and drive clinical research.

## Methods and Motivations for Their Use

### *General Geometric Framework*

The excess DOF problem posed the question of spatial consistency in the motor output. How do we achieve a uniquely consistent movement path solution, given a set of goals and context? How do we do so despite the large number of possibilities that the body affords its end effectors? Along these lines, one could also ask how to attain consistency in the temporal profile to travel along a stable spatial path. The relationship between these two aspects of a movement trajectory is not as trivial as it has been assumed. The general assumption has been that the output trajectory already contains information on the systematic path in space and also on the timing to travel along that path. The implicit assumption of many models of motor control is that the time duration of a motion to be performed is a priori known and realized in the motor trajectory (e.g., Flash and Hogan 1985; Uno et al. 1989; Biess et al. 2007, 2011, among others).

If a nervous system is trained to move within a given time constraint, it will do so with great precision. Yet the very learning process that takes place varies greatly along the two spatial and temporal components of the motor output trajectory (Torres and Andersen 2006) and recruits different classes of neurons in the posterior parietal cortex during the planning stage, before the movement execution (Torres et al. 2013c). Within the pointing, reaching and reach-to-grasp family of goal-directed motions, such variations in temporal profiles occur along a highly conserved spatial hand path (Torres 2010) but also along a conserved postural path despite the abundant DOF (Torres and Zipser 2002, 2004; Torres and Andersen 2006) (Fig. 1a). Indeed, the speed independence in goal-directed motion paths (Atkeson and Hollerbach 1985; Nishikawa et al. 1999) has been difficult to reconcile with the theoretical assumption that the timing of the motion is known a priori. Somehow one would need a ticking clock set somewhere to make that idea fit the empirical data.

Empirical data collected under unrestrained conditions, or without a priori constraining the system to move within a predefined time period does not follow such a pattern. Instead, when the system gradually learns, it seems to arrive at a notion of time duration for a movement segment in a gradual way, by adjusting different parameters as a function of task demands. For example, in obstacle avoidance the distance traveled along an initial segment of the movement that falls within 200–300 ms, before passing by the obstacle, seems to matter more than the remaining distances in the initial stages of learning (Torres and Andersen 2006). Then further adjustments along other sub-movements past the time to the firs velocity peak, are performed till the endpoint is reached (Fig. 1a). Such adjustments made at the level of the extrinsic path traveled by the end effector, correspond to intrinsic postural adjustments. These intrinsic adjustments occur along multiple DOF, recruiting and releasing them on demand. Under certain conditions (Torres

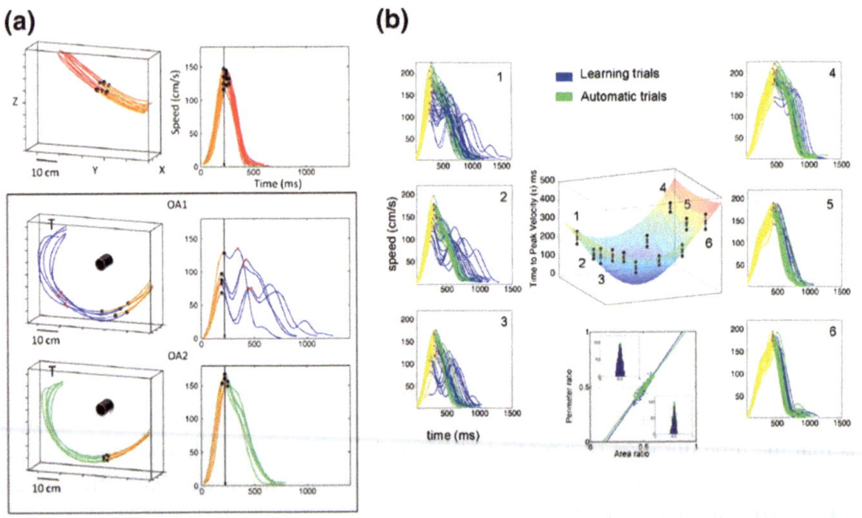


**Fig. 1** Speed invariance of some kinematics parameters during the learning of obstacle avoidance. **a** Straight reaches to a visual target are rather automatic after training. They follow a *bell-shaped* speed profile with a very systematic timing whereby the peak velocity (marked by a *black dot* on the hand trajectory and on the corresponding speed profile) is reached at a similar time and special location along the hand path despite random performance across 15 targets where this is one of them. The curved trajectories to the same target as in the straight reaches conserve their curvature but have highly variable distances traveled along the initiation of the reach, up to the peak velocity. *Blue curves* are early in the first five trials whereas *green* are later in the last five trials of the session. During the learning period the movement time was rather variable, yet the time to reach the peak was stable relative to the variations in the distance traveled during that time. **b** Six different targets marked on the surface representing the time to the peak velocity corresponding to each of 12 spatial locations, with error bars representing the deviations from the estimated mean. *Inset* shows a speed invariant symmetry. This symmetry is a geometric index conserved despite the trial-by-trial dramatic changes in movement dynamics (As in (**a**), *blue* is earlier learning and *green* is for later trials when the obstacle avoidance skill was acquired). Panel A previously published in (Torres et al. 2013c); panel B in (Torres 2013b)

and Andersen 2006), it is possible to locally linearize the forward and inverse maps connecting the intrinsic and extrinsic parameters. This in turn reveals the variable involvements of the dimensions spanning the kernel and the null subspaces of the local linear transformation. The moment-by-movement variations in the intrinsic kernel and the null subspaces have been interpreted as task-relevant and task-incidental or supplementary subspaces (Torres 2001), respectively; but also as those corresponding to the controlled or uncontrolled manifolds (Scholz and Schoner 1999; Latash et al. 2002), respectively (albeit differences in the methods used by these researchers to decompose the overall space spanned by all the DOF).

Besides the map linearization, there are other ways to examine variability in movement parameters in light of geometric relationships between desired goal-driven and internally realized displacements. It is possible to identify some parameters whose variability may be less perturbed by changes in the movement speed than other parameters (Torres and Zipser 2004; Torres 2010) (Fig. 1b). Some methods may consist of deriving law-like or invariant indexes that one can probe to learn if their invariance emerges as mere byproduct of the system's biomechanical physical constraints; or if they are subject to some level of voluntary control by the nervous system (Fig. 2). We have found that some of these dynamic-invariant quantities tend to break down in the face of brain injury due to stroke or degeneration. They also tend to recover their typical ranges in different patient populations, and do so selectively, under specific forms of sensory guidance (Fig. 2). In this sense, these invariants serve to unveil possible avenues to aid the system heal under cognizant external guidance. They may also reveal specific forms of guidance adequate to help the system internally heal itself (Torres et al. 2010, 2011, 2014).

Along the lines of variability registered at the endpoint of a task, a common practice has been to consider such variations in isolation. Considering the endpoint scatter alone, without a proper characterization of the accumulated variational path in the unfolding preceding motion, from trial to trial, implicitly assumes that intermediate points in the path are not as important (Harris and Wolpert 1998). One gets the sense that the system somehow plans a posteriori. But this notion seems counterintuitive in light of the nervous system's need to compensate for signal transduction and transmission delays in the sensory and motor nerves. More importantly, within the Bayesian estimation framework used as well to characterize the final scatter, one never learns how the nervous systems' "sensory-motor priors" developed to maturity in the first place (Hamilton and Wolpert 2002; Kording and Wolpert 2004a, b). Indeed, Bayes inference seems very suitable as a framework to pose several standing problems in motor control. It may even serve to connect key elements of decision-making research in perceptual and cognitive sciences with elements of motor performance (Kording and Wolpert 2006); but as it stands the paradigms in use do not yet question how the nervous system may acquire the various required components to realize this model during development; how it may modify them on demand; or how it may altogether change them in the face or injury. Without this information obtained from actual empirical data in a way that is unrestrained by the theory, the basic science behind such methods is difficult to apply to contemporary issues in the clinical arena.

The prevailing analyses and modeling of sensory-motor phenomena under implicit assumptions that noise in our nervous system may be Gaussian-like begs the question of the true nature of the empirical distributions characterizing movement parameters across different levels of control. We will soon see that these distributions have heavy tails and that those heavy tails ought to not be squashed with a power transformation to turn them normal and enforce linearity and normality on the data. If one truly seeks to understand the nonlinear and multiplicative nature of the processes underlying naturalistic actions of daily living, we may have to empirically validate our theoretical assumptions. We may have to preserve the inherent features of the original data. It may be the case that what we consider noise today actually contains the relevant signal.

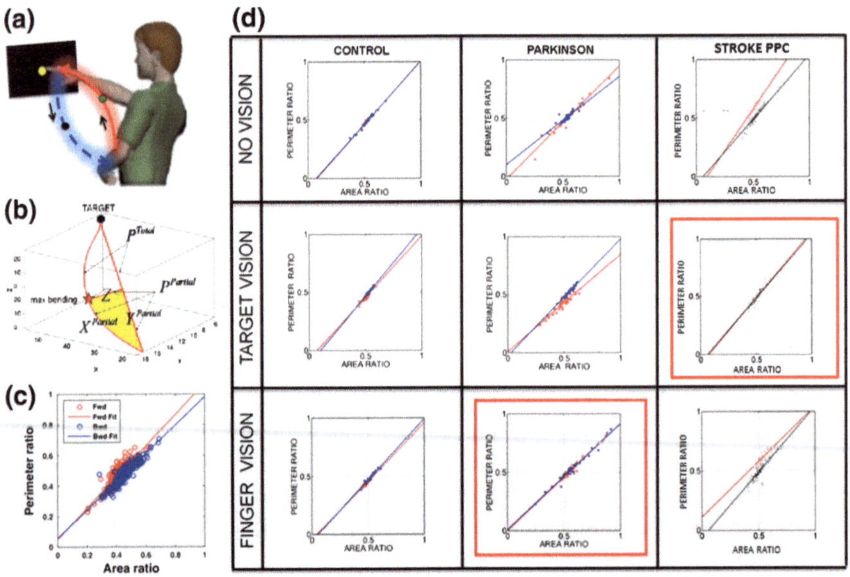


**Fig. 2** The breakdown of the geometric symmetry in different patient types along with the recovery of the symmetry using different types of sensory guidance during a pointing task. **a** Pointing task used in this experiment consisted of a continuous stroke forward and back to the target. Landmarks are the point of peak velocity in each stroke along the hand trajectory. **b** A geometric quantity (the area ratio) from the hand trajectory can be obtained in relation to the Euclidean *straight line*. The *red star* represents the point where the motion-curve maximally bends with respect to the line. The *yellow* area is then divided by the total area between the two unit speed curves. An analytical expression (partial differential equation in Torres 2001) gives geodesic curves for which this quantity is exactly 1/2. Likewise the perimeter ratio is obtained dividing the partial perimeter around the *yellow* area and the total perimeter of the two curves combined. This quantity when generated analytically is 1/2. These symmetries can then be empirically obtained to ascertain the degree to which the actual data deviates from the ideal geodesic conditions. **c** Example from one typical control showing the scatter around the ideal point (1/2, 1/2) from hundreds of trajectories forward to the target (*red*) and backwards away from it (*blue*). **d** Data from controls and patients under different forms of visual feedback while performing the task in (a). *First column* shows the performance of a representative control (each *dot* represents a trial). The tight scatter fit by the line of unity around the symmetry is conserved with similar slope for forward and back movements across all conditions (*no vision*, relying on the memory of the target; *vision of the target*, relying on the actual continuous visual feedback of the target; *finger vision*, relying on the continuous vision of a light emitting diode affixed to the moving finger while the target is off). Parkinson's disease patients (along the *second column*) broke the symmetry but recovered it with vision of the finger (Torres et al. 2011), as proprioception aligned with vision of the moving hand/arm linkage. A stroke patient with a left PPC focal lesion (*third column*) benefits instead from the vision of the target (Torres et al. 2010)

## *Wearable Sensors: Free to Move at Last*

Perhaps our ways of thinking about the neural control of movements have been shaped somehow by the instrumentation available; or perhaps we built the instruments constrained by our ways of simplifying movements that we can model within some framework that we may think is appropriate to address the problem at hand.

The use of a two-dimensional robotic linkage to probe arm movements prevails throughout behavioral and systems neuroscience studies that aim at understanding the mechanisms underlying the neural control of movements. Over the years, I have asked several questions to many of the researchers behind these studies. These have been, for the most part, intuitive questions, questions driven by common sense and seeking to build a solution to the motor control problems that stem from first principles. I was surprised by the answers I got from everyone I asked. It was as though the answers to the very questions they were asking were already there, facilitated by the methods used in data acquisition and/or by the theoretical framework of choice. In reading the methods of these papers, I knew beforehand what the answer was going to be under those conditions, and so did those who I asked, but that, in my eyes, takes away the fun of doing science.

As someone trained first in Mathematics and then in Applied Mathematics, I knew the dangers of taking a model created to solve a particular problem and applying that model (somewhat blindly) to another problem domain without properly verifying its feasibility or the main underlying assumptions. As an example to illustrate my point, I had already done some work using Pontriagin's Optimal Control framework when I started graduate school and took an interest in the field of motor control. I knew very well that one of the reasons for its invention was to be able to travel to space and back in optimal time because fuel expenditure was of the essence. It was a great solution to that problem (Pontriagin 1990).

Bringing that notion to frame the human motor control problem required solving a time-dependent and time-constrained cost functional. The solution, if one can be obtained in closed form, is an optimal trajectory providing a movement prescription in space and time. But that could not possibly be conducive of a solution that a newborn infant spontaneously came up with. This framework, so good to model the control of mechanical systems, took away the trial and error element in the whole learning process of an infant. In my eyes, the newborn using this model would have to know a priori (or be prompted to learn) how long each movement segment took. And yet, that is precisely part of what the newborn infant *spontaneously* comes to learn. It is an emergent property, rather than an a priori property of that system.

The optimal control framework was not good enough to understand the essence of the problem at hand or to capture the true nature of the empirical data obtained from unrestrained motions. Something more fundamental was missing: the missing ingredients were the variations and their rates of change at all levels of control. How can one study such variations and characterize their signatures as the person behaves throughout the day? And how can one do so as the many segments of the body spontaneously co-articulate forming different patterns? And as the ebb and flow of deliberate or automatic autonomy in our actions changes from moment to moment? I did not think that posing the problem of inverse dynamics the way I had read it in the literature was going to help me think of the essential problems the nervous system may face during development. Something else was needed.

The advent of wearable sensing technology that simultaneously registers signals from various layers of the nervous systems (Fig. 3) made a dent in the research program in my lab. From cell phone accelerometers (Torres 2013a) to research-grade wearables (Torres 2013b), we set to develop new analytics and data types that would enable the integration of multisensory data from different control levels to create new ways to track the noise-to-signal transitions across the proposed control taxonomy of the nervous systems (Fig. 3a) in a rather personalized manner. We set to also do so in social situations, such as those requiring dyadic interactions.

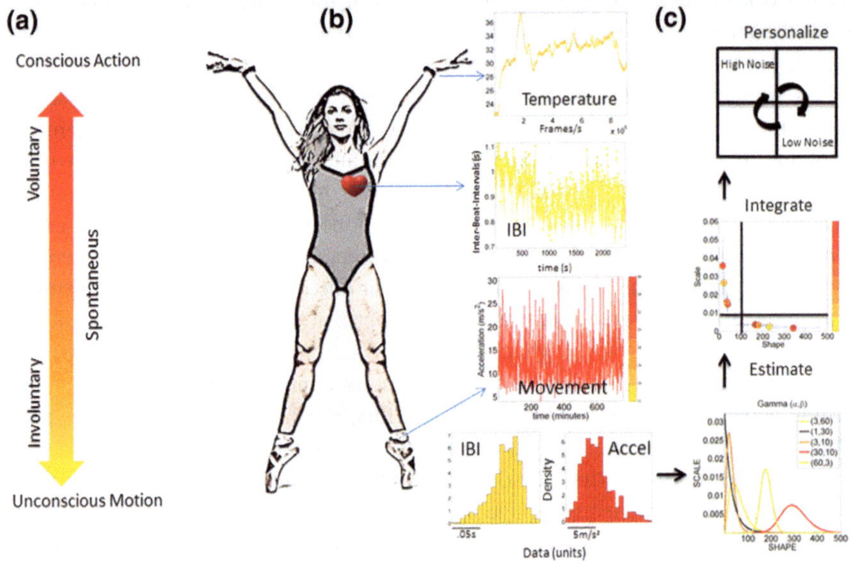


**Fig. 3** Proposed taxonomy of control levels defining volition as the proper balance in variability across the levels (Torres 2011). Different waveforms are registered using contemporary wearable sensing technology during unrestrained, naturalistic motions. This technology enables the study of the different levels of control, spanning from autonomic (e.g., heart rate, inter-heartbeat-interval IBI, temperature, etc.) to automatic (e.g., gait patterns captured with accelerometers and gyroscopes), and other motion signatures of actively produced movements under deliberate control (e.g., reaches, ballet, boxing, etc.). The fluctuations in motor performance can be extracted from each waveform and integrated (e.g., motion and temperature) to estimate in a personalized manner the transitions from spontaneous random noise to well-structured signals with high statistical regularities conducive of a predictive code (see next figure)

We wanted to be able to track those noise-to-signal transitions within the time limits of the lab experiments (Nguyen et al. 2014; 2015) and longitudinally across several months at a time (Torres and Lande 2015). We wanted to learn about the temporal dynamics of the system's inherent layers of variability without presupposing anything about it. These types of data were not available in the literature of motor control. Thus, to build our new statistical platform, we needed to build such datasets ourselves under unrestrained conditions, as people go about their lives any given day.

## *General Statistical Framework*

The variability inherently present in our movements, as pointed out by Bernstein, became the cornerstone of contemporary research in motor control. Many groups provided evidence for the importance of considering the variability in our movements (Latash et al. 2002; Scholz et al. 2003; van Beers et al. 2004, 2013; van Beers 2009). In particular, work related to noise in various areas of the nervous system (van Beers et al. 2002; van Beers 2009) was very important to turn the field around and begin to explore new empirical and theoretical questions (Harris and Wolpert 1998; Faisal et al. 2008). The stochastic feedback control model (Todorov 2005, 2009) became one of the pillars of contemporary motor control and these ideas extended as well to studies of monkey electrophysiology (Churchland et al. 2006a, b, 2010).

Despite the undeniable importance of motor variability, and the popularity of its exploration across many areas of movement research and systems neuroscience level, there was still something missing in the overall conceptualization of the problem and in the ways that theoretical assumptions were being made. Specifically a "*one-size-fits-all*" model (Fig. 4a) was being used to analyze the time series data including a variety of waveforms (ranging from spikes to local field potentials to kinematics) registered with different resolutions per instrumentation to capture brain and bodily rhythms. In essence, the data epochs gathered across trials were being averaged under the assumption of normality and homogeneity of variance. One could then speak of the "mean behavior" corresponding to the "mean mental activity." Under such assumptions, variability always referred to symmetric deviations around that assumed (theoretical) mean. Yet the very fluctuations of the waveform were lost as noise or as a nuisance (Fig. 4a). We kept speaking of an ideal expected value that we assumed, but then we never quite empirically challenged that assumption.

In a series of papers published from my lab (Torres 2011, 2012, 2013a, b; Torres et al. 2013a, b, d; Nguyen et al. 2015; Torres et al. 2016), we have shown that when empirically estimated, rather than assumed, the probability distribution functions

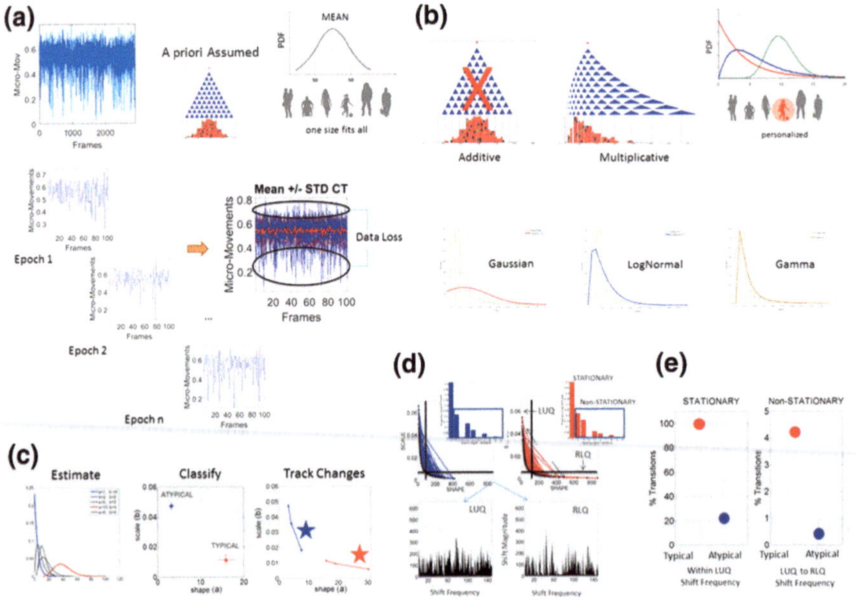


better characterizing the various degrees of fluctuations in motor performance across different levels of control, have in fact a range of skewness and dispersions that depend on many important factors (e.g., Fig. 4b). These factors range from context, fatigue and adaptation to expertise, to the type of movement the system performs (deliberate vs. spontaneous, or automatic vs. autonomic, etc.) among others. In the clinical arena, we discovered that such signatures were selectively different according to different pathologies of the nervous systems (Torres et al. 2016). Indeed, by shifting to a personalized approach (Figs. 3c and 4b) whereby the stochastic signatures of the person's fluctuations in motor performance are empirically estimated (rather than assumed) we avoided throwing away those variations as noise in performance.

One of the lessons learned over many years of research was that even when a statistical framework may be sound and adequate to study some physical phenomena, it may not be adequate to apply that framework to other phenomena; much less to do so blindly without any empirical verification, or under such restrained conditions that the derived empirical data is utterly skewed or misleading. As experts in the field of statistics have pointed out, it is important to choose the proper framework for analyses and test its validity, rather than merely assume it (Limpert et al. 2001; Limpert and Stahel 2011).

◂ **Fig. 4** The new approach to motor control is contrasted to the traditional methods. **a** Suppose that a time series of a waveform is gathered with high resolution instrumentation and some 3000 frames are gathered. The traditional model would epoch the time series (e.g., 100 frames per second each trial) and average all trials under the a priori assumption of normality. This model examines the motion data without empirically evaluating the Gaussian assumption and merely assuming the theoretical Gaussian moments. Data loss is noted by the ellipses as fluctuations are smoothed out as 'noise' when in reality they contain signal. This is the "*one size fits all approach*" currently dominating the field. **b** Suppose now that we do not assume additive statistics a priori, then the same time series is used to empirically estimate the probability distribution that the fluctuations give rise to along the continuous stream of motion. To this end one can continuously gather the peaks of the normalized waveform (values ranging from 0 to 1) and obtain a frequency histogram, then compare across different families of probability distributions which one fits the data best (e.g., using maximum likelihood estimation). This estimation can be done for each individual using the same number of frames. Since different motion types across different components of the nervous systems will span different probability distribution functions (PDF), a person is really characterized by a whole family of PDFs and the rates of change of the stochastic signatures from context to context will help empirically define truly *personalized* adaptive indexes in a stochastic sense, preserving as well the inherently nonlinear dynamic nature of the motion data. **c** Examples of estimated family of PDFs using the continuous Gamma family and classifying each individual signature on the Gamma parameter plane spanned by the shape and the scale (dispersion) of the distributions with 95 % confidence intervals. As the signatures are nonstationary they can be reestimated to track changes on the plane as a stochastic trajectory. These changes reveal the evolution of the noise-to-signal ratio which on the Gamma case is the scale parameter. **d** Computation of the noise-to-signal transitions (changes in amplitude and frequency) based on the switching between quadrants automatically defined by the median values of the estimated parameters (*black lines* are the median values across all the shape and scale empirically estimated parameters from taking blocks of $x$ number of frames and a sliding window of $y$ number of frames, where $x$ and $y$ depend on the sampling resolution of the sensors: for example at 240 Hz, with block size of 1 minute and sliding window size of 1/2 minute it is possible to provide statistical power for very tight confidence intervals for the estimates, i.e., negligible error relative to the bandwidth of the data range). Colors are coded as in **c**. The points are joined by lines in the order in which they occur to form stochastic trajectories. The *insets* show the histograms of points in the Upper Quadrant LUQ (high noise, high skewness toward the exponential limiting case with shape value of 1) versus those in the Right Lower Quadrant RLQ (low noise, toward symmetric shapes with the Gaussian distribution as the limiting case). The *insets* are histograms counting the transitions of the signatures within each quadrant (stationary transitions) and between each quadrant (nonstationary transitions). The peaks are another way to visualize the magnitude of the shifts on the Gamma plane. **d** Quantifications of the frequency of noise-to-signal transition from high noise-to-signal LUQ to low noise-to-signal RLQ. In this example, the atypical example had very low percentages of noise-to-signal transitions between the quadrants indicating stagnation in motor adaptation to physical growth and signaling poor development of neuromotor control. Note that panel **e** criteria serve as a classification index of typical versus atypical neurodevelopment. Panel D published in (Torres et al. 2016)

## Applications: Empirical Evidence

### *Recasting Thelen's Legacy and Von Holst Principle of Reafference by Considering the Maturation of Sensory-Motor Noise*

Sometime in the Spring of 1999, while I was a graduate student at the Department of Cognitive Science at UCSD, one of my Professors, the late Elizabeth Bates, introduced me to a lovely lady who did movement-related work in the context of infant development. I spent an entire afternoon engaged in one of the most memorable conversations of my career. She was Esther Thelen. I had never studied how babies learned to move, yet her ideas instantly resonated with my own way of thinking about many of the motor control problems that I was interested in. Thelen's ideas certainly left such a mark in my way of thinking about these problems, that today I cannot envision the study of motor control without considering her dynamical systems perspective on neurodevelopment as a framework to better understand the intertwined relations between motor control and the emergence and maintenance of cognition (Thelen and Smith 1994).

## The Newborn Baby: When Spontaneous Random Noise Does not Transition into Signal at Rates Congruent with the Rates of Physical Growth

The importance of considering Thelen's notion of the nonlinear dynamical nature of physical growth and motor maturation can be appreciated in Fig. 5a–c extracted from data used to build the World Health Organization (WHO) growth chart (Kuczmarski et al. 2000, 2002). This chart is the gold standard guide that Pediatricians use to follow the newborn's developmental trajectory from birth to 5 years of age. The data from this chart is currently used in combination with discrete clinical rating scales that measure (among other things) the baby's readiness to walk (e.g., the AIMS questionnaire Piper et al. 1992). They may be administered occasionally across monthly or bimonthly visits to the doctor. If instead of such sporadic discrete visitations and evaluations, parents could have a way to continuously and more frequently measure the baby's progression, there would be a higher probability of detecting risk of neurodevelopmental derailment early on, to immediately intervene. Currently the official diagnosis of a neurodevelopmental disorder is given around 3 years of age and it is only then that an "early intervention program" is provided to the parents. All that precious time prior to the official diagnosis and intervention is wasted.

The accelerated rate of change of body growth experienced in the first month of life (Fig. 5a–c) paired to the nonlinear nature of the rate of change in the noise-to-signal transitions that we captured with high-resolution sensors facilitated the very early detection of risk for neurodevelopmental stagnation. Indeed, I have been informed that several babies that were studied in this project and went on to receive months later a clinical diagnosis involving a neurodevelopmental disorder had already been flagged to be at risk several months earlier by our personalized statistical platform. This result has encouraged us to begin the validation of these biometrics in hundreds of babies, an ongoing project in my lab today.

Instead of assuming normality in our analyses, we tracked the non-stationary nature of the inherent statistics underlying the transitions from spontaneous random noise to well-structured noise, to noise with statistical regularities conducive of a predictive code. We characterized, using the changes in the amplitude and frequency of the noise-to-signal transitions of the baby's stochastic signatures of leg motions, the transition from spontaneous random movements to movements under voluntary control (Fig. 4d).

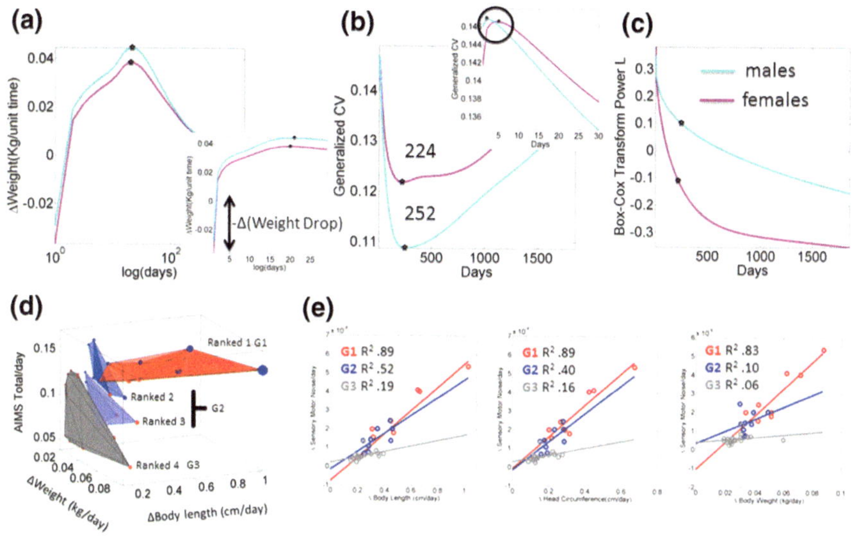


**Fig. 5** Non-linear dynamics and fast evolving stochastic signatures of early development in newborn infants shall not be treated as linear processes with a priori assumed stationary normality **a** Data from newborn infants (13,362 males and 13,623 females) extracted from the longitudinal tracking of physical growth to build the World Health Organization (WHO) growth charts (Kuczmarski et al. 2002). The changes in weight of the newborn infant follow an accelerated rate of change with initial negative (drop) followed by an accelerated (daily) increase. *Inset* zooms into the first month of life and shows the differences between males and females. The *black dot* signals a critical point in the statistical regimes of variability according to the reported generalized coefficient of variation tracked in **b**. *Inset* shows an earlier critical point in the first week of life which separates boys and girls. This separation is even more evident later on at 224 days for females and 254 days for males (a month later). Notice that this critical change in stochastic variations so evident in this rate of change data remains hidden to the pediatricians using absolute rather than incremental weight values. **c** The WHO charts is built using the percentiles of enforced normal distributions. Yet the *L* parameter of the power (Box-Cox) transformation reported in the methods to construct these charts shows the changes in the skewness of the actual weight distributions. Notice how improper it is to enforce the normal distribution in the nonlinear physical growth data of newborn infants. **d** The rates of change in physical growth of 36 newborn infants (12 typically born and 24 prelabeled clinically at risk) who were followed for 6 months. The physical growth parameters (head circumference, body length and body weight) divided by the number of days since birth are median ranked. The *red* group is composed of the typically developing infants with the highest rank in all three parameters. The *z*-axis is the Alberta Infant Motor Scale (AIMS) a metric of readiness to walk (higher is better). The size of the circle is the rate of change in head circumference (cm/day). *Blue circles* are typically born babies while *red circles* are babies at risk (premature). The other surfaces are fit to scatters of infants ranked in orderly way lower than the first group. The gray scatter is the lowest ranked. **e** The rates of change in the frequency of noise-to-signal transitions plotted as a function of the rates of change in physical growth for three groups: G1 is the Ranked 1 group (*red*), G2 is the Ranked 2 and 3 groups combined (*blue shades*), G3 is the Ranked 4 group (*gray*). Babies in the highest ranked group G1 of physical growth show a linear relation with the rate of change in stochastic signatures of neuromotor control. Babies in the group at high risk of neurodevelopmental derail are the lowest ranked (*gray*) group in **d**. They are the ones with the flat slope, denoting stagnation in the development of motor control (motions measured with inertial measurement units, accelerometers and temperature at 128 Hz, for 8 h continuously in each of 3 visits every 2 months). The intermediate group G2 falls between the two extremes (Torres et al. 2016)

Babies that developed normally had rates of physical growth congruent with the rates of change of scores indicating the readiness to walk (Fig. 5d). Their patterns of motion were further characterized by fluctuations that transitioned from noise to signal with high frequency and large amplitude of shifts between quadrants in the statistical Gamma parameter plane. As their body grew at an accelerated pace, so did the changes in neural motor control tending to more frequently organized patterns with lower rates of spontaneous random noise and higher rates of systematically predictive signals. Whenever the rates of change in such noise-to-signal patterns became stagnated in relation to those of the physical bodily growth, the relationship uncovered for typically born and typically developing babies broke down in the babies at risk (Group 3 in Fig. 5e ranked 4 in 5d).

As the babies' motions transitioned from spontaneous and random to more organized and systematic they denoted the emergence of voluntary control. Then in due time, we referred to von Holst's notion that at the motor output level it is also possible to estimate the nature of "the returning afferent stream," which these emergent voluntary movements themselves cause. We know today that this evolving relationship between noise-to-signal transitions and physical body growth must occur at congruent rates, changing in tandem if development is to occur along the typical pathway (Fig. 5e). When this does not occur it is *very likely* that in due time we will be in the presence of a neurodevelopmental disorder, perhaps even more specifically of one labeled autism.

## The Stagnation of Maturation in Motor Noise: Autism Spectrum Disorders

Autism has been described as a disorder of the mind, a cognitive disorder where the social and communication skills that typically emerge in young children by the age of 3–4 years old does not appear in a child that goes on to receive this diagnosis. The child lives in social isolation, often bullied and excluded by peers that do not understand that this child senses the world differently and as such acts differently. The child is trapped within routines and rituals that do not correspond to social expectations. Yet, very likely those odd "ritualistic behaviors" help dampen the high uncertainty of the external social world. We learned through our research that most likely such rituals also keep the child safe from the extremely high uncertainty that his/her self-generated motions contain. These motions persistently have high random and noisy signatures. The proposition by von Holst that "*voluntary movements show themselves to be dependent on the returning stream of afference which they themselves cause*" is violated in autism. Their signatures—better fit by the memoryless Exponential distribution—denote the occurrence of every event anew, with no accumulation of information conducive of a predictive code. Figure 6a shows the comparison of the empirically estimated PDF's for autism and controls. The much wider dispersion and lack of diversification of shapes across the population of different ages denote the absence of maturation in autism. Unlike typical development showing the transitions in noise-to-signal ratios and the evolution in the shapes of the distributions from highly skewed to symmetric ones, autism statistical signatures are "stuck."

Indeed, in autism the signature of well-organized, systematically predictive signal indicative of voluntary control has not appeared by 4 years of age; is absent at 10 years of age; is not there either at 25 years of age and beyond. In the context of 176 individuals whose motor output stochastic patterns we have characterized during goal-directed reaches (Torres et al. 2016), the presence of high noise and highly skewed probability distributions tending toward the memoryless (most random) exponential distribution in autism remains prevalent throughout all ages (Brincker and Torres 2013). This prevalence remains as well to some extent when there is some capacity for language, but never reaches the typical ranges of noise-to-signal ratios quantified in controls. Those children who acquire some language (some phrases, some echolalia, etc.) may show to some degree less dispersion in the distributions empirically estimated from the fluctuations in motor performance. Yet, even at young adulthood, these signatures resemble more those of a typically developing 3-year-old child than those of a young adult peer. Figure 6A shows this result in two panels (taken from Torres et al. 2013a). Something is amiss in the course of maturation of the sensory-motor systems. As in the babies at risk of neurodevelopmental derail, these individual's stochastic signatures of fluctuations in motor performance of the voluntary kind (from goal-directed motions) lack the proper frequency in the transitions from noise to signal. There is stagnation in the evolution of their noise patterns. Excess noise and randomness of those patterns is the norm in this ASD population. According to these statistical signatures, the individual with ASD must live "in the here and now."

## The Involuntary Motions of ASD Individuals in the fMRI

It has been extremely challenging to do sensory-motor research in autism. Cast as a mental disorder in the Diagnostic Statistical Manual of Mental Disorders (DSM-5 version) of the (American Psychiatric Association and DSM-5 Task Force 2013), the manual does not include criteria based on physiologically driven biometrics. Movements and their sensations are out of the question in mental illnesses (Rogers 1992).

The DSM-5 criteria for diagnoses are purely based on interpretation of the behaviors that the clinician thinks s/he observed. There is no standardized scale to be able to integrate continuous outcome measures, such as those that we use to measure behaviors by their underlying fluctuations in motor performance, and discrete numbers that a clinician may designate to one isolated aspect of the behavior. Yet, physiologists are asked to correlate continuous data from physical measurements with discrete scores from clinical inventories. Mathematically this is very inappropriate because the linear regression methods in use require multivariate normality—blatantly violated in many cases; they also require linearity which is as well violated in many bodily related signals by virtue of the redundancy and

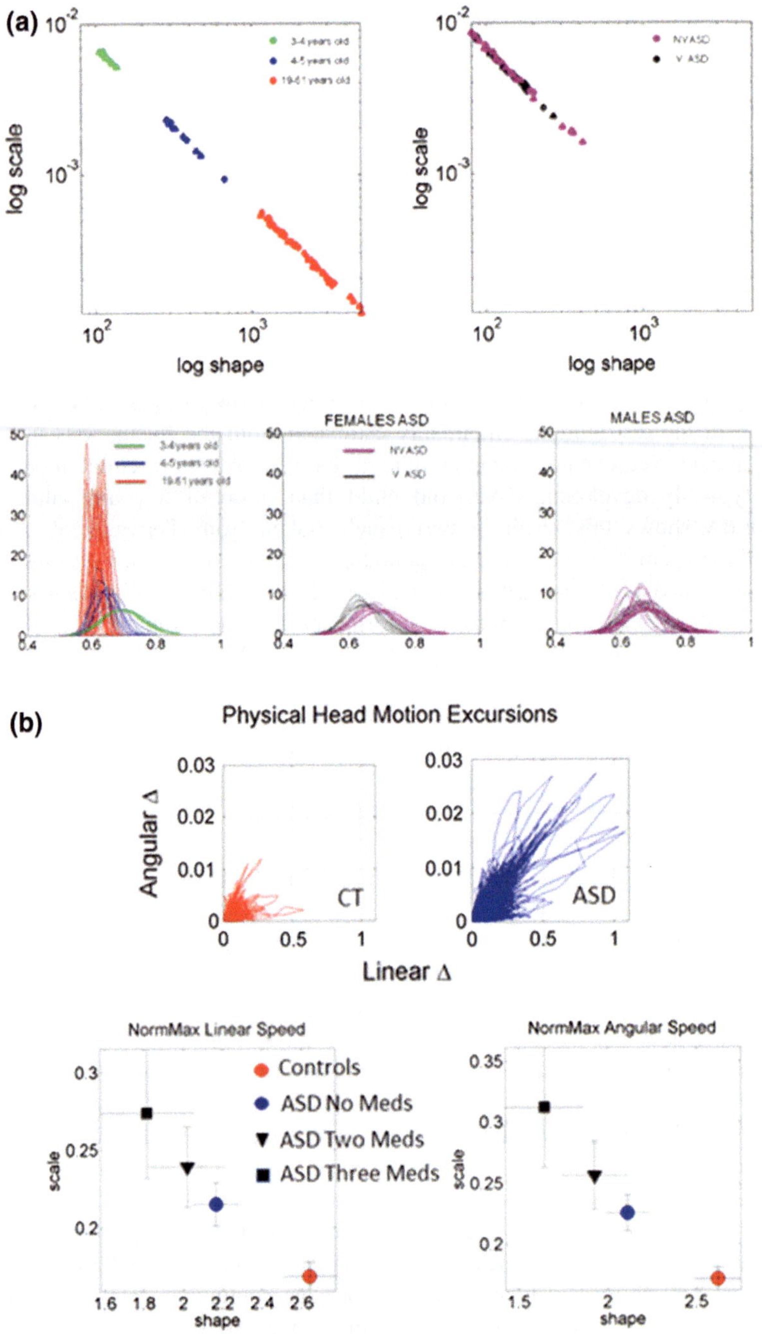
(a)
log scale
log shape
3-4 years old
4-5 years old
19-61 years old
NVASD
V ASD
FEMALES ASD
MALES ASD
(b)
Physical Head Motion Excursions
Angular Δ
Linear Δ
CT
ASD
NormMax Linear Speed
NormMax Angular Speed
Controls
ASD No Meds
ASD Two Meds
ASD Three Meds
scale
shape

◀ **Fig. 6** Stagnation in the noise levels of individuals with a diagnosis of autism spectrum disorders across life and meds influence in their motor control. **a** Typical development of the stochastic signatures derived from goal-directed behavior shows an evolution from random noise to systematic signal in typically developing individuals. In particular, the transition from 3 years old (*green*) to 4 years old (*blue*) shows a diversification in the shapes and dispersions of the PDFs that remains absent from the individuals with ASD, regardless of age, sex, or verbal capabilities. By college age the controls have very low dispersion and symmetric distributions. The presence of higher noise (broad dispersion) in ASD does not go away with age. There is no maturation as in the controls. **b** Involuntary micro-motions of the head captured in the fMRI and reported in the ABIDE databases. Actual physical excursions of the head in controls (CT) and ASD are captured by linear displacement and angular rotations throughout the scanning session. The excess involuntary head motions and the higher rates of noise are present in ASD whether or not they are on meds. Yet, the intake of psychotropic medications exacerbates the noise levels systematically with the number of meds the person is on. This effect is true for both linear displacements and angular rotations of the head. Panel A figures published in (Torres et al. 2013a); panel B Figure published in (Torres and Denisova 2016) licensed under a Creative Commons Attribution 4.0 International License

complex nonlinear dynamics present in the moving body. Financially these requirements pose very complex conflict of interests too. The amount of money needed to run a proper study with good statistical power in autism requires a very large budget to pay for such subjective discrete scores. Current NIH-NSF budgets cannot afford proper science in ASD, so the scientific goals are extremely modest and do not positively impact the lives of those affected by the disorder. We are left with largely pharma-driven criteria, criteria that do not help the affected person, or his/her supporting family unit.

Further compounding this issue is the fact that the makers of the DSM-IV and DSM-5 have significant financial ties with the pharmaceutical companies (Cosgrove et al. 2006, 2014; Cosgrove and Krimsky 2012), a fact that is not entirely unrelated to the large amount of psychotropic medications the people diagnosed under DSM-criteria are under.

As I was wondering about the effect of such drugs on motor control in ASD, I came across a piece of knowledge that I did not know: in fMRI research, people are asked to remain still while in the magnet and this is critical to the validity of each scanned frame. Motions of the head can be precisely measured from scan to scan and those linked to considerable motions (according to a set threshold) are removed, a practice called "scrubbing." These frames do not enter in the analyses of the imaging data.

Kristina Denisova, a postdoctoral fellow at Columbia University shared head motion data with my lab under Institutional Review Board Data Sharing Agreements between the two universities. The head motion data was obtained from raw fMRI scans publicly available in the Autism Brain Imaging Data Exchange (ABIDE) and ADHD-200 databases using standard open access tools available to fMRI researchers (Friston et al. 1995). In recent years, there has been some debate that individuals with ASD may move more than controls in the fMRI scanner and that perhaps some of the motor artifacts are linked to over connectivity theories in

ASD formulated using fMRI data (Deen and Pelphrey 2012). It has been difficult to assess such controversial data because each lab "scrubs" different number of scans from their sets, so the best bet to unravel the mystery of excess motions in ASD or in other disorders of the nervous systems, is to look at all the scans from the original data. Given the head motion data, the remaining problem is how to extract meaningful waveforms from these raw data so as to pose the question of possible motor control differences in ASD.

Using the new statistical platform for personalized analyses of micro-movements, I analyzed the time series of head linear displacements and head angular rotations during resting state across a number of sites. Specifically, using methods in Fig. 4b–d, I was able to extract the micro-movements waveform from the head motions of 1048 participants with ASD, ADHD, and controls. The idea was to identify and characterize stochastic signatures of involuntary micro-movements embedded in the head motions of participants within these populations.

Given the conceptualization of a necessary balance across the proposed taxonomy of control levels (Fig. 3a) to have proper volition (Torres 2011; Torres et al. 2011), I reasoned that excess involuntary motions of the head may flag impairments in the ability to remain still at will (a superb accomplishment of volitional control). But how could we distinguish between inherent issues with the sensory-motor systems and problems induced by psychotropic meds so heavily prescribed across the ASD and ADHD populations?

The ABIDE and the ADHD-200 databases list as well the psychotropic meds for a large number of participants affected by the disorders. These drugs have known side effects in the motor systems (reported in their labels and acknowledged by their makers in publicly available websites). Since motor side effects confound the issue of motor problems in ASD, it has been difficult to address motor issues in the research arena. I set to test the idea of separating motor issues in ASD and ADHD by comparing those participants under psychotropic meds and participants that were medication-naïve, both in relation to controls.

To this end, I first characterized the levels of head micro-movements noise in controls that did not take meds and were typical. Then I examined the signatures of those fellows who had a diagnosis of ASD/ADHD but took no psychotropic drugs and compared the PDF's to those also in these spectrums who did take psychotropic drugs. The questions were (1) if medication-naïve individuals with ASD/ADHD had atypical levels of noise accumulation in the micro-movements of head motions; (2) if psychotropic drug intake had a systematic impact on those estimated signatures of the fellows under psychotropic drugs.

Several interesting results emerged from these analyses. Individuals with autism have a prevalence of involuntary head motions, even when their head is padded and they are lying in a resting position. The frequency of the micro-movements and the rates of accumulation of their noise are significantly higher from those of neurotypical controls. More important yet, the probability distribution functions describing the micro-movements systematically increase the noise-to-signal ratio and the skewness with the number of psychotropic drugs that they take. This can be seen in Fig. 6b. This figure shows the profound differences in statistical signatures of involuntary head micro-motions between ASD and controls. A system with excess involuntary activity and stagnation in the noise and randomness of fluctuations in bodily rhythms (Fig. 6a) will be utterly challenged to make use of voluntarily driven reafference. That form of reafference proposed by von Holst would not emerge under such conditions. In this sense, volitional control of one's actions leading to action ownership and self-agency necessarily would require a good balance across all the levels of the proposed control taxonomy of Fig. 3a. This in turn would require mastering proper noise-to-signal ratios across multilayered signals of the nervous systems.

The DSM-makers oppose the inclusion of sensory-motor issues as biological core problems in mental illnesses. The ADOS-2 insists that the children about to be diagnosed with ASD have no issues with bodily sensations and motor control, despite reports by affected individuals that they cannot feel their body—let alone control their bodily motions at will. The objective evidence presented by our work under a new statistical lens irrevocably shows that movement (and as a corollary of the reafference principle, movement sensation) is impeded in ASD, strongly suggesting that these differences in neurodevelopment most likely fail to properly scaffold the emergence of volition. The persistent stagnation on the evolution of this form of sensory feedback since birth may prevent the CNS from forming a proper bridge with the PNS so as to—in closed loop—guide the acquisition of action ownership and self-agency. As many in the spectrum of autism would tell us, "my body has a mind of its own; I cannot control it as I want to" (see also accounts in Robledo et al. 2012).

## *Habilitate, Rehabilitate, and Enhance Volition*

The analyses of the fMRI open access databases showed us excess noise and the prevalence of randomness in the micro-movements' signatures of involuntary head motions of 1048 participants. Despite this prevalence in ASD and ADHD participants, we next asked whether such noisy and random signatures in the motor output could be reshaped. We reasoned that in the likely absence of proper kinesthetic reafference, we could close the feedback loops with external input. This form of sensory substitution had already worked for an extreme case of actual physical deafferentation: the case of Ian Waterman.

IW is a very special subject who lost his proprioception from the neck down but managed to walk and move purposefully again by replacing the lost proprioception with vision and motor imagery (Cole 1995). We had the opportunity to reexamine IW in the lab recently, as we had studied his motor performance under this statistical platform and found that his signatures of voluntary and spontaneous motor performance resembled those of the ASD participants, particularly and more importantly when IW was moving in the dark with no visual feedback (Torres et al. 2014, 2016). If IW, who had actual physical deafferentation, was able to restore volition in his actions using external (visual) feedback, there was a chance that the children with ASD could do so as well. The challenge was that these children were nonverbal, or had very limited speech. Communicating with the children was challenging for us. Instructing or prompting an experiment was not feasible.

We developed an interface to spontaneously, without instructions, evoke volitional control in these children. It had to be done gradually, first dampening the noise and randomness of the fluctuations in motor performance in supplementary (involuntary) motions, with the hopes of positively impacting—or even evoking in some cases—the voluntary level of control. This strategy would bring into play von Holst's principle of reafference as soon as they developed motions with the statistical signatures of voluntary control. To that end we needed to first evoke the self-discovery of a GOAL, and then encourage the search for a systematic pattern. If the children's visual and auditory systems were able to detect change in sensory input, we could do this.

Using Thelen's notion of goal self-discovery in infants through a process of trial and error, we designed a closed-loop interface that achieved the above-mentioned objectives. Figure 7A shows the idea (Torres et al. 2013d), which is rather simple. We embedded a region of interest (ROI) in the peripersonal space of the child (this space around the child moves with the child, as it is defined relative to the child's body). We had positional sensors (Polhemus Liberty 240 Hz, Colchester VT) on the child's hand coded relative to the child, as well as relative to a world axis. Thus, every time the child passed the hand through the ROI, even if he naturally moved around, we could trigger external media to attract his attention. Each of the children figured out the ROI within minutes. They discovered the GOAL: namely, to move the hand to the ROI so as to sustain the media playing continuously. The media could be a movie of their liking with sound of their liking, or simply their own image projected from a video camera in front of them.

The children's self-discovery of the GOAL led to trial and error till they pinpointed exactly where the hand had to be located in order for the media to play. They self-discovered cause and effect by turning random motions into GOAL-directed motions under voluntary control. Then, through that trial and error process (similar to the arm flailing that an infant would do when learning to reach) they developed the stochastic signatures of volition in their motions. Specifically we were able to quantify all the while the frequency and stationarity of noise-to-signal transitions in their voluntary reaches to the GOAL as well as in the coexisting spontaneous motion segments continuously supplementing the emergent GOAL-directed ones. Figure 7b shows snapshots of these noise-to-signal transitions using the empirically estimated probability distributions as well as the Gamma plane log–log plots, whereby each point is an estimated PDF from continuous accumulations of measurements of angular velocity with shifting signatures as the system learns over the time span of minutes. At the end of the session, when they held the arm in place to keep the hand in the ROI that played the media, their stochastic patterns had spontaneously acquired the typical signatures that we had

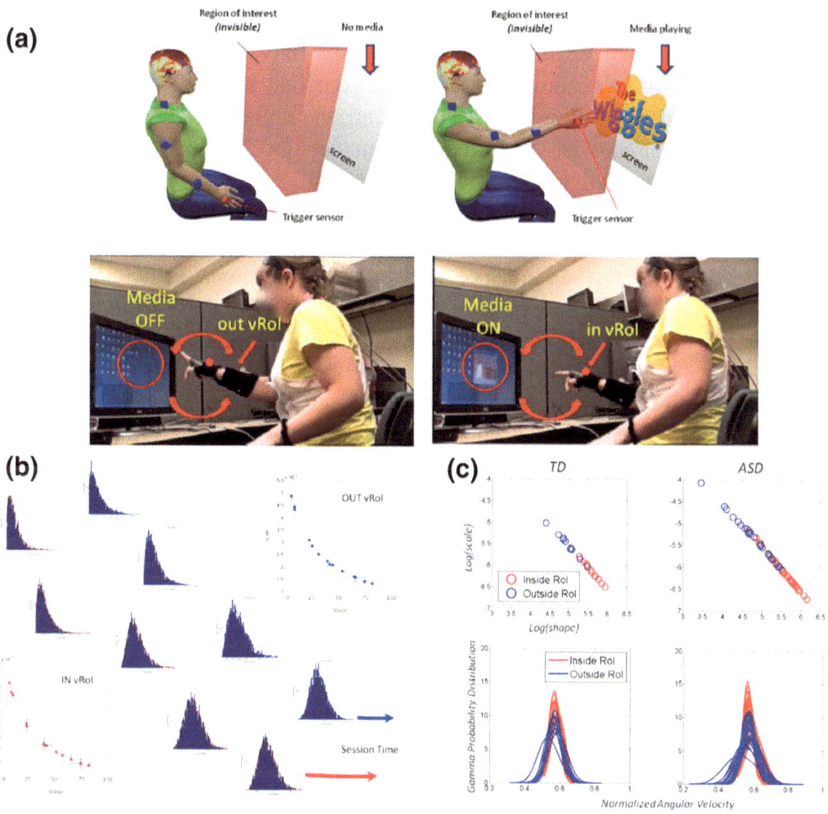


**Fig. 7** Habilitating and enhancing volition in ASD. **a** Simple paradigm using the notion of body–machine interfaces to help the child self-discover a goal and pursue it actively to evoke volition in the motions. The child wears sensors that register the position at all times and the distance to a virtual region of interest (vRoI) that we create. When the hand passes through the vRoI a movie and music is triggered on the screen. The child actively searches for that location, first through trial and error, then systematically till she discovers the goal location and holds the hand in it steadily to watch the video continuously. **b** Snapshots of the frequency histograms accumulating the fluctuations in angular velocity while the hand rotates and translates through space. Motions outside and inside the vRoI are tracked and the stochastic trajectory shown on the Gamma parameter plane. **c** The shifts from typically developing and ASD children converging toward symmetric probability distributions with low dispersion, i.e., high signal with statistical regularities characteristic of typical controls with high volition. The 25 nonverbal children with ASD mastered this task on their own and in the process learned how to control their movements at will. Figure from (Torres et al. 2013d)

quantified in typical peers at the same school. Were these changes transient? We had witnessed in those same children that prompting them during Applied Behavioral Analysis (ABA) therapy led to a somewhat robotic structure in their actions. Furthermore, over the school breaks, despite daily drilling sessions of hundreds of rote repetitions, they would come back from the break needing considerable effort to achieve the level of performance they had left at before the break. We came back to the school 4–5 weeks after our first and only session using this setup. We found that despite no practice at all, the children rapidly reacquired the ROI and the desired well-structured, systematic noise levels with the statistical regularities of a predictive code.

An important lesson from this study was that the signature of motor noise in the system of a child with a diagnosis of autism has the potential to change into typical ranges beyond the transient goals of a controlled experiment. This potential can be turned into the possibility for generalization and transfer of these motor phenomena from the lab to the activities of daily living, a future goal for investigation in our lab.

These children's CNS can spontaneously learn the bridge between intentional thoughts and well-controlled actions that follow a purpose. They can gain volition of their bodily rhythms at the level of motor performance that we measured. That is a first step into teaching them gestural communication, a form of language that they could use for social exchange.

The social deficits that the DSM-5 labels these children with at the exclusion of sensory-motor components keep this disorder as a mental deficit that can be heavily medicated. The fMRI data showing the dramatic effect that psychotropic meds have on these children seriously question the validity of a diagnosis system that has financial ties with pharmaceutical companies (Cosgrove et al. 2006, 2014; Cosgrove and Krimsky 2012). The question then remains as to whether the DSM-makers and BigPharma will ever inform the parents of the deleterious effects that those drugs have on the sensory-motor systems of their children.

There is precious potential in the individuals with autism. All we have to do is harness it and put it to the service of the neural control of movements. Our field could make a tremendous contribution to this because researchers in it have the knowledge and the tools to do it.

## Concluding Remarks

This chapter provides a new way to examine the concept of volition and use it in clinical settings. I revisited some old motor control questions using new data types and new analytics through the lens of unrestrained behaviors. Several examples of possible applications of this new statistical platform were also discussed and an invitation was extended to the Motor Control field at large to delve into the issue of volition with relation to clinical applications. The work presented here was inspired by the ever fresh ideas of Bernstein, von Holst, and Thelen. I cannot envision contemporary motor control without their influence as we renovate the body of knowledge that we have accumulated in our field and extend it to the neurodevelopmental arena.

## Take Home Message

Our field is at a stage of exploration. While we continue to guide our empirical work by sound theoretical ideas and mathematical frameworks, we also need to rethink how we gather empirical data such as to maximize "purity" in our results. We need to relax some of the constraints that we have imposed on our data collection and better verify assumptions that we have made in our analyses. If we do not do that, we risk falling into a trap of self-deception, stalling progress along the way.

The nervous system is very flexible. In particular the sensory-motor systems have evolved to adapt and learn easily to rapidly operate under many constraints. Those imposed by our experimental exercises are no exception. If we shift to naturalistic, unrestrained settings, where we continuously record all aspects of our behaviors, not just those that we render relevant to the goals of a task; we will gain new insights and will promote new ways of thinking about old problems. But even more exciting perhaps, we will be able to formulate new problems and ultimately apply our knowledge to help others in need.

I believe that if we want to truly understand the neural control of movement from first principles, we ought to let the body freely move. We ought to let the brain discover its volition and show us how it arrives at intended actions and thoughts in such a perfectly coupled way so as to seamlessly couple mental to physical activities. Acting as mere spectators of such processes, as they spontaneously unfold, rather than prompting and steering their unfolding, may lead us to a better understanding of the complex and beautiful sensory-motor phenomena that ultimately shapes our mental and social spaces.

## References

American Psychiatric Association, DSM-5 Task Force (2013) Diagnostic and statistical manual of mental disorders: DSM-5. American Psychiatric Association, Washington, DC

Atkeson CG, Hollerbach JM (1985) Kinematic features of unrestrained vertical arm movements. J Neurosci 5:2318–2330

Bernstein N (1967) The co-ordination and regulation of movements. Oxford Press, Oxford

Biess A, Liebermann DG, Flash T (2007) A computational model for redundant human three-dimensional pointing movements: integration of independent spatial and temporal motor plans simplifies movement dynamics. J Neurosci 27:13045–13064. doi:10.1523/JNEUROSCI.4334-06.2007 27/48/13045 [pii]

Biess A, Flash T, Liebermann DG (2011) Riemannian geometric approach to human arm dynamics, movement optimization, and invariance. Phys Rev E: Stat, Nonlin, Soft Matter Phys 83:031927

Brincker M, Torres EB (2013) Noise from the periphery in autism. Front Integr Neurosci 7:34. doi:10.3389/fnint.2013.00034

Churchland MM, Afshar A, Shenoy KV (2006a) A central source of movement variability. Neuron 52:1085–1096. doi:S0896-6273(06)00871-3

Churchland MM, Yu BM, Ryu SI, Santhanam G, Shenoy KV (2006b) Neural variability in premotor cortex provides a signature of motor preparation. J Neurosci 26:3697–3712. doi:26/14/3697
Churchland MM, Yu BM, Cunningham JP et al (2010) Stimulus onset quenches neural variability: a widespread cortical phenomenon. Nat Neurosci 13:369–378. doi:nn.2501
Cole J (1995) Pride and a daily marathon. MIT Press, Cambridge
Cosgrove L, Krimsky S (2012) A comparison of DSM-IV and DSM-5 panel members' financial associations with industry: a pernicious problem persists. PLoS Med 9:e1001190. doi:10.1371/journal.pmed.1001190
Cosgrove L, Krimsky S, Vijayaraghavan M, Schneider L (2006) Financial ties between DSM-IV panel members and the pharmaceutical industry. Psychother Psychosom 75:154–160. doi:10.1159/000091772
Cosgrove L, Krimsky S, Wheeler EE, Kaitz J, Greenspan SB, DiPentima NL (2014) Tripartite conflicts of interest and high stakes patent extensions in the DSM-5. Psychother Psychosom 83:106–113. doi:10.1159/000357499
Deen B, Pelphrey K (2012) Perspective: brain scans need a rethink. Nature 491:S20
Faisal AA, Selen LP, Wolpert DM (2008) Noise in the nervous system. Nat Rev Neurosci 9: 292–303. doi:nrn2258
Flash T, Hogan N (1985) The coordination of arm movements: an experimentally confirmed mathematical model. J Neurosci 5:1688–1703
Friston KJ, Holmes AP, Poline JB, Grasby PJ, Williams SC, Frackowiak RS, Turner R (1995) Analysis of fMRI time-series revisited. Neuroimage 2:45–53. doi:0.1006/nimg.1995.1007
Hamilton AF, Wolpert DM (2002) Controlling the statistics of action: obstacle avoidance. J Neurophysiol 87:2434–2440
Harris CM, Wolpert DM (1998) Signal-dependent noise determines motor planning. Nature 394:780–784. doi:10.1038/29528
Kording KP, Wolpert DM (2004a) Bayesian integration in sensorimotor learning. Nature 427: 244–247. doi:10.1038/nature02169
Kording KP, Wolpert DM (2004b) The loss function of sensorimotor learning. Proc Natl Acad Sci USA 101:9839–9842. doi:10.1073/pnas.0308394101
Kording KP, Wolpert DM (2006) Bayesian decision theory in sensorimotor control. Trends Cogn Sci 10:319–326. doi:S1364-6613(06)00127-6
Kuczmarski RJ, Ogden CL, Grummer-Strawn LM et al (2000) CDC growth charts: United States. Adv Data 1–27
Kuczmarski RJ, Ogden CL, Guo SS et al (2002) 2000 CDC growth charts for the United States: methods and development. Vital Health Stat 11:1–190
Latash ML, Scholz JP, Schoner G (2002) Motor control strategies revealed in the structure of motor variability. Exerc Sport Sci Rev 30:26–31
Limpert E, Stahel WA (2011) Problems with using the normal distribution and ways to improve quality and efficiency of data analysis. PLoS ONE 6:e21403. doi:10.1371/journal.pone.002140310-PONE-RA-16396
Limpert E, Stahel WA, Abbt M (2001) Log-normal distributions across the sciences: keys and clues. Bioscience 51:341–352
Nguyen J, Papathomas TV, Ravaliya JH, Torres EB (2014) Methods to explore the influence of top-down visual processes on motor behavior. J Vis Exp. doi:10.3791/51422
Nguyen J, Majmudar UV, Ravaliya JH, Papathomas TV, Torres EB (2015) Automatically characterizing sensory-motor patterns underlying reach-to-grasp movements on a physical depth inversion illusion. Front Hum Neurosci 9:694. doi:10.3389/fnhum.2015.00694
Nishikawa KC, Murray ST, Flanders M (1999) Do arm postures vary with the speed of reaching? J Neurophysiol 81:2582–2586
Piper MC, Pinnell LE, Darrah J, Maguire T, Byrne PJ (1992) Construction and validation of the Alberta Infant Motor Scale (AIMS). Can J Public Health 83 Suppl 2:S46–50
Pontriagin LS (1990) Optimal control and differential games: collection of papers. American Mathematical Society, Providence

Robledo J, Donnellan AM, Strandt-Conroy K (2012) An exploration of sensory and movement differences from the perspective of individuals with autism. Front Integr Neurosci 6:107. doi:10.3389/fnint.2012.00107

Rogers DM (1992) Motor disorder in psychiatry: towards a neurological psychiatry. Wiley, Chichester

Scholz JP, Schoner G (1999) The uncontrolled manifold concept: identifying control variables for a functional task. Exp Brain Res 126:289–306

Scholz JP, Kang N, Patterson D, Latash ML (2003) Uncontrolled manifold analysis of single trials during multi-finger force production by persons with and without Down syndrome. Exp Brain Res 153:45–58. doi:10.1007/s00221-003-1580-8

Smith LB, Thelen E (2003) Development as a dynamic system. Trends Cogn Sci 7:343–348. doi:S1364661303001566

Thelen E, Smith LB (1994) A dynamic systems approach to the development of cognition and action. MIT Press, Cambridge

Todorov E (2005) Stochastic optimal control and estimation methods adapted to the noise characteristics of the sensorimotor system. Neural Comput 17:1084–1108. doi:10.1162/0899766053491887

Todorov E (2009) Efficient computation of optimal actions. Proc Natl Acad Sci USA 106: 11478–11483. doi:0710743106

Torres EB, Denisova K (2016) Motor noise is rich signal in autism research and pharmacological treatments. Sci Rep 6:37422. doi:10.1038/srep37422

Torres E (2001) Theoretical framework for the study of sensory-motor integration. In: Cognitive science, PhD, University of California, San Diego, La Jolla, p 109

Torres EB (2010) New symmetry of intended curved reaches. Behav Brain Funct 6:21. doi:1744-9081-6-21

Torres EB (2011) Two classes of movements in motor control. Exp Brain Res 215:269–283. doi:10.1007/s00221-011-2892-8

Torres EB (2012) Atypical signatures of motor variability found in an individual with ASD. Neurocase Neural Basis Cogn 1:1–16. doi:10.1080/13554794.2011.654224

Torres EB (2013a) The rates of change of the stochastic trajectories of acceleration variability are a good predictor of normal aging and of the stage of Parkinson's disease. Front Integr Neurosci 7:50

Torres EB (2013b) Signatures of movement variability anticipate hand speed according to levels of intent. Behav Brain Functions 9:10. doi:10.1186/1744-9081-9-10

Torres E, Andersen R (2006) Space-time separation during obstacle-avoidance learning in monkeys. J Neurophysiol 96:2613–2632. doi:00188.2006

Torres EB, Lande B (2015) Objective and personalized longitudinal assessment of a pregnant patient with post severe brain trauma. Front Hum Neurosci 9:128. doi:10.3389/fnhum.2015.00128

Torres EB, Zipser D (2002) Reaching to grasp with a multi-jointed arm. I. Computational model. J Neurophysiol 88:2355–2367. doi:10.1152/jn.00030.2002

Torres EB, Zipser D (2004) Simultaneous control of hand displacements and rotations in orientation-matching experiments. J Appl Physiol (1985) 96:1978–1987. doi:10.1152/japplphysiol.00872.2003

Torres EB, Raymer A, Gonzalez Rothi LJ, Heilman KM, Poizner H (2010) Sensory-spatial transformations in the left posterior parietal cortex may contribute to reach timing. J Neurophysiol 104:2375–2388. doi:jn.00089.2010

Torres EB, Heilman KM, Poizner H (2011) Impaired endogenously evoked automated reaching in Parkinson's disease. J Neurosci 31:17848–17863. doi:31/49/17848

Torres EB, Brincker M, Isenhower RW et al (2013a) Autism: the micro-movement perspective. Front Integr Neurosci 7:32. doi:10.3389/fnint.2013.00032

Torres EB, Isenhower RW, Yanovich P, Rehrig G, Stigler K, Nurnberger J, Jose JV (2013b) Strategies to develop putative biomarkers to characterize the female phenotype with autism spectrum disorders. J Neurophysiol 110:1646–1662. doi:10.1152/jn.00059.2013

Torres EB, Quian Quiroga R, Cui H, Buneo CA (2013c) Neural correlates of learning and trajectory planning in the posterior parietal cortex. Front Integr Neurosci 7:39. doi:10.3389/fnint.2013.00039

Torres EB, Yanovich P, Metaxas DN (2013d) Give spontaneity and self-discovery a chance in ASD: spontaneous peripheral limb variability as a proxy to evoke centrally driven intentional acts. Front Integr Neurosci 7:46. doi:10.3389/fnint.2013.00046

Torres EB, Cole J, Poizner H (2014) Motor output variability, deafferentation, and putative deficits in kinesthetic reafference in Parkinson's disease. Front Hum Neurosci 8:823. doi:10.3389/fnhum.2014.00823

Torres EB, Isenhower RW, Nguyen J et al (2016) Toward precision psychiatry: statistical platform for the personalized characterization of natural behaviors. Front Neurol 7:8. doi:10.3389/fneur.2016.00008

Uno Y, Kawato M, Suzuki R (1989) Formation and control of optimal trajectory in human multijoint arm movement. Minimum torque-change model. Biol Cybern 61:89–101

van Beers RJ (2009) Motor learning is optimally tuned to the properties of motor noise. Neuron 63:406–417. doi:S0896-6273(09)00516-9

van Beers RJ, Baraduc P, Wolpert DM (2002) Role of uncertainty in sensorimotor control. Philos Trans R Soc Lond B Biol Sci 357:1137–1145. doi:10.1098/rstb.2002.1101

van Beers RJ, Haggard P, Wolpert DM (2004) The role of execution noise in movement variability. J Neurophysiol 91:1050–1063. doi:10.1152/jn.00652.200300652.2003

van Beers RJ, Brenner E, Smeets JB (2013) Random walk of motor planning in task-irrelevant dimensions. J Neurophysiol 109:969–977. doi:10.1152/jn.00706.2012

von Holst E, Mittelstaedt H (1950) The principle of reafference: interactions between the central nervous system and the peripheral organs. In: Dodwell PC (ed) Perceptual Processing: stimulus equivalence and pattern recognition. Appleton-Century-Crofts, New York, pp 41–72

# Part V
# Impaired Motor Control and Rehabilitation

# Motor Lateralization Provides a Foundation for Predicting and Treating Non-paretic Arm Motor Deficits in Stroke

**Robert L. Sainburg, Candice Maenza, Carolee Winstein and David Good**

**Abstract** Brain lateralization is a ubiquitous feature of neural organization across the vertebrate spectrum. We have developed a model of motor lateralization that attributes different motor control processes to each cerebral hemisphere. This bilateral hemispheric model of motor control has successfully predicted hemisphere-specific motor control and motor learning deficits in the ipsilesional, or non-paretic, arm of patients with unilateral stroke. We now show across large number and range of stroke patients that these motor performance deficits in the non-paretic arm of stroke patients vary with both the side of the lesion, as well as with the severity of contralesional impairment. This last point can be functionally devastating for patients with severe contralesional paresis because for these individuals, performance of upper extremity activities of daily living depends primarily and often exclusively on ipsilesional arm function. We present a pilot study focused on improving the speed and coordination of ipsilesional arm function in a convenience sample of three stroke patients with severe contralesional impairment. Over a three-week period, patients received a total of nine 1.5 h sessions of training that included intense practice of virtual reality and real-life tasks. Our results indicated substantial improvements in ipsilesional arm movement kinematics, functional performance, and that these improvements carried over to improve functional independence. In addition, the contralesional arm improved in our measure of contralesional impairment, which was likely due to improved participation in activities of daily living. We discuss of our findings for physical rehabilitation.

R.L. Sainburg (✉) · C. Maenza
Department of Kinesiology, The Pennsylvania State University, 29 Rec Building, Biomechanics Laboratory, University Park, PA 16802, USA
e-mail: rls45@psu.edu

R.L. Sainburg · C. Maenza · D. Good
Department of Neurology, Penn State Milton S. Hershey College of Medicine, Hershey, USA

C. Winstein
Department of Biokinesiology and Physical Therapy, University of Southern California, Los Angeles, USA



J. Laczko and M.L. Latash (eds.), *Progress in Motor Control*, Advances in Experimental Medicine and Biology 957,
DOI 10.1007/978-3-319-47313-0_14

**Keywords** Bilateral hemispheric model · Contralesional · Paresis · Ipsilesional · Rehabilitation · Dynamic dominance · Jebsen-Taylor hand function test · Interlimb transfer

## Neural Lateralization for Motor Control Processes

The seminal research of Gazzaniga (1998) on disconnection syndrome in split brain patients established neural lateralization as a fundamental feature of brain organization. They proposed that distributing different neural processes across the hemispheres was a natural consequence of developing complex functions during the course of evolution. Gazzaniga's research provided elegant support for this view of cerebral lateralization as a neural optimization process. The research of MacNeilage, Rogers, and Vallortigara (Bisazza et al. 1998; MacNeilage et al. 2009) demonstrated that brain lateralization served as a basic organizational principle over the course of vertebrate evolution and showed that neural lateralization is found throughout the vertebrate spectrum. While a great deal of research on brain lateralization has been conducted in humans, early research was largely limited to cognitive and perceptual processes, with little attention to the motor systems.

Based on asymmetries in reaching performance and adaptation to novel dynamic and visual motor environments in healthy adults, we introduced the dynamic dominance hypothesis of motor lateralization (Sainburg 2002). The dynamic dominance model proposes that the left hemisphere, in right-handers, is specialized for predictive processes that specify smooth and efficient movement trajectories under mechanically stable environmental circumstances, while the right hemisphere is specialized for impedance control mechanisms that confer robustness to movements performed under unpredictable and mechanically unstable environmental conditions. This hypothesis has been supported by empirical studies (Sainburg 2002), computational modeling studies (Yadav and Sainburg 2011, 2014) and studies in patients with unilateral brain lesions (Schaefer et al. 2007, 2012; Haaland et al. 2009; Mutha et al. 2012; Mani et al. 2013, 2014). This hypothesis of motor lateralization is consistent with the broader model of brain lateralization proposed by Rogers and colleagues (Bisazza et al. 1998). According to their model, the left hemisphere is "specialized for control of well-established patterns of behavior, under ordinary and familiar circumstances," while the right hemisphere is specialized for "detecting and responding to unexpected stimuli in the environment" (MacNeilage et al. 2009). The dynamic dominance model provides the movement analog to Roger's model, and thus places handedness in the context of a larger array of neurobehavioral asymmetries across the animal kingdom (Sainburg 2014).

## Non-paretic Arm Motor Deficits in Stroke

The dynamic dominance hypothesis is a bihemispheric model of motor control, in which each hemisphere contributes different control mechanisms to voluntary movement. One of the strongest predictions of this model is that unilateral brain damage affecting sensorimotor centers should produce hemisphere-specific deficits in the non-paretic ipsilesional arm of stroke patients. The role of contralateral motor areas in controlling limb movements has been well-established (Lawrence and Kuypers 1968; Kuypers 1978, 1982). The role the ipsilateral hemisphere has been implicated by the robust occurrence of ipsilesional motor deficits in both animal models of unilateral brain damage (Grabowski et al. 1993) as well as human stroke survivors (Wyke 1967; Fisk and Goodale 1988; Winstein and Pohl 1995; Desrosiers et al. 1996; Pohl and Winstein 1999; Yarosh et al. 2004; Schaefer et al. 2009a). In addition, both electrophysiological and neural imaging studies have shown that unilateral arm and hand movements recruit motor-regions in both cerebral hemispheres (Tanji et al. 1988; Kawashima et al. 1993, 1998; Dassonville et al. 1997). Motor deficits in the non-paretic arm of patients with unilateral stroke have been documented as early as 1967 (Wyke 1967). More recent research has shown that these deficits are functionally limiting and that they persist throughout the chronic phase of stroke (Pohl and Winstein 1999; Winstein et al. 1999; Haaland et al. 2000, 2004, 2009; Prestopnik et al. 2003a, b; Wetter et al. 2005; Haaland 2006; Schaefer et al. 2007, 2009a, b, 2012; Chestnut and Haaland 2008; Poole et al. 2009; Mutha et al. 2010, 2011a, b, 2013). In fact, studies of non-paretic arm function in chronic stroke patients have reported performance deficiencies on a number of clinical tests, including the Purdue Pegboard Test (Rapin et al. 1966), the Jebsen-Taylor Hand Function Test (Schaefer et al. 2009b), and a variety of tests that directly assess or simulate activities of daily living (Desrosiers et al. 1996; Wetter et al. 2005; Haaland et al. 2012). In addition, significant deficits in movement coordination and accuracy have been revealed by studies that have employed motion analysis (Haaland and Harrington 1989a, b, 1996; Winstein and Pohl 1995; Winstein et al. 1996; Prestopnik et al. 2003a; Haaland et al. 2004, 2009; Schaefer et al. 2009a, b, 2012; Mutha et al. 2010, 2011a, b; 2012, 2013; Mani et al. 2014).

Over the past decade, our laboratory has systematically assessed the hemisphere specificity, neural foundations, and functional implications of non-paretic arm motor deficits. In brief, our findings have indicated that non-paretic arm motor deficits in stroke result from a loss of the specific contributions of the ipsilateral hemisphere to motor control. Our research has established that right and left hemisphere damage lead to deficits in different motor control processes (Winstein and Pohl 1995; Schaefer et al. 2012; Mutha et al. 2013; Sainburg 2014).

Two important aspects of non-paretic arm deficits in stroke patients are that the quality of the motor deficits depends on the hemisphere that has been damaged by stroke, and the extent of the deficits depends on the severity of paretic arm motor impairment. Figure 1 shows data from a study of multidirection reaching movements in patients with unilateral stroke (Schaefer et al. 2009b). Figure 1a shows

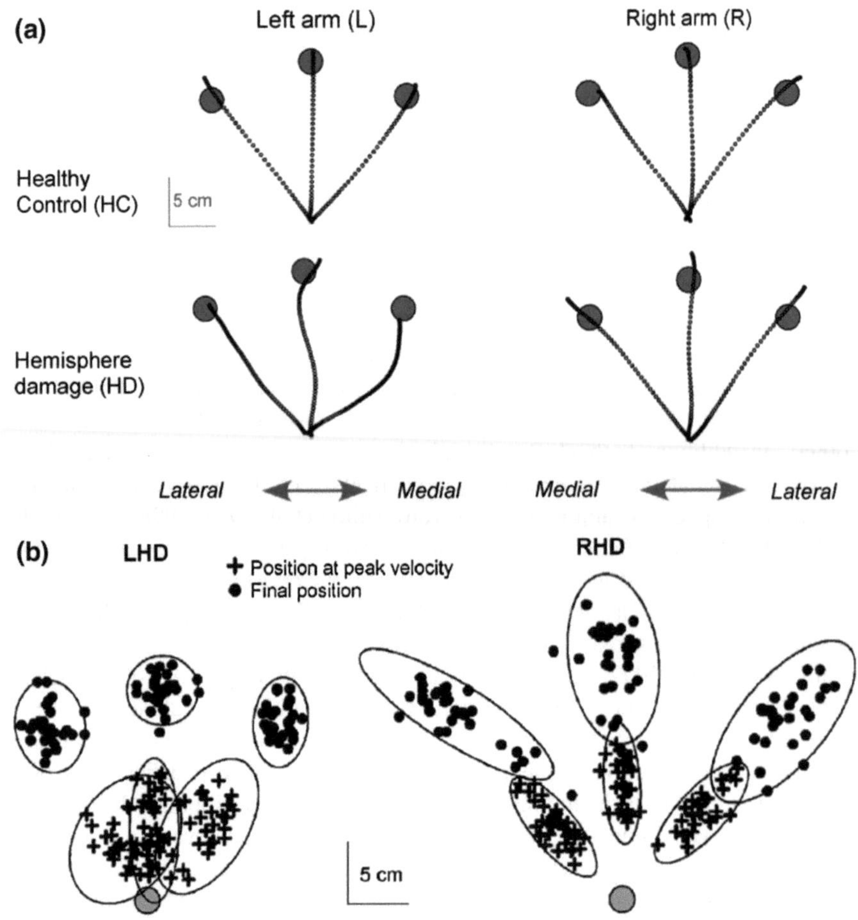


**Fig. 1** Multidirection reaching movements in patients with unilateral stroke (Schaefer et al. 2009b). **a** Reaching movements in the non-paretic arm of patients with *left* (LHD) and *right* (RHD) hemisphere damage. **b** Variance in hand positions during the initial trajectory phase (*bottom ellipses*), and in the final position (*top ellipses*) of the movement

reaching movements in the non-paretic arm of patients with left and right hemisphere damage. While patients with left hemisphere damage (LHD) made highly curved movements with accurate final positions, patients with right hemisphere damage (RHD) made straight movements with poor final position accuracies. Figure 1b shows the variance in hand positions during the initial trajectory phase (bottom ellipses), and in the final position (top ellipses) of the movement. These findings demonstrate a robust double dissociation between hemisphere of damage and motor control process effected by a unilateral hemisphere lesion: While RHD patients have deficits in stabilizing accurate final positions across trials, LHD patients have deficits in stabilizing accurate initial trajectories across trials.

These results provide evidence that each hemisphere contributes different control mechanisms to both arms, and have been corroborated by studies that examined the effects of unilateral brain damage on adaptation to novel visuomotor rotations (Mutha et al. 2011a, b), and studies that have examined responses to sudden visual perturbations (Schaefer et al. 2012). This bihemispheric model of arm control also predicts hemisphere specific deficits in the contralesional arm of stroke patients, a prediction that has been supported in patients with mild paresis (Mani et al. 2013).

## Relevance of Motor Lateralization to Stroke Rehabilitation

Regardless of the robust occurrence of ipsilesional, non-paretic arm, motor deficits in patients with unilateral stroke, Occupational and Physical Therapy motor remediation protocols continue to either completely focus on the paretic arm, or employ bilateral movements that do not challenge non-paretic arm coordination. As stated above, the non-paretic arm tends to have substantial deficits in movement coordination, speed, and accuracy that limit both performance of activities of daily living (ADL) and functional independence. Specifically, non-paretic arm motor deficits are most severe and debilitating in patients with severe paresis. In fact, the usual standard of care in rehabilitation for these patients with severe contralesional paresis is task training in essential ADL rather than intensive remediation of either arm. It is important to stress that in these patients, performance of ADL depends on non-paretic arm function. We propose that intense remediation, focused on improving the speed, coordination, and accuracy of the non-paretic arm can improve motor performance and functional independence in stroke patients with severe hemiparesis. Currently, it is not known whether remedial treatment can improve non-paretic arm function. We are beginning to address this issue by examining the effects of non-paretic arm training on functional independence.

*Non-paretic Arm Functional Deficits* We recently tested how non-paretic arm motor function might depend on the severity of paresis in the contralesional arm, and on the side of the brain that is damaged. Figure 2 shows data from 72 age and gender-matched control participants (36 using the right arm, and 36 using the left arm), 22 LHD survivors, and 29 RHD survivors. The y-axis represents the Jebsen-Taylor Hand Function Test score, taken as a percentage of right dominant arm function in our control group. The JTHFT is a clinical assessment of unilateral arm function, validated in stroke survivors (Beebe and Lang 2009), that includes a range of tasks that elicit the coordination requirements of functional daily activities, including writing, turning pages, placing large and small objects on a table, stacking small objects, and feeding (Jebsen et al. 1969, 1971). The left column (control) shows the difference between healthy participants performing with the left arm (Gray) and the right arm (Black). The data are normalized to control subjects' dominant arm performance (100 %). The data are stratified on the x-axis by hand (in the case of stroke survivors, this is always the non-paretic, ipsilesional arm), and severity of contralesional paresis, as measured by the upper extremity component of

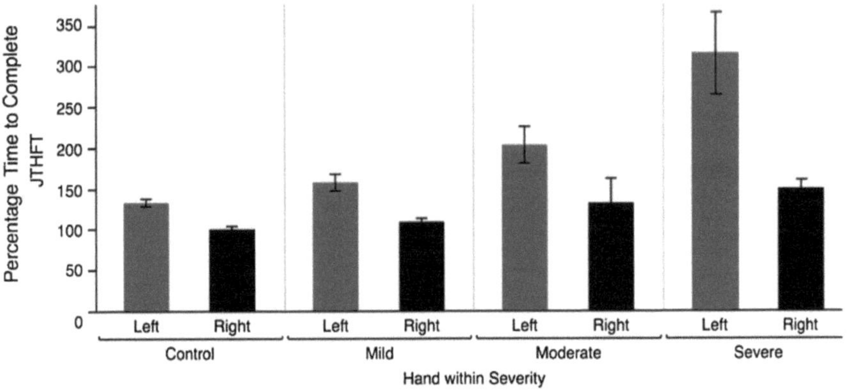


**Fig. 2** Data from 72 age and gender matched control participants (36 using the right arm, and 36 using the left arm), 22 LHD survivors, and 29 RHD survivors. Y-axis represents the Jebsen-Taylor Hand Function Test score. Data are normalized to control subjects' dominant arm performance (100 %). X-axis stratified by hand (in the case of stroke survivors, this is always the non-paretic, ipsilesional arm), and severity of contralesional deficit

the Fugl-Meyer et al. (1975) assessment of motor impairment (mild > 55, moderate > 35, Severe ≤ 35).

As illustrated in Fig. 2, in stroke survivors, JTHFT performance with the non-paretic arm was impacted substantially by both the severity of paretic arm impairment and the side of the brain that was damaged. Participants with the most severe paresis in the contralesional arm had the greatest motor deficit in the non-paretic arm. This effect depended on the side of the lesion, such that LHD was associated with greater non-paretic arm deficits than RHD. More specifically, LHD survivors took, on average, 300 % longer to complete the tasks than age matched control participants. Thus, stroke survivors who are forced to rely most extensively on their non-paretic arms for performance of ADL have the greatest deficits in coordination in the non-paretic arm. The fact that we tested these stroke survivors 1.8 years (±0.3 SE) on average after their stroke suggests that these deficits do not spontaneously improve over time.

*Remediation of the Non-paretic Arm* Remediation of the non-paretic arm is so novel that little empirical evidence exists as to whether such intervention could affect non-paretic arm control and coordination, and functional independence. We recently conducted a pilot study to test this hypothesis, which suggests that intense non-paretic arm training not only improves motor coordination and functional performance in the trained arm, but also improves functional independence and modestly improves paretic arm function. This is consistent with another recent pilot randomized clinical study that indicated that patients who received therapy that included training of the non-paretic arm improved functional independence greater than those receiving traditional therapy alone (Pandian et al. 2015). These studies provide preliminary evidence that non-paretic arm remediation might improve functional independence and even functioning in the paretic arm, possibly by increasing participation in ADL.

Our preliminary study was done with a three stroke survivor convenience sample. All participants showed significant contralesional paresis and deficits in performance of the non-paretic arm. All three participants were male. Details about age, initial Fugl-Meyer and initial JTHFT scores, and time since stroke are listed below:

SS 1: Age 58 Right MCA. Fugl-Meyer: 14, JTHFT: 69 s., TSS: 2 yr, 1 month.
SS 2: Age 66, Left MCA. Fugl-Meyer: 27, JTHFT: 125 s., TSS: 7 yrs, 0 month.
SS 3: Age 75, Left MCA. Fugl-Meyer: 45, JTHFT: 96 s., TSS: 3 yrs, 1 month.

We chose our outcome measures using the criteria that the tests can easily and quickly be administered in our environment, reflect the performance variables of interest, and have reported psychometrics in stroke survivors that include validity, reliability, and "responsiveness" or sensitivity for tracking changes in performance over time. These included the Functional Independence Measure (FIM), focusing on the components relevant to performance of ADL. In order to quantify paretic arm impairment, we used the upper extremity section of the Fugl-Meyer assessment. Lin et al. (2009) reported psychometric characteristics of the Fugl-Meyer, demonstrating substantial responsiveness (Effect Size 0.37–0.52), as well as high inter-rater reliability (ICC = 0.92–0.98). Our primary measures of non-paretic arm function include the Jebsen-Taylor Hand Function Test (JTHFT) and kinematic measurements, based on motion analysis of reaching movements. The JTHFT was discussed above, and we will describe our kinematic measures below.

Our kinematic recording and experimental setup is shown in Fig. 3. This set-up was used for both kinematic testing and virtual reality based movement training. Kinematic testing was be done for 2D, arm-supported reaching tasks, while training was done in 3-dimensions.

This custom designed apparatus allows participants to perform tasks in 2-dimensions as well as tasks that require participants to move in 3-dimensions and support their arms against gravity. For arm-supported tasks (2D), the forearms rest on air-cushion sleds that support the arms against gravity and nearly eliminate friction. For 3D tasks, the arm is held above the table top, and a cursor, representing hand position, can only be seen when the arm maintained off the table top. Task and movement feedback is displayed on a horizontal mirror positioned 35 cm above the table surface. This mirror reflects the stimuli presented on a horizontal, inverted, 60″ HDTV display. The proximal interphangeal joint of the index finger is reflected by the position of the cursor. Six Degree of freedom Trackstar® magnetic sensors are attached to the limbs, while positions of boney landmarks are used to digitize the hand, forearm, and upper arm segments, allowing calculation of 10 degrees of freedom per arm, recorded at 116 Hz. Data are low-pass filtered using a 12 Hz zero-lag Butterworth filter, prior to differentiating to yield velocity and acceleration profiles.

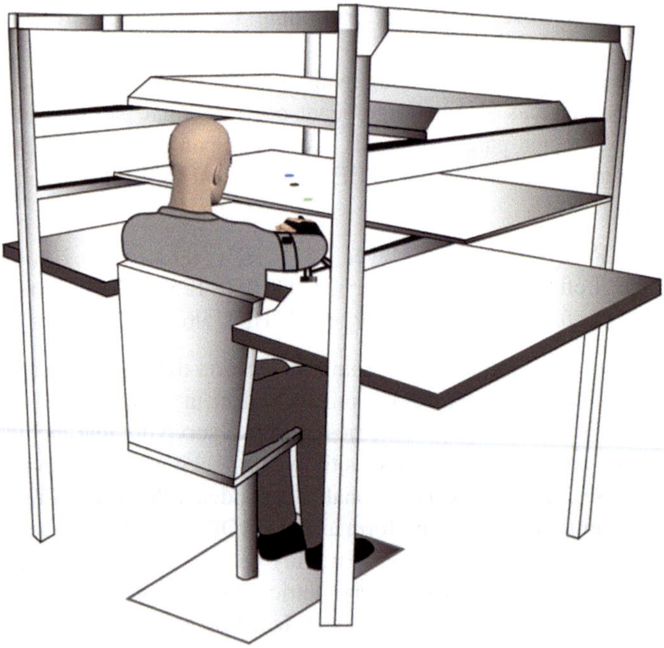

**Fig. 3** Participants were positioned facing a horizontally positioned mirror that reflected the 55″ monitor. The arm was supported over a horizontal table top, positioned just below shoulder height (adjusted to each individual's comfort), by an air-jet system, which reduced the effects of gravity and friction

Testing of non-paretic arm kinematics were done during arm-supported reaching tasks made in multiple directions, in order to be consistent with our previous characterization of non-paretic arm motor control deficits (see Fig. 1) (Schaefer et al. 2007, 2009b, 2012; Mutha et al. 2010, 2012, 2014). This study will use the same three direction reaching paradigm that was described in Fig. 1. Our primary kinematic measures included Distance Error, the vector magnitude difference between the two vectors, defined by (1) The beginning and end of the hand path, and (2) The specified start position and the target, Direction Error, the angular difference between the two vectors defined above, Hand path aspect ration, a measure of the hand path's deviation from linearity, measured as the ratio between the minor axis and the major axis of the hand path, maximum tangential velocity, and movement duration.

Figure 4 shows the timeline for the intervention. This preliminary study included a baseline period of 2 weeks, during which no intervention was done. This established stability of our dependent measures. Two baseline tests (Test 1 and Test 2) were given (2 weeks apart) prior to non-paretic arm training, and Test 3 was given immediately after the three-week training period. Subjects received 3 1.5 h sessions per week for 3 weeks. Figure 5 shows the 90 min intervention session. Participants were first exposed to virtual reality 'games' designed to focus on

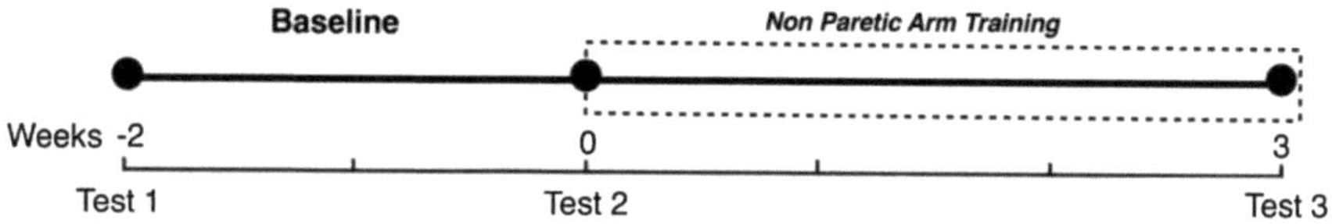


**Fig. 4** Timeline for the intervention—baseline period of 2 weeks, during which no intervention was done. Two baseline tests (Test 1 and Test 2) were given (2 weeks apart) prior to Non-Paretic arm training, and Test 3 was given immediately after the 3-week training period

specific components of control that we have previously shown to be deficient in the non-paretic arm of right or left hemisphere damaged stroke patients. For the first 20 min of training, patients practiced tasks adapted to the motor control deficits associated with each hemisphere. These included virtual shuffleboard, which focuses on predictive aspects of trajectory control, tracing games that focused on feedback-mediated position control, and targeted reaching movements that incorporate both aspects of motor control. These tasks were performed in our VR environment, but with the arm unsupported and held above the table top. For the next 40-minute phase of the session, participants practiced real-life tasks: First, they engaged in mild resistive exercises, using elastic putty and elastic bands, designed for resistive exercises of the hand and arm, respectively. After 5 min of preparation and a 5-minute break, all participants engaged in a series of six tasks, randomized in order between sessions. These tasks are shown in the flow chart in Fig. 5.

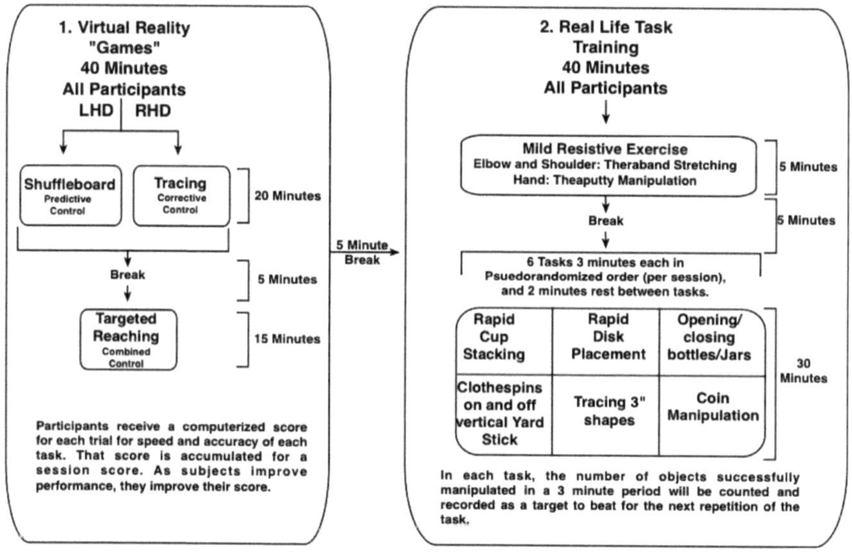


**Fig. 5** Participants were initially exposed to virtual reality games for 40 min, followed by real-life tasks for another 40 min

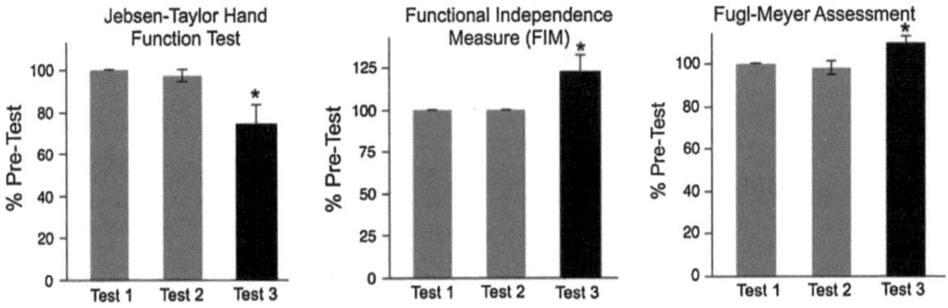


**Fig. 6** Mean ± SE for clinical assessments

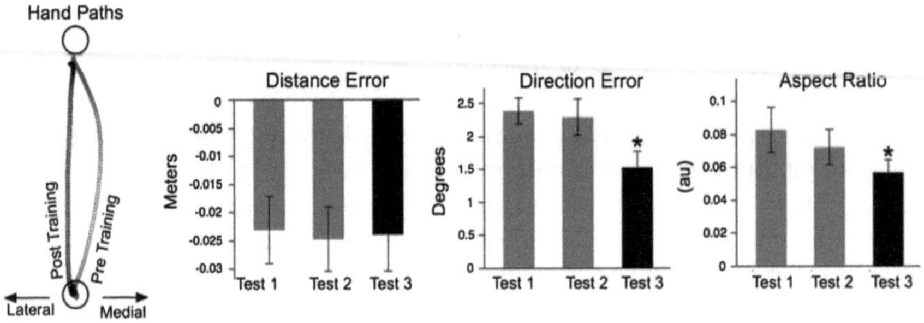


**Fig. 7** Sample Handpaths prior to and following training, and associated measures (Mean ± SE) for the two pretests (testl, test2), and the post test (test3)

The results of this pilot study are shown in Figs. 6 and 7. As predicted, there were no significant differences between our baseline measures, for Test 1 and Test 2. This demonstrates stability in baseline performance over time. The effect of training, however, was reflected by a 25.8 % reduction in performance time for the Jebsen-Taylor Hand Function Test, and a 9.8 % reduction in impairment, as measured by the Fugl-Meyer Assessement. Most importantly, the FIM score improved by 22.6 %, representing an increase of 3.3 ± (SE = 0.88) points out of 28 points on the upper extremity self-care components of the FIM.

Our kinematic measures in the non-paretic arm also showed improvements. Figure 7 shows example hand paths from a left hemisphere damaged patient using the left non-paretic arm, prior to training (Test 1) and following training (Test 2). Bar graphs show average kinematic measures (±SE), across all participants and test-sessions. Movements became straighter and more accurately directed toward the target (upper circle) following training, but the distance of the movements was accurate even in baseline movements. These effects were consistent across all subjects, as shown by the significant reduction in direction error and aspect ratio (a measure of hand path straightness), only for Test 3, post-training.

In summary, our pilot data showed promising improvements in non-paretic arm motor coordination and function. In addition, modest improvements in paretic arm impairment, measured by Fugl-Meyer score were also shown. Most importantly, these intervention related changes in performance translated to improvements in functional independence. Whether these effects are specific to the training paradigm, robust across a range of patients, and durable will need to be examined in future studies with more extensive recruitment.

## Discussion and Implications for Rehabilitation

For the most part, physical rehabilitation of upper extremity function following stroke has understandably been focused on training movements in the contralesional arm. However, the research discussed above provides compelling evidence that ipsilesional assessment and intervention also needs to be addressed. When patients have severe contralesional paresis, the ipsilesional arm is often the primary manipulator, or even the sole manipulator. Therefore, effective performance of ADL relies upon efficient coordination of this arm and hand (Haaland et al. 2012). We provided evidence that this arm shows substantial coordination deficits that limit performance of ADL and functional independence. Patients who must live with a severely paretic dominant arm, unfortunately have the most severe non-paretic arm deficits in coordination and functional performance.

We have provided preliminary evidence that intensive training of this ipsilesional, non-paretic arm could substantially improve functional independence in patients with hemiparesis. Although preliminary, this is among the first empirical evidence indicating that such intervention might lead to positive effects on motor performance and functional independence. However, our findings are consistent with a recent pilot intervention study that compared a group of patients who received therapy that included training of the non-paretic arm and the paretic arm to another group who only received "traditional" therapy for only the paretic arm (Pandian et al. 2015). The results indicated that when non-paretic intervention was combined with paretic arm training, the speed and accuracy of non-paretic arm movements improved, and the impairment level of the paretic arm modestly improved, when compared to patients who received traditional therapy alone. Combined with our current results, these findings indicate that focused non-paretic arm training might produce both improvements in non-paretic arm motor performance and in paretic arm function, both of which should facilitate improvements in functional independence. This finding appears at odds with the well-studied phenomenon of learned non-use of the paretic arm, an effect that has been successfully addressed by constraining the non-paretic arm in patients with moderate to mild paresis (Wolf et al. 1989, 2002, 2005, 2006). Constraint induced movement therapy uses constraint of the non-paretic arm in combination with intense practice of the paretic arm to facilitate recovery of paretic arm function. However, it should be stressed that this approach is most effective in patients with high-moderate to mild paresis, who are able to manipulate objects with the paretic arm, and engage in unilateral functional activities with that arm. The current study addresses recovery of function in patients with severe contralesional paresis, who generally have no voluntary movement of the wrist and fingers. Nevertheless, there is no conclusive evidence to predict whether non-paretic arm training will influence paretic arm function, either positively or negatively. Further and more extensive intervention research is necessary to address this question.

It would be inaccurate to suggest that the non-paretic arm is not addressed at all in rehabilitation. In fact, bilateral arm training has a long history in rehabilitation practice. Because most ADL are performed with both hands contributing to different aspects of the activity (Wetter et al. 2005; Haaland et al. 2012), bilateral training should be an important component to therapeutic intervention in stroke patients. It should also be stressed that bilateral training may be necessary because unilateral training may not spontaneously carry-over to bilateral activities. Recent research has indicated that in healthy young participants, adaptation to novel force fields imposed with robotic devices, and to distorted visual feedback, transfers only partially to bilateral movements, when the same arm experiences the imposed environments (Nozaki et al. 2006; Wang and Sainburg 2009). This suggests that learning a new task unilaterally (with each hand separately) may not spontaneously transfer to bilateral performance. These findings emphasize the importance of bilateral coordination in rehabilitation. Indeed, bilateral training has a long history in Occupational Therapy treatment, where manipulation of dowels and rolling pins has often been used to encourage bilateral arm use.

However, notwithstanding the importance of bilateral movement intervention, our current results indicate the importance of intense non-paretic arm remediation in patients with severe contralesional impairment. Bilateral training cannot challenge the non-paretic arm to the extent required for remediation. As Winstein and colleagues (Rose and Winstein 2005) demonstrated, when using the two arms together, the non-paretic arm slows down and adapts its movement characteristics to match that of the paretic arm. Obviously, this cannot challenge the speed and accuracy requirements of non-paretic arm coordination to the extent necessary for improvement of non-paretic function. It should also be stressed that in patients with severe contralesional paresis, many ADL must be performed unilaterally with the non-paretic arm. The fact that non-paretic arm deficits can be extensive and can persist throughout the chronic phase of stroke (see Fig. 2) indicates that forced use of the non-paretic arm does not spontaneously lead to amelioration of non-paretic arm motor deficits. We suggest that this is because the patients are not exposed to graded difficulty in challenging tasks, such as was done in the current training paradigm.

## Take Home Message

The research presented in this chapter has demonstrated that many stroke patients with severe moderate to paresis in the contralesional arm have substantial and functionally limiting motor coordination deficits in the non-paretic ipsilesional arm.

Furthermore, the combination of moderate to severe paresis with persistent motor deficits in the non-paretic arm limits performance of and participation in ADL. Based on our current results we recommend intense rehabilitation, sequentially focused on each arm, followed by practice in bilateral ADL when possible. It is critical to test this approach in a broader sample of stroke patients. A potential caveat to such a training scheme is reflected by the work of Jones et al. (1989). They showed, in an acute model of stroke in rats, that initial training of ipsilesional forelimb reaches limits the subsequent response to training in the contralesional forelimb. While this study was done in an acute lesion model in rats that bears only partial resemblance to the impairments seen in humans, it is possible that training the ipsilesional arm in the acute phase of stroke could be counter productive. To date, no studies of ipsilesional training in acute stroke are available in the literature. In direct contrast to the interpretation of the study of Jones et al., studies of motor learning in humans often demonstrate transfer of learning between the arms, (Wang and Sainburg a, b), and mirror training intervention has shown positive transfer between the arms in stroke patients (Stevens and Stoykov 2004; Oujamaa et al. 2009; Selles et al. 2014). We recommend further studies of ipsilesional arm intervention in stroke survivors with moderate to severe contralesional paresis to determine whether such training can positively affect functional outcomes and participation in human stroke survivors.

## References


Beebe JA, Lang CE (2009) Relationships and responsiveness of six upper extremity function tests during the first six months of recovery after stroke. J Neurol Phys Ther 33:96–103

Bisazza A, Rogers LJ, Vallortigara G (1998) The origins of cerebral asymmetry: a review of evidence of behavioural and brain lateralization in fishes, reptiles and amphibians. Neurosci Biobehav Rev 22:411–426

Chestnut C, Haaland KY (2008) Functional significance of ipsilesional motor deficits after unilateral stroke. Arch Phys Med Rehabil 89:62–68

Dassonville P, Zhu XH, Uurbil K, Kim SG, Ashe J (1997) Functional activation in motor cortex reflects the direction and the degree of handedness. Proc Natl Acad Sci USA 94:14015–14018

Desrosiers J, Bourbonnais D, Bravo G, Roy PM, Guay M (1996) Performance of the 'unaffected' upper extremity of elderly stroke patients. Stroke 27:1564–1570

Fisk JD, Goodale MA (1988) The effects of unilateral brain damage on visually guided reaching: hemispheric differences in the nature of the deficit. Exp Brain Res 72:425–435

Fugl-Meyer AR, Jaasko L, Leyman I, Olsson S, Steglind S (1975) The post-stroke hemiplegic patient. 1. A method for evaluation of physical performance. Scand J Rehabil Med 7:13–31

Gazzaniga MS (1998) The split brain revisited. Sci Am 279:50–55

Grabowski M, Brundin P, Johansson BB (1993) Paw-reaching, sensorimotor, and rotational behavior after brain infarction in rats. Stroke 24:889–895

Haaland KY (2006) Left hemisphere dominance for movement. Clin Neuropsychol 20:609–622

Haaland KY, Harrington D (1989a) The role of the hemispheres in closed loop movements. Brain & Cogn 9:158–180

Haaland KY, Harrington DL (1989b) Hemispheric control of the initial and corrective components of aiming movements. Neuropsychologia 27:961–969

Haaland KY, Harrington DL (1996) Hemispheric asymmetry of movement. Curr Opin Neurobiol 6:796–800

Haaland KY, Harrington DL, Knight RT (2000) Neural representations of skilled movement. Brain 123(Pt 11):2306–2313

Haaland KY, Mutha PK, Rinehart JK, Daniels M, Cushnyr B, Adair JC (2012) Relationship between arm usage and instrumental activities of daily living after unilateral stroke. Arch Phys Med Rehabil 93:1957–1962

Haaland KY, Prestopnik JL, Knight RT, Lee RR (2004) Hemispheric asymmetries for kinematic and positional aspects of reaching. Brain 127:1145–1158

Haaland KY, Schaefer SY, Knight RT, Adair J, Magalhaes A, Sadek J, Sainburg RL (2009) Ipsilesional trajectory control is related to contralesional arm paralysis after left hemisphere damage. Exp Brain Res 196:195–204

Jebsen RH, Griffith ER, Long EW, Fowler R (1971) Function of "normal" hand in stroke patients. Arch Phys Med Rehabil 52:170–174 (passim)

Jebsen RH, Taylor N, Trieschmann RB, Trotter MJ, Howard LA (1969) An objective and standardized test of hand function. Arch Phys Med Rehabil 50:311–319

Jones RD, Donaldson IM, Parkin PJ (1989) Impairment and recovery of ipsilateral sensory-motor function following unilateral cerebral infarction. Brain 112(Pt 1):113–132

Kawashima R, Matsumura M, Sadato N, Naito E, Waki A, Nakamura S, Matsunami K, Fukuda H, Yonekura Y (1998) Regional cerebral blood flow changes in human brain related to ipsilateral and contralateral complex hand movements—a PET study. Eur J Neurosci 10:2254–2260

Kawashima R, Yamada K, Kinomura S, Yamaguchi T, Matsui H, Yoshioka S, Fukuda H (1993) Regional cerebral blood flow changes of cortical motor areas and prefrontal areas in humans related to ipsilateral and contralateral hand movement. Brain Res 623:33–40

Kuypers HG (1978) The motor system and the capacity to execute highly fractionated distal extremity movements. Electroencephalogr Clin Neurophysiol (Suppl): 429–431

Kuypers HG (1982) A new look at the organization of the motor system. Prog Brain Res 57:381–403

Lawrence DG, Kuypers HG (1968) The functional organization of the motor system in the monkey. I. The effects of bilateral pyramidal lesions. Brain 91:1–14

Lin JH, Hsu MJ, Sheu CF, Wu TS, Lin RT, Chen CH, Hsieh CL (2009) Psychometric comparisons of 4 measures for assessing upper-extremity function in people with stroke. Phys Ther 89: 840–850

MacNeilage PF, Rogers LJ, Vallortigara G (2009) Origins of the left & right brain. Sci Am 301:60–67

Mani S, Mutha PK, Przybyla A, Haaland KY, Good DC, Sainburg RL (2013) Contralesional motor deficits after unilateral stroke reflect hemisphere-specific control mechanisms. Brain 136:1288–1303

Mani S, Przybyla A, Good DC, Haaland KY, Sainburg RL (2014) Contralesional arm preference depends on hemisphere of damage and target location in unilateral stroke patients. Neurorehabil Neural Repair 28:584–593

Mutha PK, Haaland KY, Sainburg RL (2012) The effects of brain lateralization on motor control and adaptation. J Mot Behav 44:455–469

Mutha PK, Haaland KY, Sainburg RL (2013) Rethinking motor lateralization: specialized but complementary mechanisms for motor control of each arm. PLoS ONE 8:e58582

Mutha PK, Sainburg RL, Haaland KY (2010) Coordination deficits in ideomotor apraxia during visually targeted reaching reflect impaired visuomotor transformations. Neuropsychologia 48:3855–3867

Mutha PK, Sainburg RL, Haaland KY (2011a) Critical neural substrates for correcting unexpected trajectory errors and learning from them. Brain 134:3647–3661

Mutha PK, Sainburg RL, Haaland KY (2011b) Left parietal regions are critical for adaptive visuomotor control. J Neurosci 31:6972–6981

Mutha PK, Stapp LH, Sainburg RL, Haaland KY (2014) Frontal and parietal cortex contributions to action modification. Cortex 57:38–50

Nozaki D, Kurtzer I, Scott S (2006) Limited transfer of learning between unimanual and bimanual skills within the same limb. Nat Neurosci 9:3

Oujamaa L, Relave I, Froger J, Mottet D, Pelissier JY (2009) Rehabilitation of arm function after stroke. Literature review. Ann Phys Rehabil Med 52:269–293

Pandian S, Arya KN, Kumar D (2015) Effect of motor training involving the less-affected side (MTLA) in post-stroke subjects: a pilot randomized controlled trial. Topics in stroke rehabilitation. doi:10.1179/1074935714Z.0000000022

Pohl PS, Winstein CJ (1999) Practice effects on the less-affected upper extremity after stroke. Arch Phys Med Rehabil 80:668–675

Poole JL, Sadek J, Haaland KY (2009) Ipsilateral deficits in 1-handed shoe tying after left or right hemisphere stroke. Arch Phys Med Rehabil 90:1800–1805

Prestopnik J, Haaland K, Knight R, Lee R (2003a) Hemispheric dominance for open and closed loop movements. Soc Neurosci Abstr 30

Prestopnik J, Haaland K, Knight R, Lee R (2003b) Hemispheric dominance in the parietal lobe for open and closed loop movements. J Int Neuropsychol Soc 9:1–2

Rapin I, Tourk LM, Costa LD (1966) Evaluation of the Purdue Pegboard as a screening test for brain damage. Dev Med Child Neurol 8:45–54

Rose DK, Winstein CJ (2005) The co-ordination of bimanual rapid aiming movements following stroke. Clin Rehabil 19:452–462

Sainburg RL (2002) Evidence for a dynamic-dominance hypothesis of handedness. Exp Brain Res 142:241–258

Sainburg RL (2014) Convergent models of handedness and brain lateralization. Front Psychol 5:1092

Schaefer SY, Haaland KY, Sainburg RL (2007) Ipsilesional motor deficits following stroke reflect hemispheric specializations for movement control. Brain 130:2146–2158

Schaefer SY, Haaland KY, Sainburg RL (2009a) Dissociation of initial trajectory and final position errors during visuomotor adaptation following unilateral stroke. Brain Res 1298:78–91

Schaefer SY, Haaland KY, Sainburg RL (2009b) Hemispheric specialization and functional impact of ipsilesional deficits in movement coordination and accuracy. Neuropsychologia 47: 2953–2966

Schaefer SY, Mutha PK, Haaland KY, Sainburg RL (2012) Hemispheric specialization for movement control produces dissociable differences in online corrections after stroke. Cereb Cortex 22:1407–1419

Selles RW, Michielsen ME, Bussmann JB, Stam HJ, Hurkmans HL, Heijnen I, de Groot D, Ribbers GM (2014) Effects of a mirror-induced visual illusion on a reaching task in stroke patients: implications for mirror therapy training. Neurorehabil Neural Repair 28:652–659

Stevens JA, Stoykov ME (2004) Simulation of bilateral movement training through mirror reflection: a case report demonstrating an occupational therapy technique for hemiparesis. Topics Stroke Rehabil 11:59–66

Tanji J, Okano K, Sato KC (1988) Neuronal activity in cortical motor areas related to ipsilateral, contralateral, and bilateral digit movements of the monkey. J Neurophysiol 60:325–343

Wang J, Sainburg RL (2009) Generalization of visuomotor learning between bilateral and unilateral conditions. J Neurophysiol 102:2790–2799

Wetter S, Poole JL, Haaland KY (2005) Functional implications of ipsilesional motor deficits after unilateral stroke. Arch Phys Med Rehabil 86:776–781

Winstein CJ, Merians AS, Sullivan KJ (1999) Motor learning after unilateral brain damage. Neuropsychologia 37:975–987

Winstein CJ, Pohl PS (1995) Effects of unilateral brain damage on the control of goal-directed hand movements. Exp Brain Res 105:163–174

Winstein CJ, Pohl PS, Cardinale C, Green A, Scholtz L, Waters CS (1996) Learning a partial-weight-bearing skill: effectiveness of two forms of feedback [published erratum appears in Phys Ther 1997 Mar; 77(3):328]. Phys Ther 76:985–993

Wolf SL, Blanton S, Baer H, Breshears J, Butler AJ (2002) Repetitive task practice: a critical review of constraint-induced movement therapy in stroke. Neurologist 8:325–338

Wolf SL, Lecraw DE, Barton LA, Jann BB (1989) Forced use of hemiplegic upper extremities to reverse the effect of learned nonuse among chronic stroke and head-injured patients. Exp Neurol 104:125–132

Wolf SL, Thompson PA, Morris DM, Rose DK, Winstein CJ, Taub E, Giuliani C, Pearson SL (2005) The EXCITE trial: attributes of the Wolf motor function test in patients with subacute stroke. Neurorehabil Neural Repair 19:194–205

Wolf SL, Winstein CJ, Miller JP, Taub E, Uswatte G, Morris D, Giuliani C, Light KE, Nichols-Larsen D (2006) Effect of constraint-induced movement therapy on upper extremity function 3 to 9 months after stroke: the EXCITE randomized clinical trial. JAMA 296:2095–2104

Wyke M (1967) Effect of brain lesions on the rapidity of arm movement. Neurology 17:1113–1120

Yadav V, Sainburg RL (2011) Motor lateralization is characterized by a serial hybrid control scheme. Neuroscience 196:153–167

Yadav V, Sainburg RL (2014) Handedness can be explained by a serial hybrid control scheme. Neuroscience 278:385–396

Yarosh CA, Hoffman DS, Strick PL (2004) Deficits in movements of the wrist ipsilateral to a stroke in hemiparetic subjects. J Neurophysiol 92:3276–3285

# Control of Cycling Limb Movements: Aspects for Rehabilitation

**Jozsef Laczko, Mariann Mravcsik and Peter Katona**

**Abstract** Walking, swimming, cycling, and running are cyclic movements that are often performed in training programs or rehabilitation protocols by athletes or people with neuromotor disorders. The muscular and kinematic activities that are acquired during cyclic movements reveal control principles, especially for the optimization and stabilization of motor performance, for a given criterion in rehabilitation processes. The influence of external loads and resistive forces on limb movements should be considered in rehabilitation protocols and when assessing physical activity levels or defining activity patterns for the artificial control of limb movements. This chapter focuses on special cyclic limb movements: lower and upper limb cycling. Two aspects of this research and applications are discussed. First, variances of movement patterns are examined at different levels of the motor system (endpoint, joint configuration, muscle) during unimanual right and left arm cycling and bimanual cycling movements. Second, it is shown that the muscle activity patterns that are acquired during lower and upper limb cycling in able-bodied people may be used to define and improve stimulation patterns for functional electrical stimulation-driven cycling movements in spinal cord-injured individuals. This report also discusses the advantages of the application and control of these types of movements for the rehabilitation of people with paralyzed limbs.

J. Laczko (✉)
Department of Information Technology and Biorobotics, Faculty of Science, University of Pécs, 6, Ifjusag St, Pécs H-7624, Hungary
e-mail: laczkoj@gamma.ttk.pte.hu

J. Laczko · M. Mravcsik
Laboratory of Rehabilitation-Technology, Wigner Research Centre for Physics, Hungarian Academy of Sciences, 29–33, Konkoly Thege M. St, Budapest H-1121, Hungary
e-mail: mravcsik.mariann@gmail.com

P. Katona
University of Physical Education, 44 Alkotas St, Budapest H-1123, Hungary
e-mail: katona@tf.hu

M. Mravcsik
Faculty of Science, University of Pécs, 6, Ifjusag St, Pécs H-7624, Hungary



J. Laczko and M.L. Latash (eds.), *Progress in Motor Control*, Advances in Experimental Medicine and Biology 957,
DOI 10.1007/978-3-319-47313-0_15

**Keywords** Unimanual · Bimanual · Crank resistance · Upper limb · Lower limb · Power production · Electromyography · Functional electrical stimulation (FES)

## Introduction

Cyclic limb movements are often used for the rehabilitation of people with neural-based motor dysfunctions. Different types of cyclic movements were investigated. Differences between discrete and cyclic limb movements were analyzed in healthy people (Hogan and Sternad 2007; Smits-Engelsman et al. 2002, 2006; van Mourik and Beek 2004) and in survivors of stroke (Massie et al. 2014). Cycling upper or lower limb movements are a special type of movement that is clearly distinguishable from other rhythmic movements and sequences of discrete movements. For example, cycling arm movements occur when a circle is continuously and repetitively drawn or tracked by a pen held in the hand. This task may reveal motor control properties in people with Parkinson's disease (Keresztenyi et al. 2009). However, the most often performed cycling limb movement is cycling on a bicycle, tricycle or an ergometer, which are frequently available at many rehabilitation centers and in the homes of people with neural-based movement disorders; but, these types of movements have received less attention from a motor control perspective. Cycling movements on an ergometer or bicycle constrain the path of the endpoint of the limb to a plane and a circle within the plane. This circle is the path on which the foot (placed on the pedal in lower limb cycling) or the hand (grasping the handlebar during upper limb cycling) moves. The endpoint of the limb must move on the same path during these movements, even if external conditions, such as resistive forces, are imposed. This constraint makes cycling movements different than other types of cyclic limb movements, such as walking, swimming, or repetitively executed unimanual and bimanual reaching movements, in which the endpoint paths are not restricted (Laczko et al. 2001; Domkin et al. 2002) and may vary accross trials. However, studies of repetitively executed reaching and object transporting arm movements (Tibold et al. 2011; Tibold and Laczko 2012) have demonstrated that the effect of load was not significant on the variances of endpoint trajectory. Furthermore, the load did not affect torque profiles in the elbow and shoulder joints (Tibold and Laczko 2012). Participants grasped and held objects of various weights in their hand and placed the object from an initial to a target position, vertically upward or downward. Movements were executed repetitively in a series. The task was performed using two objects, a light, empty compact disc case, and a load of 20 N, and it was found that the effect of load was not significant on the variances of endpoint trajectory. The arm configuration variances (defined by the variances of intersegmental joint angles in the shoulder, elbow, and wrist) when transporting the heavy object upward (against gravity) did not differ significantly from those observed for movements executed with the light object. The arm configuration variance was significantly higher when the object was transported downward for movements executed with heavy object compared to movements with light object. The load had a significant effect on the variances of muscle activity patterns (EMG amplitudes) in both movement directions. A larger load was associated with increased variances in flexor and extensor activities. Beside external load, neural disorders also influence motor variances, but not in all motor tasks.

Cyclic reaching movements were examined in stroke survivors (Massie et al. 2012), and variances in the arm movements were compared for the affected and non-affected arms. Participants were asked to reach continuously between the center of two targets in a sagittal plane as quickly and accurately as possible. Variance was a measure of the consistency of fingertip contact points in relation to the average contact point. The variance of the target contact did not differ for the affected and non-affected arms. Endpoint position in these cyclic object transporting and reaching movements was stabilized by neural control, and this control was reflected in unaffected endpoint variances.

The variance in cycling movements at the endpoint level (foot or hand) is expected to be small and practically independent from crank resistance if cycling cadence is constant because of the fixed path of the endpoint. We hypothesized that the effect of crank resistance on arm configuration variance and muscle activity variance is significant. The central nervous system may compensate for external forces by employing altered and more variable motor patterns at joint and muscle levels to ensure successful execution of the motor task. This effect was previously proposed by investigating swimming arm movements (Martens et al. 2015).

Here we present results related to the observed variances of movement patterns obtained in upper limb cycling. We investigated arm configuration variance in a joint space defined by intersegmental joint angles in the shoulder, elbow and wrist joints and muscle activity variances in a muscle space defined by elbow and shoulder muscles. We compared arm configuration variances and muscle activity variances for the dominant and non-dominant arms and for unimanual and bimanual cycling. The effect of crank resistance and pedaling cadence on muscle activities for lower limb cycling are also presented. The application of functional electrical stimulation-driven cycling movements of spinal cord-injured individuals is also discussed.

## Arm Configuration and Muscle Activity Variances in Upper Limb Cycling

Arm cycling on an ergometer is often used in medical rehabilitation, but there is much less literature available on kinematic and muscle activity data on these movements than there is on lower limb cycling on a bicycle. External load in cycling movement is manifested in the crank resistance of a bicycle or ergometer. The effect of crank resistance on arm cranking movements of people with differing levels of spinal cord injury was investigated and appropriate resistance loads were determined for C5, C6, and C7 levels of injury (Jacobs et al. 2003, 2004).

We investigate arm cycling movements of 12 right-handed, able-bodied participants (26 ± 4.9 years) when cycling under 2 resistance conditions (RC) on a MEYRA (Kalletal, Germany) ergometer. Cycling was performed bimanually and unimanually by the right and left arms in a random order with a cadence of 60 revolutions per minute (rpm) and directed by a metronome. Participants were cycling for 30 s in each condition. Joint coordinates and muscle activities (surface EMG) of biceps, triceps, and delta anterior and posterior muscles were measured and analyzed. We paid special attention to the variances of arm movements and the dependence of variances on cycling conditions.

The data presented were obtained when cycling under crank resistances of 1.16 and 3.09 Nm. Time courses of joint coordinates and muscle activities (EMG amplitude) were segmented to cycles based on the number of cycles completed by the subjects (30). Separate cycles were time normalized, and endpoint (hand position) variances, arm configuration (shoulder, elbow, wrist angles), and muscle activity (biceps, triceps, ddelta anterior, delta posterior) variances across cycles were computed and averaged for each participant and cycling condition (for each RC, for bimanual and unimanual cycling, for the left and right arm) separately. The difference between variances obtained in the two RCs was assessed using Student's T test ($p = 0.05$), and the differences between variances in the left and right arm and between variances obtained in unimanual and bimanual cycling was also compared ($p = 0.05$). The endpoint variance did not depend on crank resistance, as expected, because of the fixed endpoint path and constant pedaling cadence. The question was whether the angular variances (arm configuration variance) and muscle activity variances were significantly different under various cycling conditions. Preliminary data showed that the angular variances did not differ significantly by crank resistance (Mravcsik and Laczko 2015) for the right or left arm and for bimanual or unimanual cycling. Table 1 presents the mean (across subjects) arm configuration variances per degrees of freedom and standard errors.

**Table 1** Mean arm configuration variances (and SE) in various cycling conditions

| Crank resistance (Nm) | Variance ($^{o2}$) Bimanual | | Variance ($^{o2}$) Unimanual | |
|---|---|---|---|---|
| | Left | Right | Left | Right |
| 1.16 | 18.4 (0.9) | 15.4 (1.9) | 16.4 (1.6) | 16.5 (1.9) |
| 3.09 | 15.3 (1.3) | 16.7 (2.0) | 17.7 (1.3) | 15.2 (1.5) |

## *Comparison of Arm Configuration Variances Obtained in Unimanual and Bimanual Cycling*

There were no significant differences between arm configuration variances obtained in unimanual and bimanual cycling (Fig. 1).

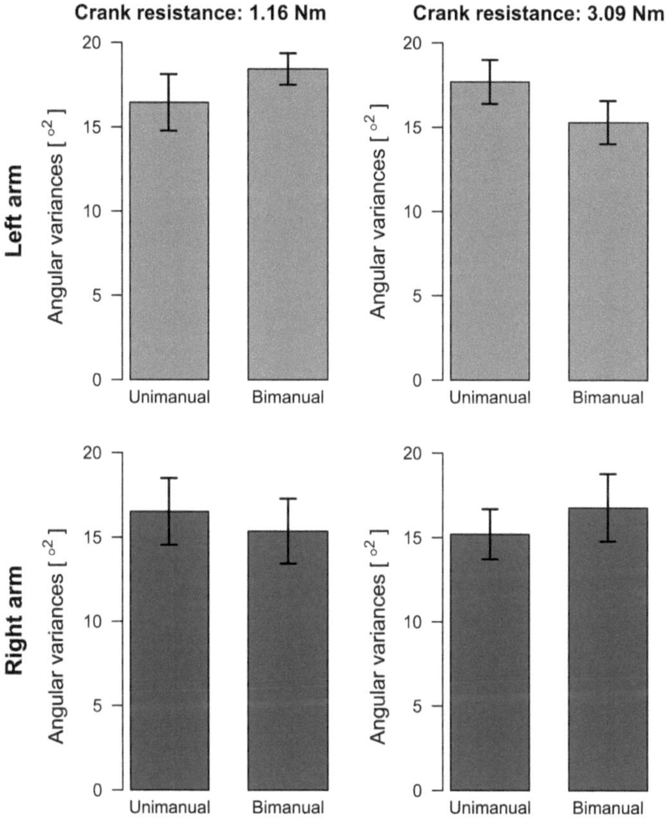


**Fig. 1** Comparison of arm configuration variances in bimanual and unimnual cyling. Average (across subjects) arm configuration variances in the left arm (*upper diagrams*) and right arm (*lower diagrams*) when cycling was performed against a crank resistance of 1.16 and 3.09 Nm. *Vertical lines* above the bars represent standard errors

## *Comparison of Arm Configuration Variances Obtained in the Two Arms*

There were no significant differences in arm configuration variances obtained in the dominant and non-dominant arms (Fig. 2).

These results suggest that arm configuration variance was not affected significantly by the arm (left or right) used or the number of arms (one or two) involved in the cycling.

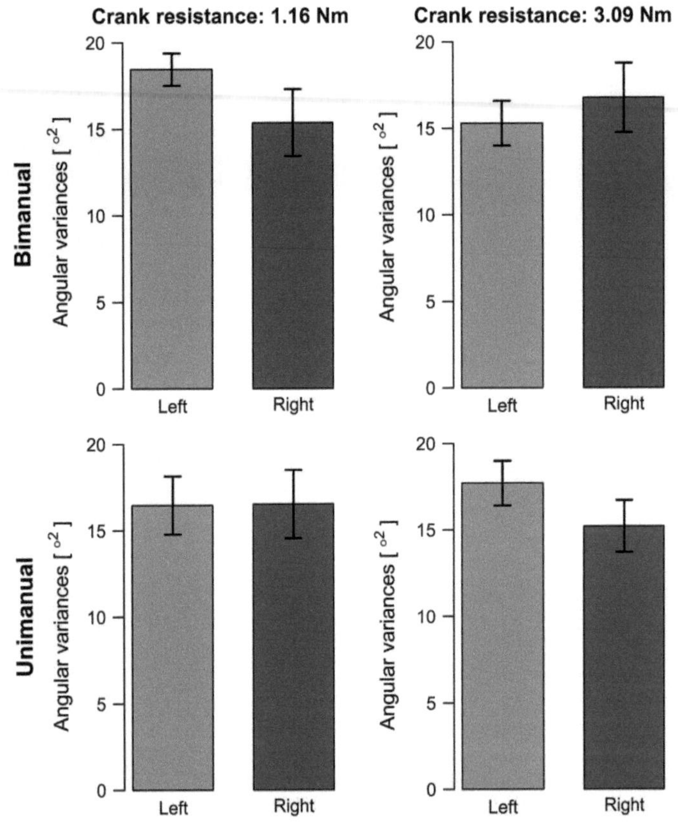


**Fig. 2** Comparison of arm configuration variances in the *left* and *right* arms. *Upper diagrams* average arm configuration variances (and SE) obtained in bimanual cycling *Lower diagrams* average arm configuration variances (and SE) in unimanual cycling

## *Comparison of Muscle Activity Variances Obtained in Unimanual and Bimanual Cycling*

We found significant differences in the muscle activity variances when unimanual and bimanual cycling were compared (Mravcsik et al. 2016a) (Fig. 3). The muscle activity variances were significantly higher in unimanual compared to bimanual cycling in the left and right arms.

Thus muscle activity variances were affected by crank resistance and also by the number of arms participated in the cycling. The question remained whether these effects were different for the dominant and non-dominant arm or not?

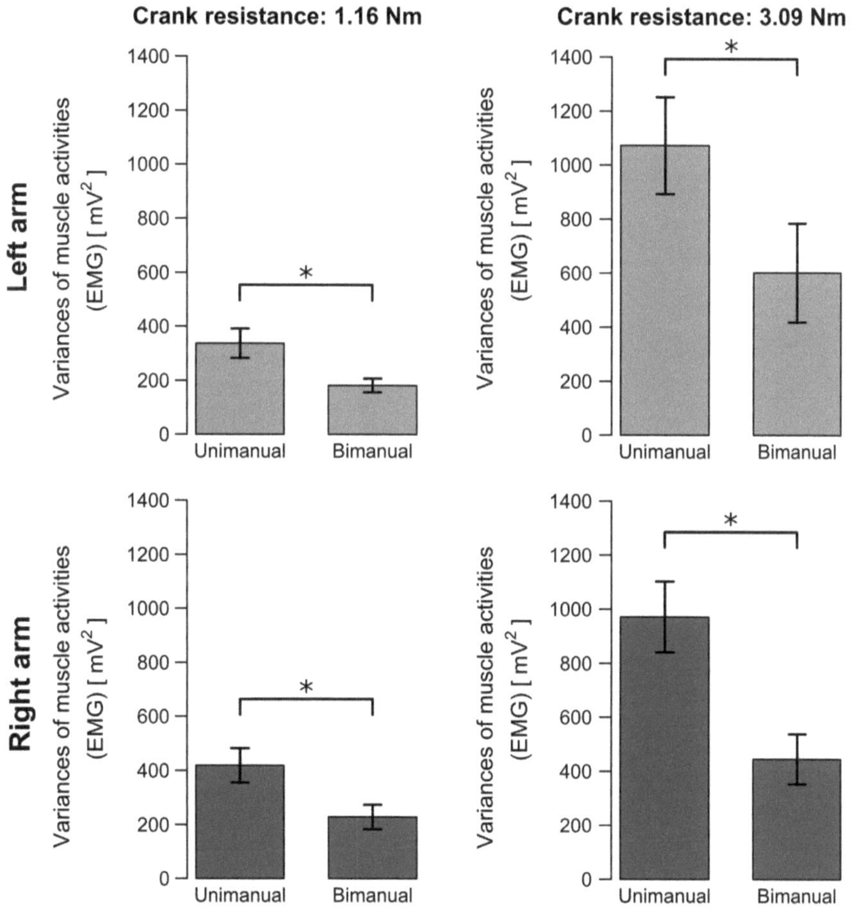


**Fig. 3** Average muscle activity variances obtained during unimanual versus bimanual cycling in various cycling conditions. *Upper panels* present results for the left arm *lower panels* for the right arm; *Left panels* for crank resistance of 1.16, *Right panels* for crank resistance of 3.09

## *Comparison of Muscle Activity Variances Obtained in the Two Arms*

There were no significant differences between muscle activity variances obtained in the left and right arms (Fig. 4).

Therefore, during arm cycling neither arm configuration variances nor muscle activity variances differed significantly for the dominant and non-dominant arm. This result was obtained for low (1.16 Nm) and high (3.09 Nm) crank resistances and for unimanual and bimanual cycling too. This result was surprising because there are significant differences in the movement characteristics of the dominant and non-dominant arms in most motor tasks. Movement pattern variances did not differ significantly for the two arms during cycling movements, which may reflect an equally stable neural control of the two arms during arm cycling.

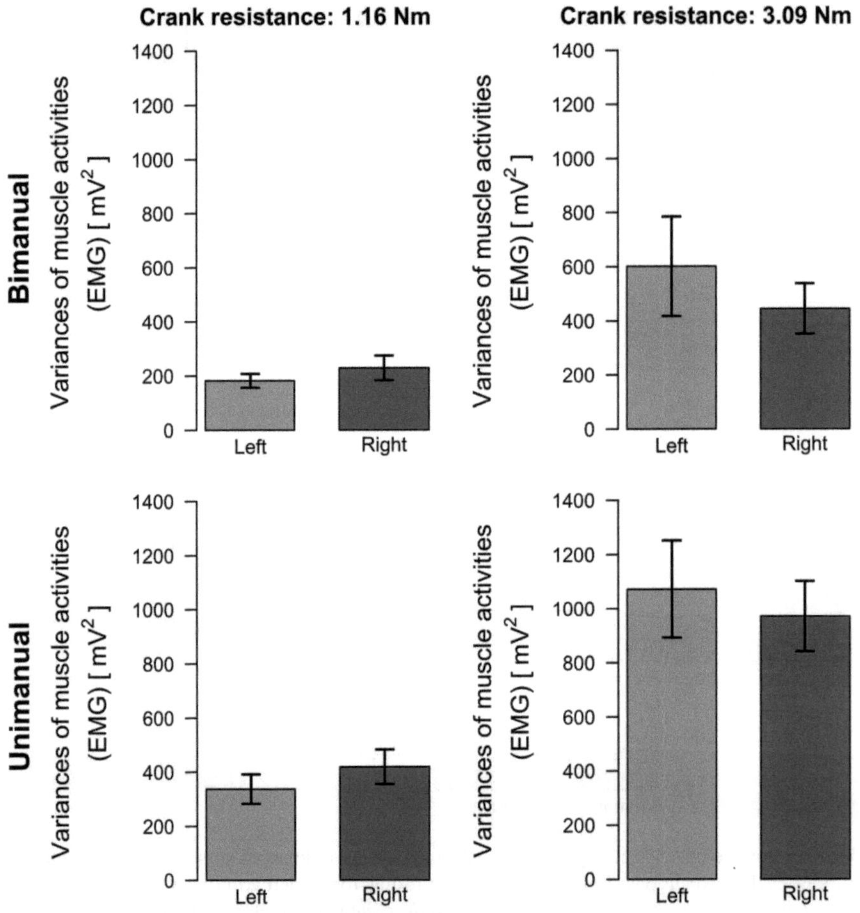


**Fig. 4** Comparison of the left and right arms muscle activity variances in bimanual and unimanual cycling. Results obtained when cycling was performed against crank resistance of 1.16 Nm (*left diagrams*) and 3.09 Nm (*right diagrams*)

We note that arm configuration variances in the bimanual reaching movements of right-handed participants were compared between arms (Domkin et al. 2002). Arm configuration variance before practice was higher in the right arm (which used more available joint angle combinations) than in the left arm, but the arm configuration variances became equal in the two arms after practicing because the variance in the right arm decreased more than the variance in the left arm. Participants in our pilot studies of arm cycling did not practice the movement, and even so there were no significant differences in arm configuration variances between the arms.

Asymmetry of performance is known in human limb movements (Sainburg 2002). Sainburg and colleagues developed the dynamic dominance model, which is presented in another chapter of this book. This model proposes that dominant arm control in right-handed individuals is optimized for the planning of effective movement trajectories and intersegmental coordination under stable environmental conditions, whereas, the non-dominant left arm control is optimized for impedance control and responses to unexpected changes in environmental conditions. This model was nicely proven for reaching, goal directed arm movements. The left hemisphere in right-handed people is specialized for controlling well-established patterns of behavior, and the right hemisphere is responsible for responding to changes in the environment. We assume that arm cycling is a special motor task in that respect that movement stability (quantified by variability) is comparable and did not differ significantly between the arms. This manual symmetry in variances of arm cycling movements may be interesting from a rehabilitation point of view. Arm cycling is an ideal rehabilitation intervention when muscle strengthening in both arms is the goal and motor stability is vital. Motor stability was quantified here by joint configuration (joint angles) variances and muscle activity (EMG amplitudes) variances.

The role of lateralization in arm cycling is under investigation. Future studies may reveal how stroke survivors perform cycling movements with the paretic and non-paretic arms. Bimanual and unimanual arm cycling should be included in training protocols.

Keeping in mind that controlling lower limb cycling is different than upper limb cycling we note a result about symmetry found in lower limb cycling and relates to rehabilitation intervention (Ambrosini et al. 2012). It has been shown that there was no significant difference between the paretic and non-paretic leg in hemiparetic patients in terms of joint torque and muscle activity, after having functional electrical simulation driven cycling trainings. The symmetry in variances of activity patterns may be an object of further research, and it should also be examined for upper limb cycling.

Differences or similarities in the control of arm cycling in various cycling conditions may be reflected not only in motor variances but in coactivation of flexor-extensor muscle pairs too. We analyzed the coactivation intervals of flexor-extensor muscle pairs (biceps-triceps and delta anterior-delta posterior muscle pairs) in both arms. According to our pilot study coactivation ranges of flexor-extensor muscles do not differ significantly in various resistance conditions neither in the dominant arm, nor in the non-dominant arm (Mravcsik and Laczkó 2014). These results indicate, that without significant decreases in muscle coactivation, only greater individual muscle forces ensure the increased power output that is required to cycle against higher resistance, and this holds for both arms.

## Lower Limb Cycling

Lower limb cycling was studied and analyzed to a greater extent than arm cycling. This emphasis is partially due to sport applications, and lower limb cycling is more widely used in rehabilitation protocols when reaching higher power output is the goal.

Power output while cycling on an ergometer may be changed by altering pedaling cadence or crank resistance. We investigated these effects in lower limb cycling and hypothesized that changes in crank resistance and pedaling cadence would have a significant effect on the stability of neural control of the movement. When cycling faster and/or against higher crank resistance, then not only the amplitude of muscle activities (EMG amplitude) increases, but the activity patterns of individual muscles and muscle coactivation may also change. The effects of cadence change (Neptune et al. 1997; Brown and Kautz 1999; Chapman et al. 2007; Katona et al. 2014), and the effect of load (Baum and Li 2003; Hug et al. 2011; Smits-Engesman 2006) on muscle activity patterns were investigated. The combined effect of cadence change and crank resistance change on the amplitude of EMG signals and the size of range of muscle coactivation was analyzed (Katona et al. 2013, 2014). Able-bodied participants performed cycling movements on an ergometer at 2 pedaling cadences, 60 and 45 rpm, and 3 levels of crank resistances. Surface EMG signals were recorded on the left quadriceps and hamstrings at a 1000 Hz sampling frequency.

Figure 5 presents EMG signals from the quadriceps and hamstrings muscle groups during cycling with slow (45 rpm) and fast (60 rpm) cadences. The quadriceps exhibited a periodic activity in both cadence conditions, which is related to its role in this movement as the main power output provider. The EMG of hamstrings exhibited a rather calm activity with a small periodic increase in the amplitude (Fig. 5).

Raw sEMG signals were filtered (fourth-order Butterworth filter) and smoothed (RMS) before further data processing and analyses. The average EMG amplitudes for each participant and cycling conditions were computed.

The ratio of average EMG amplitudes obtained in the two cadence (fast/slow) conditions was higher when cycling against a higher resistance. This holds for knee flexors and extensors (Table 2). This result indicates that if resistance is increased, then not only muscle activities (EMG) are increased, but their rates of change with respect to changes in cadence are higher when cycling is performed against higher crank resistance. Notably, the ratio of cadence (fast versus slow cycling) and the ratio of power output was independent of resistance conditions in this study, but the ratio of average muscle activities (EMG amplitudes) depended on crank resistance and was significantly higher in higher RCs.

Comparing of the ratios of EMG amplitudes of flexors and extensors during fast versus slow cycling demonstrated that hamstring activity increased generally at a higher rate than quadriceps, but this difference was not significant.

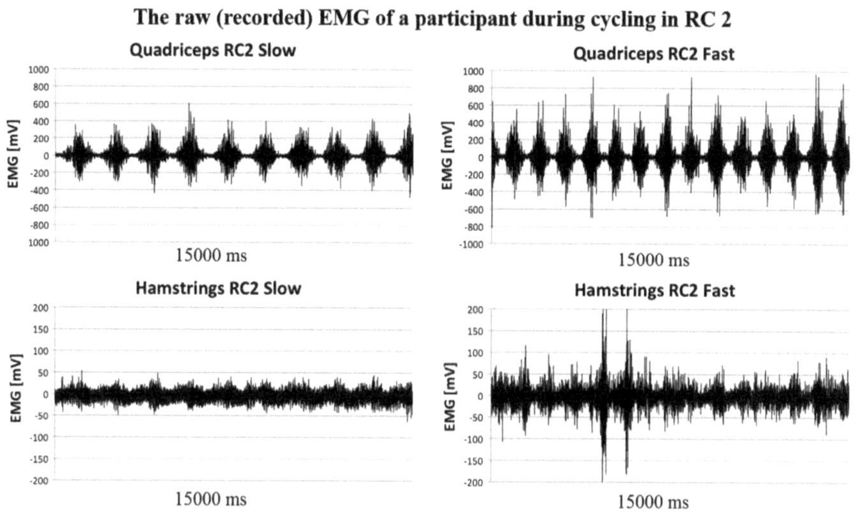


**Fig. 5** Raw (recorded) sEMG data of a participant cycled for 15 s in moderate resistance conditions (RC2). The graphs on the *left* and *right* present data obtained in slow (45 rpm) and fast (60 rpm) cycling respectively. (Adapted from Katona et al. 2014, Fig. 1)

**Table 2** Crank resistance, power output, ratio of power outputs and the ratio of average EMG amplitudes obtained in the two cadence conditions are presented for the three resistance conditions (low—RC1, moderate—RC2, high—RC3)

| Dimension | Resistance (at 60 rpm) [Nm] | Resistance (at 45 rpm) [Nm] | Power (fast) [W] | Power (slow) [W] | Ratio of power [ratio] | Ratio of $EMG_{mean}$_Q [fast/slow] | Ratio of $EMG_{mean}$_H [fast/slow] |
|---|---|---|---|---|---|---|---|
| RC1 | 2.51 | 2.12 | 15.76 | 9.99 | 1.58 | 1.23 | 1.22 |
| RC2 | 3.76 | 3.18 | 23.61 | 14.98 | 1.58 | 1.31 | 1.49 |
| RC3 | 5.02 | 4.24 | 31.53 | 19.97 | 1.58 | 1.46 | 1.51 |

Fast represents cycling with a cadence of 60 rpm, and slow represents cycling at 45 rpm. The mean (across participants) ratio of average EMG amplitudes with the standard deviations is provided for the two muscle groups (Q—quadriceps and H—hamstrings) separately (Katona et al. 2014)

The hamstrings generate relatively small power compared to the quadriceps, and when the cadence increases it may co-contract with the quadriceps. Muscle co-contraction may modify and regulate power output. We established angular positions of the crank in which each muscle in a pair (Quadriceps-Hamstrings) were active simultaneously. These angular positions defined the range of coactivation of the muscle pair. A muscle was considered active when its EMG amplitude exceeded 30 % of its mean EMG amplitude, as assessed in the actual RC (Ozgünen et al. 2010). The size of coactivation range of hamstring and quadriceps muscle groups increased with increased cycling cadence (Katona et al. 2013). In contrast, coactivation decreased with increased crank resistance. These findings reflect the adaptation of neural control as a response to altered external forces. One question in medical rehabilitation is whether to increase cycling cadence or crank resistance for individual trainings. Increased power output may be achieved by increasing pedaling cadence or crank resistance or both factors. When cycling cadence increased, than the size of coactivation range of hamstrings-quadriceps muscle groups increased (Katona et al. 2013). When the effect of cadence was investigated by acquiring sEMG on parts of the quadriceps (vastus medialis, vastus lateralis and rectus femoris) separately, then no significant change was observed in the size of coactivation ranges for the following muscle pairs: vastus medialis—biceps femoris; vastus medialis—biceps femoris; rectus femoris—biceps femoris (Laczko et al. 2014). However, if cycling is executed against higher resistances, then muscle coactivation is decreased for these muscle pairs. Thus, neural control may help generate sufficient power output by reducing the ranges where flexor-extensor muscles coactivate and the interval in which flexor-extensor muscles generate torque (knee torque) in opposite directions. If power output is increased by higher cadence, then maintaining or increasing the range of coactivation may be important for regulating and smoothing the movement by flexor activities and "Breaking" the speed caused primarily by extensor activity. It is another question that how cycling people are able to control and keep a desired constant speed in various cycling conditions. Preliminary study shows that with increased resistance this ability decreases (Valy et al. 2015). Our studies are based on electromyographic patterns of two or four muscles per limbs. A higher number of muscles were considered to reveal muscle synergies and their dependence on crank torque during lower limb cycling (Hug et al. 2011). Using nonnegative matrix factorization it was shown that three synergies accounted for the majority of variances of EMG signals of 11 muscles. These synergies may reflect a robust neural control strategy as they remained consistent when pedaling with altering crank torques.

## Functional Stimulation-Driven Limb Cycling

Upper limb and lower limb cycling is a special motor task that is used in functional electrical stimulation-based rehabilitation protocols. Cycling has an advantage compared to other types of locomotion, such as walking, because the participant does not have to hold the weight of the entire body. There are several centers where FES-driven cycling is used in rehabilitation protocols. The aim of these rehabilitation training programs is not only muscle strengthening by active muscle force generation, but also to improve cardiovascular and respiratory functions. The psychological affect is also very important because the SCI participants generate active forces using their own muscles. The participants use their own muscles to actively cycle for tens of minutes continuously, even if the motor command is coming not from the brain but from an artificial device (stimulator and controller). The physiological benefits of functional electrical stimulated cycling are presented in many papers (Frotzler et al. 2009; Lai et al. 2010; Perret et al. 2010; Dolbow et al. 2013; Brurok et al. 2013) and there are several studies about the effects of electrical stimulation (Theisen et al. 2002; Szecsi et al. 2014) and stimulation parameters (as current amplitude, pulse width and stimulation frequency) on energy and power output of FES-driven lower limb cycling (Eser et al. 2003; Hunt et al. 2006). We investigated the effect of various stimulation frequencies on power output (Laczko et al. 2012) and found that higher stimulation frequency did not necessarily result in higher performance and the dependence of performance on stimulation frequency varied among individuals.

Other crucial questions, from a motor control view, are the timing of muscle activations and muscle co-contractions. One possible method to establish proper artificial stimulation patterns is based on measured muscle activity patterns in able-bodied cycling people. The stimulation pattern derived from EMG data acquired from hamstrings and quadriceps muscle groups of 41 able-bodied persons lead to a stimulation pattern for our SCI participants (Pilissy et al. 2008; Laczko 2011). Two pairs of surface electrodes on each leg, one for the quadriceps, and one for the hamstring muscle groups were applied.

Recently we established a new stimulation pattern when three parts of the quadriceps (vastus medialis, vastus lateralis, and rectus femoris) were stimulated separately by three pairs of electrodes and the hamstrings with one pair of electrodes. In this case the power output increased significantly and this was achieved applying lower stimulation current amplitude per muscles (Mravcsik et al. 2016b).

Our observations on upper and lower limb cycling performed by able-bodied people may be used to improve and adjust rehabilitation protocols and artificial stimulation patterns to various cycling conditions to attain the desired power outputs when SCI people or stroke survivors are cycling and their movements are generated by FES.

## Translation to Rehabilitation

The presented features of upper limb cycling suggest some mechanisms to define rehabilitation protocols depending on the goals of the rehabilitation. The larger variability of muscle activities across cycles during unimanual upper limb cycling compared to bimanual cycling supports that it is more difficult to maintain consistent, periodic muscle activity patterns via central control in unimanual cycling than it is in bimanual cycling. This result is not surprising, but it is remarkable that the control of the non-dominant arm seems to be as stable as the control of the dominant arm in arm cycling, even without training. Arm cycling offers an ideal rehabilitation procedure when stability is important and strengthening the muscles in both arms is the aim. Joint angular variances as well as muscle activity variances are similar and stable control of the movement is maintained equally well in the dominant and non-dominant arm. This result is obtained from able-bodied unimpaired participants, but this idea should be translated to arm cycling in stroke survivors in future studies. Whether arm cycling is a motor task in which the affected and non-affected arms are not as different as in the performing of other motor tasks is a question for future research.

Functional electrical stimulation (FES)-driven cycling is in demand for lower limb muscle strengthening, and additionally it is advantageous for the cardiovascular and respiratory systems. Analyses of cycling limb movements of able-bodied people are profoundly useful for establishing stimulation patterns for FES-controlled movements of spinal cord-injured (SCI) individuals. Adequate stimulation patterns are required in SCI individuals to allow them to drive the crank of an ergometer using their own (electrically stimulated) muscles at various cycling cadences and crank resistances. The above-mentioned investigations impact the potential stimulation patterns that are used in FES-assisted cycling, namely, stronger individual muscle stimulation is needed to increase cycling cadence. If the aim is to cycle against higher crank resistance, then the time duration should be decreased in which hamstring and quadriceps muscles are stimulated in parallel. We suggest increasing quadriceps stimulation at a higher rate compared to the hamstrings when cycling against a higher resistance. If the aim is to cycle with a higher cadence, then the hamstrings activity should be increased at a higher rate compared to the quadriceps. FES-driven limb cycling apparatuses are commercially available and used mainly by SCI individuals, but stroke survivors may also use this method. Stimulation patterns varying between apparatuses and the user can define stimulation patterns in most cases. Results obtained from investigations of the motor control of limb cycling facilitate the efficient use of these resources.

Beside FES applications, lower limb cycling may also be used in the rehabilitation of patients with cerebral palsy (CP). Near-maximal adaptation of CP bodies toward healthy levels was demonstrated as electromyographic mean amplitude and frequency changed similarly both in the affected and non-affected legs in athletes with CP and able-bodied athletes who performed cycling training (Runciman et al. 2015).

## Take Home Message

The use of cycling limb movements in a larger extent of rehabilitation protocols is suggested for people with impaired limb movement control (e.g., stroke or spinal cord injury). The proper selection of rehabilitation training depends on the goal of the training. Upper or lower limb cycling is not a usual movement type in everyday behavior, but it is beneficial in rehabilitation paradigms for muscle strengthening and for improvements in cardiovascular and respiratory functions while maintaining stable motor control. Arm cycling is special because its neural control seems equally robust for the dominant and non-dominant arms in unimpaired individuals. Future studies may reveal differences in cycling pattern variances in the affected and non-affected limbs of stroke survivors. Control principles from upper and lower limb cycling movements of able-bodied people should be further identified to improve rehabilitation protocols and stimulation patterns for the efficient control of functional electrical stimulation-driven cycling training in spinal cord-injured people and stroke survivors with paralyzed limbs.

**Acknowledgements** The authors would like to thank the National Institute for Medical Rehabilitation and the Pázmány Péter Catholic University in Budapest for the possibility to perform measurements reported in this paper.

## References

Ambrosini E, Ferrante S, Ferrigno G, Molteni F, Pedrocchi A (2012) Cycling Induced by Electrical Stimulation Improves Muscle Activation and Symmetry During Pedaling in Hemiparetic Patients. IEEE Trans Neural Syst Rehabil Eng 20(3):320–330

Baum BS, Li L (2003) Lower extremity muscle activities during cycling are influenced by load and frequency. J Electromyogr Kinesiology 13(2003):181–190

Brown DA, Kautz SA (1999) Speed-dependent reductions of force output in people with poststroke hemiparesis. Phys Ther 79(10):919–30

Brurok B, Tørhau T, Karlsen T, Leivseth G, Helgerud J, Hoff J (2013) Effect of lower extremity functional electrical stimulation pulsed isometric contractions on arm cycling peak oxygen uptake in spinal cord injured individuals. J Rehabilitation Med 45:254–259

Chapman AR, Vicenzino B, Blanch P, Hodges PW (2007) Leg muscle recruitment during cycling is less developed in triathletes than cyclists despite matched cycling training loads. Exp Brain Res 181:503–518

Dolbow DR, Gorgey AS, Dolbow JD, Gater DR (2013) Seat pressure changes after eight weeks of functional electrical stimulation cycling: a pilot study. Top Spinal Cord Inj Rehabil 19(3): 222–228

Domkin D, Laczko J, Jaric S, Johansson H, Latash ML (2002) Structure of joint variability in bimanual pointing tasks. Exp Brain Res 143:11–23

Eser PC, Donaldson N, Knecht H, Stussi E (2003) Influence of different stimulation frequencies on power output and fatigue during FES-cycling in recently injured SCI people. Neural Syst Rehabil IEEE Trans Eng 11(3):236–240

Frotzler A, Coupaud S, Perret C, Kakebeek TH, Hunt KJ, Eser P (2009) Effect of detraining on bone and muscle tissue in subjects with chronic spinal cord injury after a period of electrically-stimulated cycling: a small cohort study. J Rehabil Med 41:282–285

Hogan N, Sternad D (2007) On rhythmic and discrete movements: reflections, definitions and implications for motor control. Exp Brain Res 181:13–30

Hug F, Turpin NA, Couturier A, Dorel S (2011) Consistency of muscle synergies during pedaling across different mechanical constraints. J Neurophysiol 106:91–103

Hunt KJ, Ferrario C, Grant S, Stone B, McLeanb AN, Fraser MH, Allan DB (2006) Comparison of stimulation patterns for FES-cycling using measures of oxygen cost and stimulation cost. Med Eng Phys 28:710–718

Jacobs PL, Mahoney ET, JohnsonB (2003) Reliability of arm wingate anaerobic testing in persons with complete paraplegia. J Spinal Cord Med 26(2):141–144

Jacobs PL, Johnson BM, Mahoney ET, Carter AB, Somarriba GA (2004) Effect of variable loading in the determination of upper-limb anaerobic power in persons with tetraplegia. J Rehabil Res Dev 41(1):9–14

Katona P, Pilissy T, Laczko J (2013) Speed and crank resistance affects coactivation of knee muscles during cycling movements. In: San Juan, Puerto Rico, society for the neural control of movement, 23rd annual meeting abstracts, p 1

Katona P, Pilissy T, Tihanyi A, Laczko J (2014) The combined effect of cycling cadence and crank resistance on hamstrings and quadriceps muscle activities during cycling. Acta Physiol Hung 101(4):505–516

Keresztenyi Z, Cesari P, Fazekas G, Laczkó J (2009) The relation of hand and arm configuration variances while tracking geometric figures in Parkinson's disease—"aspects for rehabilitation". Int J Rehabil Res 32(1):53–63

Laczko J (2011) Modeling of human movements, neuroprostheses. Clin Neurosci/Ideggyogy Szle 64(7–8):162–167. IF 0.488

Laczko J, Jaric S, Domkin D, Johansson H, Latash ML (2001) Stabilization of kinematic variables in the control of bimanual pointing movements. In: International joint conference on neural networks, Washinghton, DC. July 14–19 2001, ISBN 0-7803-7044-9/ ©2001 IEEE, pp 1256–1260

Laczko J, Katona P, Waszlavik E, Klauber A (2012) Dependence of cycling performance on training time and stimulation frequency during FES driven cycling. In: IFESS 2012, 17th annual meeting—International functional electrical stimulation society, pp 448–451

Laczko J, Katona P, Valy A (2014) Muscle co-activation as function of crank angle when cycling on an ergometer with altered power output. Program No. 630.06, 2014 Neuroscience Meeting Planner. Society for Neuroscience

Lai CH, Chan WH, Chan WP, Peng CW, Shen LK, Chen JJ, Chen SC (2010) Effects of functional electrical stimulation cycling exercise on bone mineral density loss in the early stages of spinal cord injury. J Rehabil Med 42:150–154

Martens J, Daly D, Deschamps K, Fernandes RJP, Staes F (2015) Intra-individual variability of surface electromyography in front crawl swimming. PLoS ONE 10(12):e0144998. doi:10.1371/journal.pone.0144998

Massie CL, Malcolm MP, Greene DP, Browning RC (2012) Kinematic motion analysis and muscle activation patterns of continuous reaching in survivors of stroke. J Motor Behav 44(3):213–222

Massie CL, Malcolm MP, Greene DP, Browning RC (2014) Biomechanical contributions of the trunk and upper extremity in discrete versus cyclic reaching in survivors of stroke. Top Stroke Rehabil 21(1):23–32

Mravcsik M, Laczkó J (2014) Muscle co-activation during arm cycling against altering crank resistances. In: 7th World Congress of Biomechanics, Control Number 3864

Mravcsik M, Laczko J (2015) Variances of joint configuration and muscle activity patterns during arm cycling against external resistances. Program Number 609.06, 2015 Neuroscience Meeting Planner. Society for Neuroscience, Chicago IL, USA (Online)

Mravcsik M, Zentai N, Botzheim L, Laczko J (2016a) Unimanual versus bimanual arm cycling movements - muscle activity variances. Program Number 886.01, 2016 Neuroscience Meeting Planner. Society for Neuroscience, San Diego CA, USA (Online)

Mravcsik M, Klauber A, Laczko J (2016b) FES driven lower limb cycling by four and eight channel stimulations—a comparison in a case study. In: Proceedings of the 12th Vienna International Workshop on Functional Electrical Stimulation, pp. 89–93, ISBN: 978-3-900928-12-4

Neptune RR, Kautz SA, Hull ML (1997) The effect of pedaling rate on coordination in cycling. J Biomech 30(10):1051–1058

Ozgünen KT, Celik U, Kurdak SS (2010) Determination of an Optimal Threshold Value for Muscle Activity Detection in EMG Analysis. J Sports Sci Med 9(4):620–628. eCollection 2010

Perret C, Berry H, Hunt KJ, Donaldson N, Kakebeeke TH (2010) Feasibility of functional electrical stimulated cycling in subjects with spinal cord injury: an energetics assessment. J Rehabil Med 42:873–875

Pilissy T, Klauber A, Fazekas G, Laczkó J, Szécsi J (2008) Improving functional electrical stimulation driven cycling by proper synchronization of the muscles. Ideggyogy Sz 61(5–6):162–167
Runciman P, Derman W, Ferreira S, Albertus-Kajee Y, Tucker R (2015) A descriptive comparison of sprint cycling performance and neuromuscular characteristics in able-bodied athletes and paralympic athletes with cerebral palsy. Am J Phys Med Rehabil 94:28–37
Sainburg RL (2002) Evidence for a dynamic-dominance hypothesis of handedness. Exp Brain Res 142:241–258
Smits-Engelsman BC, Swinnen SP, Duysens J (2006) The advantage of cyclic over discrete movements remains evident following changes in load and amplitude. Neurosci Lett 396(1):28–32. Epub 2005 Dec 2
Smits-Engelsman BCM, Van Galen GP, Duysens J (2002) The breakdown of Fitts' law in rapid, reciprocal aiming movements. Exp Brain Res 145:222–230
Smits-Engelsman BCM, Swinnen S, Duysens J (2006) The advantage of cyclic over discrete movements remains evident following changes in load and amplitude. Neurosci Lett 396(1):28–32
Szecsi J, Straube A, Fornusek C (2014) A biomechanical cause of low power production during FES cycling of subjects with SCI. J NeuroEng Rehabil 11:123
Theisen D, Fornusek C, Raymond J, Davis GM (2002) External power output changes during prolonged cycling with electrical stimulation. J Rehabil Med 34:171–175
Tibold R, Laczko J (2012) The effect of load on torques in point-to-point arm movements: a 3D model. J Mot Behav 44(5):341–350
Tibold R, Fazekas G, Laczko J (2011) Three-dimensional model to predict muscle forces and their relation to motor variances in reaching arm movements. J Appl Biomech 27:362–374
Valy A, Laczko J (2015) Timing errors during lower limb cycling under various conditions. Program Number 241.18, 2015 Neuroscience Meeting Planner. Society for Neuroscience, Chicago IL, USA, (Online)
van Mourik AM, Beek PJ (2004) Discrete and cyclical movements: unified dynamics or separate control? Acta Psychol 117(2):121–138

# Impaired Voluntary Movement Control and Its Rehabilitation in Cerebral Palsy

**Andrew M. Gordon**

**Abstract** Cerebral palsy is caused by early damage to the developing brain, as the most common pediatric neurological disorder. Hemiplegia (unilateral spastic cerebral palsy) is the most common subtype, and the resulting impairments, lateralized to one body side, especially affect the upper extremity, limiting daily function. This chapter first describes the pathophysiology and mechanisms underlying impaired upper extremity control of cerebral palsy. It will be shown that the severity of impaired hand function closely relates to the integrity of the corticospinal tract innervating the affected hand. It will also shown that the developing corticospinal tract can reorganize its connectivity depending on the timing and location of CNS injury, which also has implications for the severity of hand impairments and rehabilitation. The mechanisms underlying impaired motor function will be highlighted, including deficits in movement execution and planning and sensorimotor integration. It will be shown that despite having unimanual hand impairments, bimanual movement control deficits and mirror movements also impact function. Evidence for motor learning-based therapies including Constraint-Induced Movement Therapy and Bimanual Training, and the possible pathophysiological predictors of treatment outcome and plasticity will be described. Finally, future directions for rehabilitations will be presented.

**Keywords** Sensorimotor integration · Constraint-induced movement therapy · Bimanual · Jense-Taylor test · Hand function · Corticospinal pathways (CST) · Object-release coordination · Motor planning · Thalamocortical · Stereognosis · Mirror movements

A.M. Gordon (✉)
Department of Biobehavioral Sciences, Teachers College, Columbia University, 525 West 120th Street, New York, NY Box 93, 10027, USA
e-mail: ag275@columbia.edu



J. Laczko and M.L. Latash (eds.), *Progress in Motor Control*,
Advances in Experimental Medicine and Biology 957,
DOI 10.1007/978-3-319-47313-0_16

## Introduction

Cerebral palsy (CP) is the most common pediatric neurological disorder. CP is caused by damage to the brain during early development. Unilateral spastic cerebral palsy (USCP) (often referred to as hemiplegic CP), with motor impairments mainly lateralized to one body side, is the most common subtype, accounting for 30–40 % of newly diagnosed cases (Himmelmann et al. 2005). In this chapter, the pathophysiology and mechanisms underlying impaired upper extremity (UE) control of USCP will be reviewed. The specific pathophysiology will be related to the severity of UE impairments, where clear relationships between the type and the extent of early brain damage and/or the resulting reorganization of the developing nervous system will be demonstrated. The evidence for motor-learning based therapies will be discussed, including possible pathophysiological predictors of treatment outcome and the associated neuroplasticity.

## Pathophysiology of Unilateral Spastic CP

Corticospinal (CST) pathways from motor areas, especially the primary motor cortex (M1), develop in a corticofugal manner, approaching the spinal cord by the 20th week of gestation during typical development (see Staudt 2010). Thereafter these pathways undergo synaptogenesis, with targeted projections to the spinal–segmental level. The motor cortices are initially believed to develop bilateral projections (i.e., projections to both the contralateral and ipsilateral UEs) (Eyre et al. 2001). There is a gradual weakening of ipsilateral projections, and strengthening of contralateral projections via synaptic competition during continuing development (Staudt 2010), which is driven by M1 activity (see Friel et al. 2013). This is evidenced by studies of the developing CST in kittens, whereby blockade of M1 activity either by injection of a GABA-agonist or restraint of a forelimb, prevents the CST from establishing spinal connectivity, and results in permanent motor impairments (Friel et al. 2013). In humans, this intricate process is susceptible to prenatal and perinatal brain damage. Since the CST directly innervate hand motoneurons, which provide the capacity for selective UE movement control (Lawrence and Kuypers 1968), permanent impairments in manual dexterity may occur following damage to this developing system.

USCP is often the result of periventricular lesions, middle cerebral artery infarct, hemi-brain atrophy, brain malformation or posthemorrhagic porencephaly (e.g., Uvebrandt 1988; Cioni et al. 1999). As described above, motor areas and CST integrity are often compromised (e.g., Staudt et al. 2004; Bleyenheuft et al. 2007). The specific etiology may drastically influence subsequent development of CST pathways (Staudt 2010). The severity of upper extremity impairments largely depends on the extent of damage to the CST (e.g., Bleyenheuft et al. 2007; Scheck et al. 2012). The damage can be approximated using both conventional MRIs and diffusion tensor imaging, by measuring the asymmetry in the cross-sectional area of the cerebral peduncles through which the CST passes through (e.g., Bleyenheuft et al. 2007; Friel et al. 2014), or measuring the fractional anisotropy (Bleyenheuft et al. 2007). The asymmetry in the CST innervating each hand is strongly related to severity of manual dexterity impairments. This can be seen in Fig. 1A, where the index of asymmetry (ratio of the affected to unaffected peduncle size) is highly correlated to manual dexterity, as measured by the time to complete the Jebsen-Taylor Test of Hand Function (Friel et al. 2014). The timing of CST damage is also predictive of outcome. Cortical malformation in the first two trimesters typically results in less severe upper extremity impairments than periventricular lesions in the early, and MCA later, in the third trimester (Fig. 1b, Staudt et al. 2004). Disruptions during the intrauterine period can reduce or abolish neuronal activity in CST projections originating in the affected hemisphere. The altered activity can result in ipsilateral being maintained and strengthened during further development, with contralateral projections being partly or completely eliminated (Staudt et al. 2004; Eyre et al. 2007; see Staudt 2010) (Fig. 1c). On average, individuals who undergo such ipsilateral reorganization have more severely affected upper extremity function Fig. 1d, Holmström et al. (2010) and often the persistence of mirror movements. This reorganization may represent functional compensation of the affected hemisphere by the unaffected hemisphere. The extent to which this ipsilateral reorganization is able to compensate for the absent contralateral projections decreases with age at the time of damage (Staudt et al. 2004). CST projections from M1 to spinal interneurons, thus indirectly innervating hand muscles, are also impaired, further affecting coordination and reflex control (Harrison 1988). The net result of the early brain damage is impairments in movement execution, movement planning, sensorimotor control and the ability to coordinate the two hands together (bimanual control). These impairments are described below in relation to the pathophysiology of CP.

## *Movement Execution Impairments*

As a result of damage to the developing motor pathways, there are impairments in the execution of upper extremity movements. For example, the upper extremity is often weak, and the lack of selective finger movements, limits the ability to achieve many grasp forms (e.g., precision grip). Thus movements are unskilled, slow and clumsy (Brown et al. 1987).

Manual motor control has been quantified for more than three decades via the examination of fingertip forces during precision grasping. Fingertip coordination in typically developing (TD) children reaches near adult-like coordination by 6–8 years of age (see Gordon 2001). Conversely, children with CP at this age often have infantile-like force coordination, with prolonged delays between movement phases (e.g., between grasp contact and force initiation) and reduced coupling of grip and load force (Eliasson et al. 1991; Prabhu et al. 2011). While most children

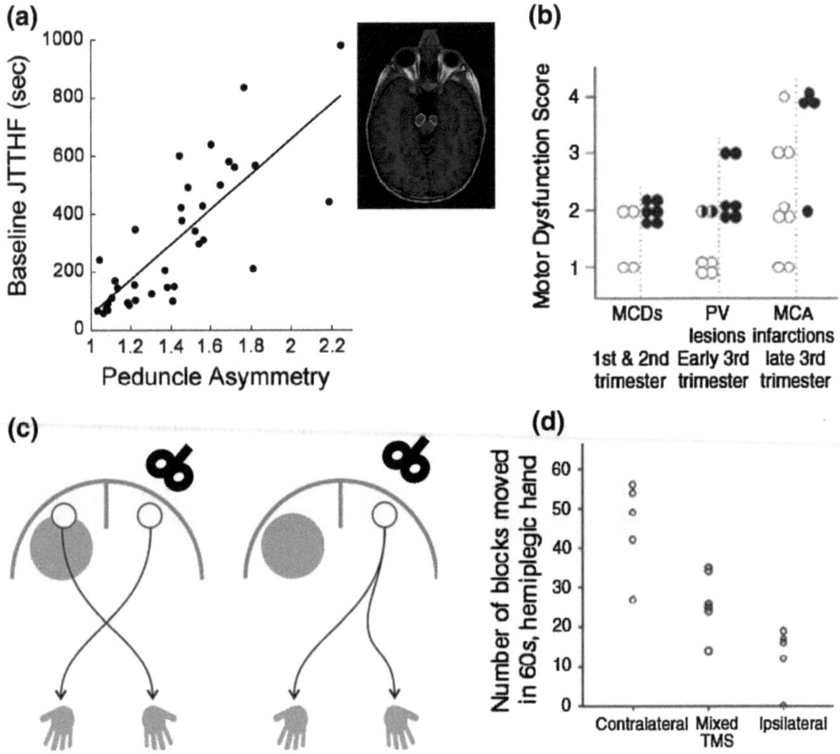


**Fig. 1** **a** Peduncle asymmetry predicts baseline unimanual dexterity (Jebsen-Taylor Test). Inset t1-weighted MRI showing cerebral peduncles, outlined. Modified from Friel et al. (2014). **b** Comparison of hand motor dysfunction scores among patients with congenital brain injuries acquired during the trimesters of gestation (malformations of cortical development [MCD], first and second trimester of pregnancy; periventricular [PV] lesions, early third trimester of pregnancy; middle cerebral artery [MCA] infarctions, late third trimester of pregnancy). *Open circles*, individuals with preserved crossed corticospinal projections from the affected hemispheres; filled circles, individuals with reorganized ipsilateral projections from the contralesional hemispheres; *half-filled circles*, individuals with both preserved contralateral and reorganized ipsilateral corticospinal projections to the more affected hand. Modified from Staudt et al. (2004). **c** Relation between box and blocks score (higher score represents greater dexterity) using the more affected hand and the CST motor projection pattern to the hand, assessed using transcranial magnetic stimulation (TMS). Modified from Holmström et al. (2010)

with CP are capable of adjusting their fingertip forces to the object's weight and texture, their forces are often excessive and variable, with reduced adaptation to the object's weight and texture compared to TD children (Eliasson et al. 1992). Object release coordination is also impaired (Eliasson and Gordon 2000), which is exacerbated when speed and accuracy constraints are imposed (Gordon et al. 2003). Fortunately, precision grip in children with CP does improve with development (Eliasson et al. 2006) and extensive practice (Gordon and Duff 1999a), which has helped inspired the development of intensive rehabilitation protocols (e.g., constraint-induced movement therapy; CIMT) (Gordon 2011, 2014).

## *Motor Planning Impairments*

Perhaps an unappreciated aspect of the motor control deficits is that individuals with unilateral spastic CP also have higher level impairments in motor planning (see Steenbergen et al. 2013). These impairments can impact precision grasping. Due to delays in feedback, the development of fingertip forces during grasp must be planned before initiation. Furthermore, sensory information about some object properties (e.g., weight) is not available until after lift off. Motor planning for grasp control involves the formation and utilization of internal models of objects based on previous experience manipulating the object (see Gordon and Duff 1999a). Children with unilateral spastic CP have an impaired ability to scale the amplitude of the force development to the known object properties in their more affected hand (Gordon et al. 1999). The impairments in force planning are reduced following extensive practice (Gordon and Duff 1999a), providing a template for intensive rehabilitation training. The planning deficits may partially reflect impaired motor learning, involving extracting appropriate sensory information to form internal models, and integrating it with motor commands during subsequent actions.

Children with unilateral spastic CP often demonstrate 'global planning' impairments that are independent of the effector used (see Steenbergen et al. 2013). Oddly, the force scaling deficits during object lifting (Eliasson et al. 1992; Gordon and Duff 1999a) and grip force coupling (Gordon et al. 2006a, b) are effector dependent, where deficits are solely seen in the more affected, and not the less affected UE. A similar finding during load force perturbations has been observed, with participants unable to anticipate the consequences of a dynamic perturbations (Bleyenheuft and Thonnard 2010). An interesting, and potentially transformative, finding is that following several lifts with the less affected hand, anticipatory planning, reflected by the rate of force development matching the object's weight, is immediately present in the more affected hand; i.e., there is a transfer of sensorimotor information and improvement in motor function (Gordon et al. 1999). In a subsequent study (Gordon et al. 2006a, b), we found that despite the normal of anticipatory force scaling during successive lifts with the more affected hand, anticipatory force scaling is transferred from the more affected to the less affected hand. Thus, the lack of anticipatory control during normal use of the more affected hand is not due to impaired sensation, yielding insufficient internal models. Instead, these findings suggest that the impaired planning may be due to an inability to integrate sensory information with the motor output in the more affected hand. Simultaneous grasp and lift of an object in each hand improved some aspects of grip performance as well (Steenbergen et al. 2008), although the grip-lift movements became slower, likely due to the requirement to divide attention between the two hands. These findings helped motivate the development of intensive bimanual training approaches (e.g., Hand-arm bimanual intensive therapy; HABIT) described below (Gordon 2011, 2014).

## Sensorimotor Impairments

Thalamocortical somatosensory projections achieve connectivity in their cortical termination sites during the third trimester. These connections normally are not damaged by PVL, or they may circumvent the lesion to terminate in the postcentral gyrus (see Staudt 2010). In contrast, MCA lesions, occurring later and in many instances affecting the postcentral gyrus, are more likely to affect the somatosensory system (Staudt 2010). Thus, children with USCP, especially of MCA origin, often have accompanying sensory impairments, which may further compromise fine motor coordination (Brown et al. 1987; Moberg 1962). Tactile perception and discrimination, stereognosis and proprioception are often disturbed (e.g., Gordon and Duff 1999b, Bleyenheuft and Thonnard 2011) with the amount of impairment related to the integrity of afferent pathways (see Scheck et al. 2012). These sensory impairments may contribute to precision grip impairments found in CP (Gordon and Duff 1999b) as they resemble precision grip impairments following digital anesthesia in healthy adults; i.e., higher and more variable grasping forces (e.g., Johansson and Westling 1984). Although there appears to be improvements in both sensory acuity and motor function following training, the relationship between sensory and motor abilities is not fixed. Such relationships have been established between stereognosis and motor function (Gordon and Duff 1999b), but conflicting results are reported for other modalities.

## Bimanual Coordination Impairments

Children with unilateral spastic CP have deficits in bimanual coordination above and beyond unimanual dexterity deficits (e.g., Steenbergen et al. 1996, 2000; see Gordon and Steenbergen 2008). Similar to motor planning, the relation between the type and timing of the lesion and bimanual coordination impairments are not known. During symmetrical, bimanual reaching tasks, children with unilateral spastic CP do coordinate their bimanual movements by compensating with their noninvolved hand, as long as the accuracy demands or task complexity are not increased. In one recent study, participants were instructed to hold a grip device in each hand and place one device on top of the other while the grip and load force were recorded in each hand (Islam et al. 2011). Children with USCP initiated the task by decreasing grip force in the releasing hand before increasing the grip force in the holding hand during the preparation phase, with the subsequent grip force increase in the holding hand being reduced and occurring later (transition phase) than that of TD children. The impairments were unrelated to presence of mirror movements, and the impairment was greater when the less affected hand served as the holding hand.

In another series of studies (Hung et al. 2004, 2010), children were asked to open a drawer with one hand and manipulate its contents with the other hand. Children with USCP were less coordinated, with reduced movement overlap of the drawer

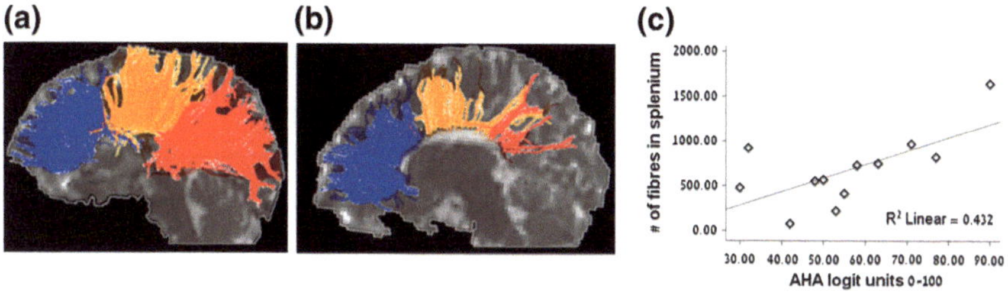


**Fig. 2** Tractography of the corpus callosum (*genu blue*, *midbody orange*, *splenium red*) in **a** a TD child and **b** a child with unilateral CP. **c** Relationship between # of fibres in the splenium of the corpus callosum and Assisting Hand Assessment scores. Modified from Weinstein et al. (2013)

opening hand and manipulating hand (and sequential completion of opening the drawer and manipulating its contents (Hung et al. 2010). Interestingly, as described later in this chapter, bimanual training improved some aspects of this coordination more than unimanual training (Hung et al. 2011).

While emphasis has been placed on CST connectivity to the affected hands in relation hand function (e.g., Bleyenheuft et al. 2007; Friel et al. 2014). However, the cause of hand impairments may not be as simple. For example, one study showed that reduced structural integrity of the corpus callosum in USCP was associated with reduced function of both the more- and less affected hands (Green et al. 2013). Interestingly, the corpus callosum integrity was also related to the quality of bimanual hand use (Fig. 2), which as described below, may be functionally more important.

## *Mirror Movements*

Many individuals with USCP have unintentional imitation of voluntary movement of the contralateral limb, known as mirror movements (Woods and Teuber, 1978). Conceivably, mirror movements could negatively impact effective performance of asymmetrical bimanual activities (e.g., tying shoes). They are believed to be caused by the ipsilateral CST projecting from the less affected M1 that innervates bilateral UEs (e.g., Carr et al. 1993; Farmer et al. 1991; Staudt et al. 2004) as described above, and/or impaired interhemispheric inhibition between the two M1s (Koerte et al. 2011).

Early studies examined the relationship between the timing of the brain injury, CST and mirror movements, quantifying mirror movements using various measures (Woods and Teuber 1978; Carr et al. 1993; Farmer et al. 1991). More recent studies demonstrated a negative impact on unimanual and bimanual hand function (Islam et al. 2011; Sukal-Moulton et al. 2013; Klingels et al. 2016; Kuhtz-Buschbeck et al. 2000; Adler et al. 2015). While individuals are able to voluntarily reduce the amplitude of mirror movements with attention (Kuhtz-Buschbeck et al. 2000), it is unknown whether mirror movements impact the ability to treat impaired hand function, or whether treatments can diminish their impact on function.

## Rehabilitation of Upper Extremity Function

### *Motor Learning Approaches to Rehabilitation*

A recent review of evidence supporting the clinical implementation of various therapeutic approaches identified only six approaches with sufficient evidence of efficacy (Novak et al. 2013). These were Constraint-induced movement therapy (CIMT), Intensive bimanual training, Goal-directed training, Context-specific training, Occupational therapy post botoxulinum toxin injections, and Home training. There were two commonalities across these approaches: all involved active movement of the participant using principles of motor learning, and all involved intensive practice.

Motor learning approaches to rehabilitation (often referred to as task-oriented or goal training) focuses on the effectiveness and efficiency of motor performance in specific actions, with activity limitations being an important aspect of CP (Bax et al. 2005) rather than on normalization of movement patterns or prevention of compensations (Carr and Shepherd 1989; Trombly 1995; Winstein and Wolf 2009). This approach is based on integrated models of motor learning and control and behavioral neuroscience. The focus is on participation and acquisition of skills and involves targeted physical and mental activity. The late eminent scholar, Ann Gentile, noted that: "The behaviour that dominates our daily lives is directed toward the accomplishment of goals. It is aimed at a specific purpose or end that we are trying to achieve" (Gentile 2000; p. 112). The approach recognizes task performance as being influenced by the person, the task and the environment (Gordon and Magill 2012). Problem solving is a key component, with the notion that the brain selects movement patterns that are available based on the constraints of the patient (Latash and Anson 1996). The associated behavioral demands of the tasks and motor skill training may be used to "shape" the desired movement pattern, and results in cortical reorganization (Plautz et al. 2000) underlying concurrent functional outcomes. For maximal efficacy, the training must be challenging, with progressively increasing behavioral demands, and involve active participation and problem solving. Skilled training in animals shows increased plasticity of UE cortical representations, whereas unskilled training did not (Kleim et al. 1998; Friel et al. 2012).

### *Constraint-Induced Movement Therapy*

CIMT is an obvious choice for eliciting intensive unimanual practice. There was a strong scientific rationale based on animal models underlying the application of intensive practice-based approaches to human upper extremity (UE) rehabilitation

(Tower 1940; Taub and Shee 1980). Starting with 'forced use' in adult stroke patients, in which the less affected UE was physically restrained to incentivize practice of the more affected UE (Wolf et al. 1989), subsequent efforts incorporated principles of behavioral psychology (shaping) and motor learning to elicit active practice of the affected UE. This active training approach along with the restraint become known as 'Constraint-induced movement therapy' (CIMT) (Taub and Wolf 1997). The first adaptation of the approach for the pediatric population leading to a case study was published more 15 years ago (Charles et al. 2001). Since then there have been many studies of CIMT to date, including more than 30 RCTs, with nearly all studies providing evidence of efficacy (see Dong et al. 2013; Eliasson et al. 2014).

The best age to conduct CIMT is unknown. Most studies include children from 2 to 7 years, although CIMT studies have included infants as young as 7 months (e.g., Lowes et al. 2014) to adolescence (Eliasson et al. 2003; Gordon et al. 2006a, b; Sakzewski et al. 2011a). Given the very early age in which brain damage occurs in children with CP, we could surmise that there is tremendous potential for recovery (Kennard 1936) and that "earlier treatment is better". However, the relationship is not as clear as one might hope (Eliasson et al. 2005; Hoare et al. 2013; Gordon et al. 2006a, b; Sakzewski et al. 2011a). Nevertheless, the studies to date indicate that children of all ages may benefit.

Animal models suggest that the best time to start treatments eliciting movements of the more affected UE, such as CIMT, may well be during the first few months of life. The as described above (Martin et al. 2011; Eyre 2007). Development of the CST requires activity-dependent competition between the two sides of the developing motor system. Transcranial magnetic stimulation studies in human infants (Eyre 2003, 2007; Eyre et al. 2001; Staudt et al. 2004) are consistent with these findings. Pharmacologically inhibiting activity of the contralesional side, thus balancing activity of the two hemispheres, immediately after unilateral brain injury in the kitten, restores motor function, normal anatomical connectivity of the CST and the motor representational map in primary motor cortex (Martin et al. 2011). A feline model of CIMT (Friel et al. 2012) examined the effects of CIMT at age 8 to 13 weeks and 20-24 weeks (Fig. 3). The earlier training restored motor function, CST connections in the spinal cord and the motor cortical representation of the forelimb, and increased cholinergic spinal interneuron density on the contralateral side. The later training (age 20–24 weeks) did not restore motor function or contralateral spinal cholinergic neuronal density, although it did increase CST connectivity and the motor cortical representation of the forelimb. Restraint alone (forced use without specific, skilled training) only restored CST connectivity (Friel et al. 2012). This alludes to the importance of increasing skilled activity of the involved UE, a principle embedded in both CIMT and bimanual training (see below), to balance the neural activity between the two hemispheres before the less affected CST "outcompetes" the affected CST (i.e., at an early age).

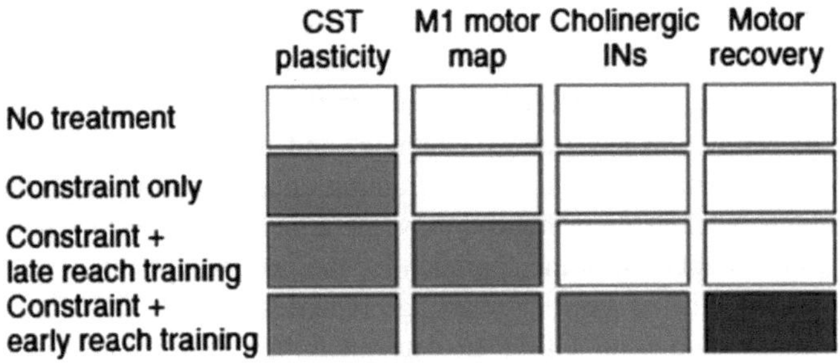


**Fig. 3** Summary of effects constraint only (forced use), constraint plus motor training in a in a feline model of unilateral spastic cerebral palsy induced by inactivation of the motor cortex. Early training was in a kitten age 8–13 weeks, late training 20–24 weeks. *Gray rectangles*, conditions in which an effect was noted. Corticospinal tract (CST) plasticity is defined as the presence of axons/varicosities within the spinal intermediate gray matter or more ventrally. An effect of treatment on the M1 motor map is defined as an increase in the number of sites from which stimulation evoked a limb motor response. A filled *light gray rectangle* in the cholinergic interneuron (INs) column indicates the presence of robust increases in the ratio of spinal cholinergic interneurons on the affected and unaffected sides. The *dark rectangle* in the motor recovery column indicates am improvement in stepping accuracy while walking over a *horizontal ladder*. From Friel et al. (2013)

## *Intensive Bimanual Training*

The efficacy CIMT lead to the obvious question of whether similar improvements could be achieved without physically restraining the less affected upper extremity. Bimanual training is developmentally focused and takes into account principles of motor learning (Thorndike 1914) (practice specificity) to train use of the more affected hand as a functional (nondominant) assist rather than as a dominant hand as done in CIMT. It uses the environment (task) to "force use the more affected hand". One child-friendly form of intensive bimanual training was developed, "Hand-arm bimanual intensive therapy" (HABIT) (Charles and Gordon 2006), with the focus improving the amount and quality of involved UE use during bimanual tasks. HABIT retains the *intensive structured practice* of CIMT and typically it is provided in a day camp environment 6 h/day for 10–15 days (i.e., 60–90 h). A small randomized control trial of HABIT was initially conducted (Gordon et al. 2007) demonstrating that children who received HABIT had improved quality of affected hand use during bimanual activities and frequency of UE use. HABIT has also been shown to be effective using a themed (learning magic tricks) approach (Green et al. 2013).

Bimanual training and CIMT result in similar improvements in both unimanual capacity and bimanual performance outcomes in CP populations (Sakzewski et al. 2011a, b; Hoare et al. 2013; Gordon et al. 2008, 2011; Facchin et al. 2011; Deppe et al. 2013; Dong et al. 2013) (Fig. 4). Bimanual training has been shown to result in better improvement in functional activities and play goals meaningful to participants and caregivers (Gordon et al. 2011; Brandão et al. 2012) largely since most goals are bimanual (Gordon et al. 2011). During a bimanual task where children are asked to open a drawer with one hand and manipulate its contents with the other hand, improvements in both the CIMT and HABIT groups were observed despite not being practiced during the training (Hung et al. 2011). However greater improvements in movement overlap (the percentage of time with both hands engaged in the task) and goal synchronization (reduced time between each hand completing the task goals) were found for the HABIT group. Together these findings are in agreement with the principle of practice specificity.

Despite the advantages and disadvantages of each approach, CIMT and HABIT are not mutually exclusive. They can be performed concurrently with sufficient intensity (Aarts et al. 2010; Cohen-Holzer et al. 2011; Taub et al. 2007) or CIMT could be followed with a bimanual training program (i.e., sequentially) where gains in unimanual capacity are translated into bimanual activities (Case-Smith et al. 2012; Taub et al. 2007). These might combine the benefits of both approaches, although comparison studies to either isolated approach have not been conducted (but see Gordon 2011).

## *Neurophysiological Predictors and Responses to Training*

Although the above intensive training protocols have demonstrated great promise, a major limitation is that the neurological damage and subsequent recovery in cerebral palsy is extremely diverse. Studies of these approaches have largely ignored

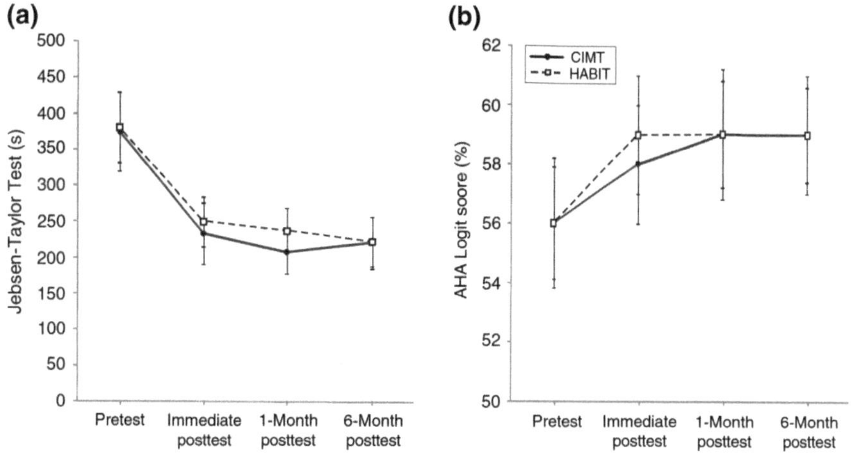


**Fig. 4** **a** Mean ± SEM time to complete the 6 timed items (writing excluded) of the Jebsen-Taylor Test of Hand Function. Faster times correspond to better performance. **b**. Mean + SEM scaled logit scores (AHA units) on the Assisting Hand Assessment; higher scores represent better performance. Abbreviations: *CIMT* Constraint-induced movement therapy; *HABIT* Hand-arm intensive bimanual therapy; *AHA* Assisting hand assessment. From Gordon et al. (2011)

this issue and tend to lump participants together and look at the group effects. There has been large variability in the response to treatment as well (e.g., Sakzewski et al. 2011b). It may be that depending on the specific lesions and brain reorganization, one approach may be more efficacious than another. For example, Staudt and colleagues (Kuhnke et al. 2008) showed that children with a contralateral CST increased manual dexterity after CIMT, whereas children with an ipsilateral CST did not. The two groups also showed differences in neuroplasticity associated with the training (Juenger et al. 2013). In contrast, Islam et al. (2014) concluded that improvements in hand use after CIMT were present in all participants, irrespective of their CST connectivity pattern. Only one case study of two children with USCP has examined the effect of CST connectivity pattern on efficacy of bimanual therapy, with improvements seen regardless of CST organization (Bleyenheuft et al. 2015) (Fig. 5).

A recent study of the neural predictors of bimanual training showed that children with greater overall severity of brain injury at baseline (higher radiological score) and decreased cortical activation of the lesional hemisphere on fMRI, showed greater improvement on bilateral hand function but less improvement on unimanual function (Schertz et al. 2016). Similarly, greater white matter damage (DTI of the

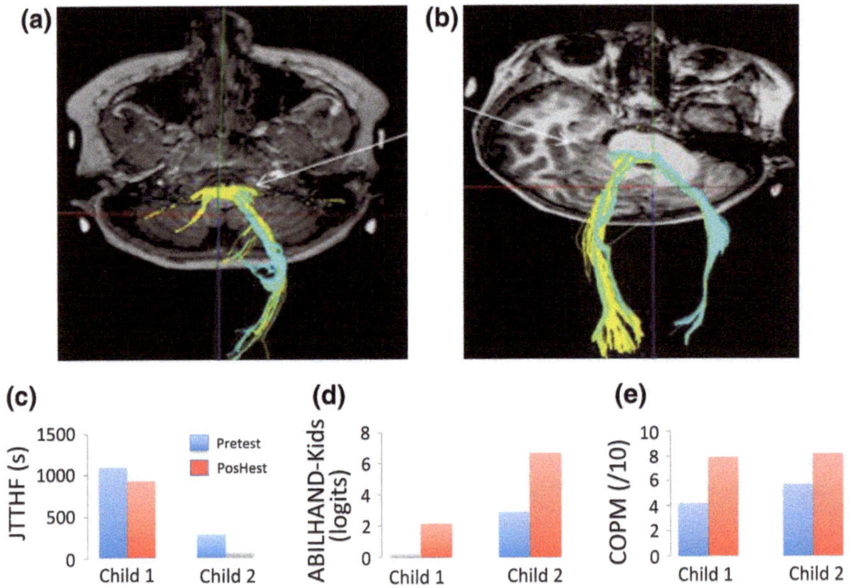


**Fig. 5** Determinist diffusion tensor imaging tracking was made from the 3 mm spheres (123 voxels) created symmetrically in both CST tracts, with their middle centered on the CST fibers as visualized in a transversal plane passing through the middle cerebellar peduncle. **a** child 1 with ipsilateral connectivity, **b** child 2 with contralateral connectivity. Pretest and posttest data for each child for the **c** Jebsen-Taylor test of hand function (JTTHF), **d** ABILHAND-Kids and **e** Canadian occupational performance measure (COPM) performance rating. Modified from Bleyenheuft et al. (2015)

posterior limb of the internal capsule and the CST were also associated with greater improvement on bimanual function but no improvement on unimanual assessment. However, the amount of damage to the CSTs alone did not relate improvement after CIMT (Rickards et al. 2014) or bimanual training (Friel et al. 2014).

One study showed increases in activation and size of the motor areas controlling the affected hand using TMS, DTI and fMRI (Bleyenheuft et al. 2015). The later study found increased activation in cortical areas involved in reward circuits, suggesting an intriguing possibility that the training may engage circuits involved with achieving self-efficacy. Another study showed that changes in DTI and fMRI parameters were seen when comparing pre- and post intervention following bimanual training. Despite variable patterns of brain plasticity, a change toward a more unilateral brain activation pattern was consistently associated with greater motor improvements. Another (TMS) study showed that the map size and amplitude of motor evoked potential (MEP) amplitudes of the affected hand motor map increased significantly after bimanual training regardless of whether the representation was in the lesional or contralesional hemisphere (Friel et al. 2016) (Fig. 6). The amount of change in these measures correlated with improvements in hand function. Finally, one study examined differences in cortical reorganization following CIMT depending on CST organization (Juenger et al. 2013). Both groups exhibited increases in sensorimotor cortical activation, which is consistent with a report of better sensory function after intensive training (Kuo et al. 2016). Interestingly TMS showed a decrease in M1 excitability in children with an ipsilateral CST pattern, whereas it showed an increase in M1 excitability in children with preserved contralateral CST organization.

Recently the question of training ingredients, specifically progressive skill training, was addressed. Skill progression has been shown to be important for inducing plasticity in animal models of learning (Kleim et al. 2002; Nudo 2003). Furthermore, in a feline model of unilateral CP, restriction of the nonparetic forelimb alone (analogous to forced use) did not drive CST development, cortical reorganization or improvements in stepping accuracy, whereas the addition to active training (analogous to CIMT) did. Children with USCP were either provided intensive bimanual training (HABIT) that included progressive increases in the difficulty of activities as hand function improved, or practice of the same activities without skill progression (bimanual play). At first glance, it appeared that both approaches lead to equal improvements in hand function as assessed with standard clinical assessments (Brandao et al. 2014). However, upon examination of progress on functional and play goals identified by caregivers, children in the skilled practice group had greater improvements (Brandao et al. 2014) and improved their movement coordination patterns as seen using kinematic analysis on the functional drawer opening task described above. Furthermore, we used TMS to probe the hand representation plasticity (Friel et al. 2016). It was shown that children in the skill progression group that the size and motor evoked potential (MEP) amplitudes of the affected hand motor map increased significantly after therapy, whereas training of the same activities without skill progression did not (Fig. 6). Thus, skill progression appears to be an important ingredient for changing motor patterns and inducing plasticity in M1.

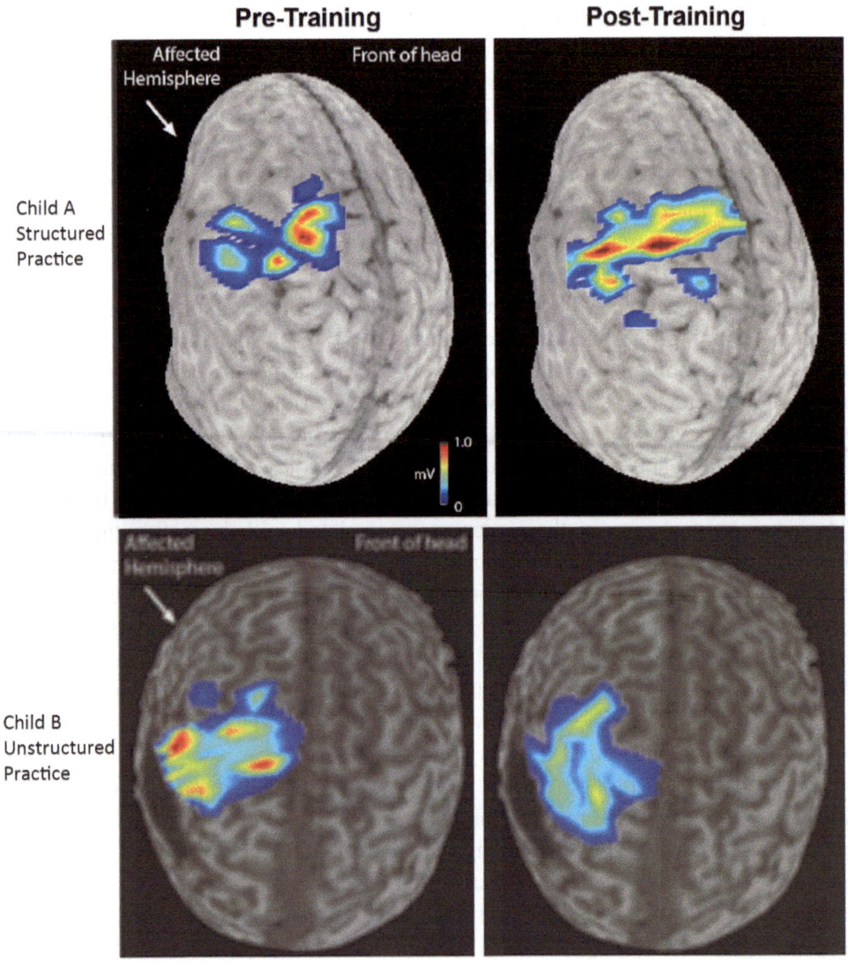


**Fig. 6** Changes in magnitude of motor evoked potentials in TMS maps after structured training in child A and unstructured practice in child B. Maps are of the affected hand located contralateral to the affected hemisphere. *Red color* indicates stronger MEP response. Not that the maps indicate a larger hand representation with stronger MEPS after structured practice only. Plotted from data in Friel et al. 2016

Together these studies highlight the variations in CNS damage underlying CP. It is essential to understand the underlying behavioral and neural mechanisms as to better target them with interventions. However, the variability complicates story and may limit the extent to which new treatments may be generalized across individuals. In fact, they highlight the notion that there cannot be a "one-size-fits-all approach" to rehabilitation.

## *Brain Stimulation*

As we begin to understand the activity-dependent reorganization of the brain following intensive training, the exciting prospect of stimulating the brain directly using rTMS or transcranial direct current stimulation (tDCS). Stimulating the CST after M1 inactivation in a feline model of hemiparesis has been shown stimulate its normal development and improve motor function (Salimi et al. 2008). Stimulating the brain using tDCS has been shown to enhance motor learning (Reis et al. 2009). In hemiplegia due to adult stroke, short bouts of inhibitory, contralesional rTMS improves hand function (Hsu et al. 2012; Lefaucheur et al. 2014). A preliminary trial of contralesional inhibitory rTMS found the procedure to be safe and feasible for patients with hemiplegia due to arterial ischaemic stroke (Kirton et al. 2008). A recent trial of contralesional rTMS in participants with unilateral CP suggested favorable tolerability and functional improvements as well (Gillick et al. 2014). In a RCT of 45 participants with USCP due to perinatal stroke examined inhibitory contralesional rTMS on hand function (Kirton et al. 2016). They found that participating in intensive rehabilitation programs alone can result in sustained functional gains. However the addition of CIMT and rTMS increases the chances of significant improvement. An important caveat is that this study did not examine CST organization in relation to the outcomes (see Staudt and Gordon 2016).

It should be noted that since almost all studies of intensive training or brain stimulation have been conducted in children in children age 2 and above. This is thought to be beyond the age in which normal contralateral CST organization is established (i.e., within the first 6 months, Eyre et al. 2007). In fact while most studies have demonstrated improvements in hand function, none have reported "curing" the impaired hand. An important remaining question is whether early treatment during this "critical period" can alter the development of CST connectivity, and possibly thwart the emergence of CP. In this regards, given the limited attention spans and movement repertoires of infants, brain stimulation may eventually hold considerable promise to this effect. Furthermore, given the potential impact of tDCS on motor learning, a key question is whether the addition of this technique in conjunction with intensive training can decrease the required dose, and thus cost, associated with these programs.

## Take Home Message

The pathophysiology and mechanisms underlying impaired upper extremity control of USCP will be reviewed. It was shown that the integrity of the CST tract closely relates to the extent of upper extremity impairments. It was also shown that the developing CST tract can reorganize its connectivity depending on the timing and locating of CNS injury. The reorganization is also predictive of the resulting severity of impairments. Mechanisms underlying impaired motor function were highlighted, including movement execution, movement planning, sensorimotor integration, bimanual movement control and mirror movements. Evidence for motor-learning based therapies was presented, including Constraint-Induced Movement Therapy and Bimanual Training, and possible pathophysiological predictors of treatment outcome were presented. Finally future possibilities of stimulating the development of the nervous system were presented.

## References


Aarts PB, Jongerius PH, Geerdink YA et al (2010) Effectiveness of modified constraint-induced movement therapy in children with unilateral spastic cerebral palsy: a randomized controlled trial. Neurorehabil Neural Repair 24:509–518

Adler C, Berweck S, Lidzba K, Becher T, Staudt M (2015) Mirror movements in unilateral spastic cerebral palsy: Specific negative impact on bimanual activities of daily living. Eur J Paediatr Neurol 19:504–509

Bax M, Goldstein M, Rosenbaum P, Leviton A, Paneth N, Dan B, Jacobsson B, Damiano D (2005) Proposed definition and classification of cerebral palsy. Dev Med Child Neurol 47: 571–576

Bleyenheuft Y, Grandin CB, Cosnard G, Olivier E, Thonnard JL (2007) Corticospinal dysgenesis and upper-limb deficits in congenital hemiplegia: a diffusion tensor imaging study. Pediatrics 120:e1502–e1511

Bleyenheuft Y, Thonnard JL (2010) Predictive and reactive control of precision grip in children with congenital hemiplegia. Neurorehabil Neural Repair 24:318–327

Bleyenheuft Y, Thonnard JL (2011) Tactile spatial resolution in unilateral brain lesions and its correlation with digital dexterity. J Rehabil Med 43:251–256

Bleyenheuft Y, Dricot L, Gilis N, Kuo HC, Grandin C, Bleyenheuft C, Gordon AM, Friel KM (2015) Capturing neuroplastic changes after bimanual intensive rehabilitation in children with unilateral spastic cerebral palsy: a combined DTI, TMS and fMRI pilot study. Res Dev Dis 43–44:136–149

Brandão M, Gordon AM, Mancini MC (2012) Functional impact of constraint therapy and bimanual training in children with cerebral palsy: a randomized controlled trial. Am J Occup Ther 66:672–681

Brandao M, Ferre CL, Kuo H-C, Rameckers E, Bleyenheuft Y, Hung Y-C, Friel K, Gordon AM (2014) Comparison of structured skill and unstructured practice during intensive bimanual training in children with unilateral spastic cerebral palsy. Neural Rehab Neur Repair 28: 452–461

Brown JK, van Rensburg E, Walsh G, Lakie M, Wright GW (1987) A neurological study of hand function of hemiplegic children. Dev Med Child Neurol 29:287–304

Carr LJ, Harrison LM, Evans AL, Stephens JA (1993) Patterns of central motor reorganization in hemiplegic cerebral palsy. Brain 116:1223–1247

Carr J, Shepherd RB (1989) A motor learning model for stroke rehabilitation. Physiotherapy 75:372–380

Case-Smith J, DeLuca SC, Stevenson R, Ramey SL (2012) Multicenter randomized controlled trial of pediatric constraint-induced movement therapy: 6-month follow-up. Am J Occup Ther 66:15–23

Charles J, Lavinder G, Gordon AM (2001) The effects of constraint induced therapy on hand function in children with hemiplegic cerebral palsy. Ped Phys Ther 13:68–76

Charles J, Gordon AM (2006) Development of hand-arm bimanual intensive therapy (HABIT) for improving bimanual coordination in children with hemiplegic cerebral palsy. Dev Med Child Neurol 48:931–936

Cioni G, Sales B, Paolicelli PB, Petacchi E, Scusa MF, Canapicchi R (1999) MRI and clinical characteristics of children with hemiplegic cerebral palsy. Neuroped 30:249–255

Cohen-Holzer M, Katz-Leurer M, Reinstein R et al (2011) The effect of combining daily restraint with bimanual intensive therapy in children with hemiparetic cerebral palsy: a self-control study. NeuroRehab 29:29–36

Deppe W, Thuemmler K, Fleischer J, Berger C, Meyer S, Wiedemann B (2013) Modified constraint-induced movement therapy versus intensive bimanual training for children with hemiplegia—a randomized controlled trial. Clin Rehabil 10:909–920

Dong VA-Q, Tung IH-H, Siu HW-Y et al (2013) Studies comparing the efficacy of constraint-induced movement therapy and bimanual training in children with unilateral cerebral palsy: a systematic review. Dev Neurorehabil 16:133–143

Eliasson AC, Bonnier B, Krumlinde-Sundholm L (2003) Clinical experience of constraint induced movement therapy in adolescents with hemiplegic cerebral palsy—a day camp model. Dev Med Child Neurol 45:357–359

Eliasson AC, Forssberg H, Hung YC, Gordon AM (2006) Development of hand function and precision grip control in individuals with cerebral palsy: a 13-year follow up study. Pediatrics 118:e1226–e1236

Eliasson AC, Gordon AM, Forssberg H (1991) Basic coordination of manipulative forces of children with cerebral palsy. Dev Med Child Neurol 33:659–668

Eliasson AC, Gordon AM, Forssberg H (1992) Impaired anticipatory control of isometric forces during grasping by children with cerebral palsy. Dev Med Child Neurol 34:216–225

Eliasson AC, Gordon AM (2000) Impaired force coordination during object release in children with hemiplegic cerebral palsy. Dev Med Child Neurol 42:228–234

Eliasson A-C, Krumlinde-sundholm L, Shaw K et al (2005) Effects of constraint-induced movement therapy in young children with hemiplegic cerebral palsy: an adapted model. Dev Med Child Neurol 47:266–275

Eliasson AC, Krumlinde-Sundholm L, Gordon AM, Feys H, Klingels K, Aarts P, Rameckers E, Autti-Rämö I, Hoare B (2014) Guidelines for future research in Constraint-induced movement therapy for children with unilateral cerebral palsy: an expert consensus. Dev Med Child Neurol 56:125–137

Eyre JA (2003) Development and plasticity of the corticospinal system in man. J Neural Transplant Plast 10:93–106

Eyre JA (2007) Corticospinal tract development and its plasticity after perinatal injury. Neurosci Biobehav Rev 31:1136–1149

Eyre JA, Smith M, Dabydeen L et al (2007) Is hemiplegic cerebral palsy equivalent to amblyopia of the corticospinal system? Ann Neurol 62:493–503

Eyre JA, Taylor JP, Villagra F et al (2001) Evidence of activity-dependent withdrawal of corticospinal projections during human development. Neurol 57:1543–1554

Facchin P, Rosa-Rizzotto M, Dalla Pozza LV et al (2011) Multisite trial comparing the efficacy of constraint-induced movement therapy with that of bimanual intensive training in children with hemiplegic cerebral palsy: Postintervention results. Am J Phys Med Rehabil 90:539–553

Farmer SF, Harrison LM, Ingram DA, Stephens JA (1991) Plasticity of central motor pathways in children with hemiplegic cerebral palsy. Neurology 41:1505–1510

Friel K, Chakrabarty S, Kuo H-C et al (2012) Using motor behavior during an early critical period to restore skilled limb movement after damage to the corticospinal system during development. J Neurosci 32:9265–9276

Friel KM, Chakrabarty S, Martin JH (2013) Pathophysiological mechanisms of impaired limb use and motor systems repair strategies after unilateral injury of the developing brain. Dev Med Child Neurol Suppl 4:27–31

Friel KM, Kuo HC, Carmel JB, Gordon AM (2014) Improvements in hand function after intensive bimanual training are not associated with corticospinal tract dysgenesis in children with hemiplegic cerebral palsy. Exp Brain Res 232:2001–2009

Friel K, Kuo H-C M, Ferre CL, Brandao M, Carmel J, Bleyenheuft Y, Gowatsky JL, Stanford AD, Rowny SB, Luber B, Bassi B, Murphy DLK, Lisanby SH, Gordon AM (2016) Skilled bimanual training drives motor cortex plasticity in children with unilateral cerebral palsy. Neural Rehab Neural Repair 30(9):834–844

Gentile AM (2000) Skill acquisition: action, movement, and neuromotor processes. In: Carr JH, Shepherd RD (eds) Movement science: foundations for physical therapy, 2nd edn. Aspen, Rockville, MD, pp 111–187

Gillick BT, Krach LE, Feyma T et al (2014) Primed low-frequency repetitive transcranial magnetic stimulation and constraint-induced movement therapy in pediatric hemiparesis: a randomized controlled trial. Dev Med Child Neurol 56:44–52

Gordon AM (2001) In: Kalverboer AF, Gramsbergen A (eds) Handbook of brain and behaviour in human development. Kluwer, Dordrecht, pp 513–537

Gordon AM (2011) To constrain or not to constrain, and other stories of intensive upper extremity training for children with unilateral cerebral palsy. Dev Med Child Neurol 53(Suppl. 4):56–61

Gordon AM (2014) In: Shepherd R (Ed) Cerebral Palsy in infancy. targeted activity to optimize early growth and development. Elsevier, pp 305–322

Gordon AM, Charles J, Schneider JA, Chinnan A (2007) Efficacy of a hand-arm bimanual intensive therapy (HABIT) for children with hemiplegic cerebral palsy: a randomized control trial. Dev Med Child Neurol 49:830–838

Gordon AM, Charles J, Duff SV (1999) Fingertip forces during object manipulation in children with hemiplegic cerebral palsy. II: Bilateral coordination. Dev Med Child Neurol Neurol 41:176–185

Gordon AM, Charles J, Steenbergen B (2006a) Movement planning is disrupted by impaired sensorimotor integration in children with hemiplegic cerebral palsy. Ped Res 60:587–591

Gordon AM, Charles J, Wolf SL (2006b) Efficacy of constraint-induced movement therapy on involved-upper extremity use in children with hemiplegic cerebral palsy is not age-dependent. Pediatrics 117:e363–e373

Gordon AM, Chinnan A, Gill S, Petr E, Hung YC, Charles J (2008) Both constraint-induced movement therapy and bimanual training lead to improved performance of upper extremity function in children with hemiplegia. Dev Med Child Neurol 50:957–958

Gordon AM, Duff SV (1999a) Fingertip forces during object manipulation in children with hemiplegic cerebral palsy. I: Anticipatory scaling. Dev Med Child Neurol 41:166–175

Gordon AM, Duff SV (1999b) Relation between clinical measures and fine manipulative control in children with hemiplegic cerebral palsy. Dev Med Child Neurol 41:586–591

Gordon AM, Hung YC, Brandao M et al (2011) Bimanual training and constraint-induced movement therapy in children with hemiplegic cerebral palsy: a randomized trial. Neuro Rehab Neural Repair 25:692–702

Gordon AM, Lewis S, Eliasson AC, Duff SV (2003) Object release under varying task constraints in children with hemiplegic cerebral palsy. Dev Med Child Neurol 45:240–248

Gordon AM, Magill, RA (2012) Motor learning: Application of principles to pediatric rehabilitation. In: Campbell SK, Palisano RJ, Orlin MN (eds) Physical therapy for children. Elsevier (Saunders), St. Louis, pp 151–174

Gordon AM, Steenbergen B (2008) In: Eliasson AC and Burtner P (eds) Improving hand function in children with cerebral palsy: theory, evidence and intervention. Clinics in developmental medicine. Mac Keith Press, London, pp 160–75. https://www.ncbi.nlm.nih.gov/pubmed/24237277

Green D, Schertz M, Gordon AM, Moore A, Schejter M, Farquharson Y, Ben Bashat D, Winstein M, Lin J-P, Fattal-Valevski A (2013) A multi-site study of functional outcomes following a themed approach to hand-arm bimanual intensive therapy (HABIT) for children with hemiplegia. Dev Med Child Neurol 55:527–533

Harrison A (1988) Spastic cerebral palsy: possible spinal interneuronal contributions. Dev Med Child Neurol 30:769–780

Himmelmann K, Hagberg G, Beckung E, Hagberg B, Uvebrant P (2005) The changing panorama of cerebral palsy in Sweden. IX. Prevalence and origin in the birth-year period 1995–1998. Acta Paediatr 94:287–294

Hoare B, Imms C, Villanueva E et al (2013) Intensive therapy following upper limb botulinum toxin A injection in young children with unilateral cerebral palsy: a randomized trial. Dev Med Child Neurol 55:238–247

Holmström L, Vollmer B, Tedroff K et al (2010) Hand function in relation to brain lesions and corticomotor-projection pattern in children with unilateral cerebral palsy. Dev Med Child Neurol 52:145–152

Hsu WY, Cheng CH, Liao KK, Lee IH, Lin YY (2012) Effects of repetitive transcranial magnetic stimulation on motor functions in patients with stroke: a meta-analysis. Stroke 43:1849–1857

Hung Y-C, Casertano L, Hillman A, Gordon AM (2011) The effect of training specificity on bimanual coordination in children with hemiplegia. Res Dev Dis 32:2724–2731

Hung Y-C, Charles J, Gordon AM (2004) Bimanual coordination of arm movements in children with hemiplegia. Dev Med Child Neurol 46:746–753

Hung YC, Charles J, Gordon AM (2010) Influence of accuracy constraints on bimanual coordination during a goal-directed task in children with hemiplegic cerebral palsy. Exp Brain Res 201:421–428

Islam M, Gordon AM, Forssberg H, Sköld A, Eliasson AC (2011) Grip force coordination during a bimanual task in unilateral cerebral palsy. Dev Med Child Neurol 53:920–926

Islam M, Nordstrand L, Holmström L, Kits A, Forssberg H, Eliasson AC (2014) Is outcome of constraint-induced movement therapy in unilateral cerebral palsy dependent on corticomotor projection pattern and brain lesion characteristics? Dev Med Child Neurol 56:252–258

Johansson RS, Westling G (1984) Roles of glabrous skin receptors and sensorimotor memory in automatic control of precision grip when lifting rougher or more slippery objects. Exp Brain Res 56:550–564

Juenger H, Kuhnke N, Braun C, Ummenhofer F, Wilke M, Walther M, Koerte I, Delvendahl I, Berweck S, Mall V, Staudt M (2013) Two types of exercise-induced neuroplasticity in congenital hemiparesis: a TMS, fMRI and MEG study. Dev Med Child Neurol 55:941–951

Kennard MA (1936) Age and other factors in motor recovery from precentral lesions in monkeys. Am J Physiol 115:138–146

Kirton A, Chen R, Friefeld S, Gunraj C, Pontigon AM, deVeber G (2008) Contralesional repetitive transcranial magnetic stimulation for chronic hemiparesis in subcortical paediatric stroke: a randomised trial. Lancet Neurol 7:507–513

Kirton A, Andersen J, Herrero M, Nettel-Aguirre A, Carsolio L, Damji O, Keess J, Mineyko A, Hodge J, Hill MD (2016) Brain stimulation and constraint for perinatal stroke hemiparesis: The PLASTIC CHAMPS trial. Neurology 3(86):1659–1667

Kleim JA, Barbay S, Nudo RJ (1998) Functional reorganization of the rat motor cortex following motor skill learning. J Neurophysiol 80:3321–3325

Kleim JA, Barbay S, Cooper NR, Hogg TM, Reidel CN, Remple MS, Nudo RJ (2002) Motor learning-dependent synaptogenesis is localized to functionally reorganized motor cortex. Neurobiol Learn Mem 77:63–77

Klingels K, Jaspers E, Staudt M, Guzzetta A, Mailleux L, Ortibus E, Feys H (2016) Do mirror movements relate to hand function and timing of the brain lesion in children with unilateral cerebral palsy? Dev Med Child Neurol 58:735–742

Koert I, Pelavin P, Kirmess B, Fuchs T, Berweck S, Laubender RP, Heinen F (2011) Anisotropy of transcallosal motor fibres indicates functional impairment in children with periventricular leukomalacia. Dev Med Child Neurol 53:179–186

Kuhnke N, Juenger H, Walther M, Berweck S, Mall V, Staudt M (2008) Do patients with congenital hemiparesis and ipsilateral corticospinal projections respond differently to constraint-induced movement therapy? Dev Med Child Neurol 50:898–903

Kuo H-C, Gordon AM, Henrionnet A, Hautfenne S, Friel K, Bleyenheuft Y (2016) The effects of intensive bimanual training with and without tactile training on tactile function in children with unilateral spastic cerebral palsy. Res Dev Disabil 49–50:129–139

Kuo H-C, Ferre CL, Carmel JB, Gowatsky JL, Stanford AD, Rowny SB, Lisanby SH, Gordon AM, Friel KM (2016) Using diffusion tensor imaging to identify corticospinal tract projection patterns in children with unilateral spastic cerebral palsy. Dev Med Child Neurol. doi: 10.1111/dmcn.13192

Kuhtz-Buschbeck JP, Sundholm LK, Eliasson AC, Forssberg H (2000) Quantitative assessment of mirror movements in children and adolescents with hemiplegic cerebral palsy. Dev Med Child Neurol 42:728–736

Latash ML, Anson JG (1996) What are "normal movements" in atypical populations? Behav Brain Sci 19:55–106

Lawrence DG, Kuypers HGJM (1968) The functional organization of the motor system in the monkey. I. The effects of bilateral pyramidal lesions. Brain 91:1–14

Lefaucheur J-P, André-Obadia N, Antal A et al (2014) Evidence-based guidelines on the therapeutic use of repetitive transcranial magnetic stimulation (rTMS). Clin Neurophysiol 125:2150–2206

Lowes LP, Mayhan M, Orr T, Batterson N, Tonneman JA, Meyer A, Alfano L, Wang W, Whalen CN, Nelin MA, Lo WD, Case-Smith J (2014) Pilot study of the efficacy of constraint-induced movement therapy for infants and toddlers with cerebral palsy. Phys Occup Ther Pediatr 34:4–21

Martin JH, Chakrabarty SS, Friel KM (2011) Harnessing activity-dependent plasticity to repair the damaged corticospinal tract in an animal model of cerebral palsy. Dev Med Child Neurol 53:9–13

Moberg E (1962) Criticism and study of methods for examining sensibility in the hand. Neurol 12:8–19

Novak I, McIntyre S, Morgan C, Campbell L, Dark L, Morton N, Stumbles E, Wilson SA, Goldsmith S (2013) A systematic review of interventions for children with cerebral palsy: state of the evidence. Dev Med Child Neurol 55:885–910

Nudo RJ (2003) Functional and structural plasticity in motor cortex: implications for stroke recovery. Phys Med Rehabil Clin N Am 14(Suppl):S57–S76

Plautz EJ, Milliken GW, Nudo RJ (2000) Effects of repetitive motor training on movement representations in adult squirrel monkeys: role of use versus learning. Neurobiol Learn Mem 74:27–55

Prabhu S, Diermayr G, Gysin P, Gordon AM (2011) Impaired anticipatory control of grasp during gait in children with hemiplegic cerebral palsy. Dev Med Child Neurol 53:865–869

Reis J, Schambra HM, Cohen LG, Buch ER, Fritsch B, Zarahn E, Celnik PA, Krakauer JW (2009) Noninvasive cortical stimulation enhances motor skill acquisition over multiple days through an effect on consolidation. Proc Natl Acad Sci USA 106:1590–1595

Rickards T, Sterling C, Taub E, Perkins-Hu C, Gauthier L, Graham M, Griffin A, Davis D, Mark VW, Uswatte G (2014) Diffusion tensor imaging study of the response to constraint-induced movement therapy of children with hemiparetic cerebral palsy and adults with chronic stroke. Arch Phys Med Rehabil 95:506–514

Sakzewski L, Ziviani J, Abbott DF et al (2011a) Randomized trial of constraint-induced movement therapy and bimanual training on activity outcomes for children with congenital hemiplegia. Dev Med Child Neurol 53:313–320

Sakzewski L, Ziviani J, Boyd RN (2011b) Best responders after intensive upper-limb training for children with unilateral cerebral palsy. Arch Phys Med Rehabil 92:578–584

Salimi I, Friel KM, Martin JH (2008) Pyramidal tract stimulation restores normal corticospinal tract connections and visuomotor skill after early postnatal motor cortex activity blockade. J Neurosci 28:7426–7434

Scheck SM, Boyd RN, Rose SE (2012) New insights into the pathology of white matter tracts in cerebral palsy from diffusion magnetic resonance imaging: a systematic review. Dev Med Child Neurol 54:684–696

Schertz M, Shiran S, Myers V, Weinstein M, Fattal-Valevski A, Artzi M, Ben Bashat D, Gordon AM, Green D (2016) Advanced imaging predictors of improvement from a motor learning based intervention for children with hemiplegia. Neural Rehab Neural Repair 30: 647–660

Staudt M (2010) Brain plasticity following early life brain injury: insights from neuroimaging. Semin Perinatol 34:87–92

Staudt M, Gerloff C, Grodd W et al (2004) Reorganization in congenital hemiparesis acquired at different gestational ages. Ann Neurol 56:854–863

Staudt M, Gordon AM (2016) Combining rTMS and CIMT: A “one-size-fits-all” therapy for congenital hemiparesis? Neurology 3(86):1652–1654

Steenbergen B, Charles J, Gordon AM (2008) Fingertip force control during bimanual object lifting in hemiplegic cerebral palsy. Exp Brain Res 186:191–201

Steenbergen B, Hulstijn W, De Vries A, Berger M (1996) Bimanual movement coordination in spastic hemiparesis. Exp Brain Res 110:91–98

Steenbergen B, Jongbloed-Pereboom M, Spruijt S, Gordon AM (2013) Impaired motor planning and motor imagery in children with unilateral spastic cerebral palsy: challenges for the future of pediatric rehabilitation. Dev Med Child Neurol Suppl 4:43–46

Sukal-Moulton T, Murray TM, Dewald JP (2013) Loss of independent limb control in childhood hemiparesis is related to time of brain injury onset. Exp Brain Res 225(3):455–463

Steenbergen B, Van Thiel E, Hulstijn W, Meulenbroek RGJ (2000) The coordination of reaching and grasping in spastic hemiparesis. Hum Mov Sci 19:75–105

Taub E, Shee LP (1980) Somatosensory deafferentation research with monkeys: implications for rehabilitation medicine. In: Ince LP (ed) Behavioral psychology in rehabilitation medicine: clinical applications. Williams and Wilkins, Baltimore and London, pp 371–401

Taub E, Wolf SL (1997) Constraint induced movement techniques to facilitate upper extremity use in stroke patients. Top Stroke Rehabil 3:38–61

Taub E, Griffin A, Nick J et al (2007) Pediatric CI therapy for stroke-induced hemiparesis in young children. Dev Neurorehabil 10:3–18

Thorndike EL (1914) Educational psychology: briefer course. Columbia University Press, New York

Tower S (1940) Pyramidal lesion in the monkey. Brain 63:36–90

Trombly C (1995) Clinical practice guidelines for post-stroke rehabilitation and occupational therapy practice. Am J Occup Ther 49:711–714

Uvebrandt P (1988) Hemiplegic cerebral palsy. Aetiology and outcome. Acta Paediatr Scand Suppl 345:1–100

Weinstein M, Green D, Geva R, Schertz M, Fattal-Valevski A, Artzi M, Myers V, Shiran S, Gordon AM, Gross-Tsur V, Ben Bashat D (2013) Interhemispheric and intrahemispheric connectivity and manual skills in children with unilateral cerebral palsy. Brain Struct Funct 219:1025–1040

Weinstein M, Myers V, Green D, Schertz M, Shiran A, Geva R, Artzi M, Gordon AM, Fattal-Valevski A, Ben Bashat D (2015) Brain plasticity following intensive bimanual therapy in children with hemiplegia: preliminary evidence. Neural Plast :798481. doi: 10.1155/2015/798481

Winstein CJ, Wolf SL (2009) Task-oriented training to promote upper extremity recovery. In: Stein J, Harvey R, Macko R, Winstein CJ, Zorowitz R (eds) Stroke recovery and rehabilitation. Demos Medical Publishing, New York, NY

Wolf SL, Lecraw DE, Barton LA et al (1989) Forced use of hemiplegic upper extremities to reverse the effect of learned nonuse among chronic stroke and head-injured patients. Exp Neurol 104:125–132

Woods BT, Teuber HL (1978) Mirror movements after childhood hemiparesis. Neurol 28: 1152–1157

# Can Motor Recovery in Stroke Be Improved by Non-invasive Brain Stimulation?

**John C. Rothwell**

**Abstract** At the present time, there is enormous interest in methods of non-invasive brain stimulation. These interact with ongoing neural activity, mainly in cerebral cortex, and have measureable effects on behaviours in healthy people. More intriguingly, they appear to have effects on synaptic plasticity that persist even after stimulation has ceased. This has led, as might be expected, to the proposal that brain stimulation methods might be therapeutically useful in rehabilitation. The rationale is that physical therapy involves learning new patterns of activity to compensate for those lost to the stroke. Enhanced "plasticity" produced by brain stimulation might increase the ability to learn and enhance therapy. However, if things really were as simple as this, brain stimulation would be on its way to becoming a standard addition to treatment in all departments of rehabilitation. The fact that this has not happened means that something is not quite correct. Is the theory untenable, or are the methods of stimulation suboptimal?

**Keywords** Paired associative stimulation (PAS) · Synaptic plasticity (STDP) · Transcranial direct current stimulation (TDCS) · Magnetic stimulation · Cortical

## Transcranial Magnetic Stimulation (TMS)

TMS employs a rapidly changing magnetic field to induce a very short pulse of electrical current in the underlying brain that is designed to be very similar to the electrical pulse delivered by conventional peripheral nerve stimulators (Barker et al. 1987). It is thought that this activates the axons of neurones near the surface of the brain, producing an action potential that travels to the synaptic terminal where it releases transmitter that could excite or inhibit further neurones. In the motor cortex, several lines of evidence suggest that a single TMS pulse causes repetitive

J.C. Rothwell (✉)
Institute of Neurology, University College London, Queen Square,
London WC1N 3BG, UK
e-mail: j.rothwell@ucl.ac.uk



J. Laczko and M.L. Latash (eds.), *Progress in Motor Control*,
Advances in Experimental Medicine and Biology 957,
DOI 10.1007/978-3-319-47313-0_17

firing of neurones that lasts several milliseconds before being "quenched" by a long-lasting IPSP (Rothwell et al. 1991).

Most of the early TMS stimulators could only deliver 1 stimulus every 3–4 s. However, from the therapeutic viewpoint, the interesting development came with stimulators that could give multiple stimuli at frequencies up to 50–100 Hz (repetitive TMS, or rTMS). Many experiments in animal brain, particularly in slices of hippocampus, have shown that repeated patterns of activity can change the effectiveness of synaptic connections for hours, days or longer (Malenka and Bear 2004). Such changes are induced in two different ways: repetitive activation of the presynaptic input (e.g., 100 Hz for 10 s) or repeated inputs that are carefully timed to precede/follow the discharge of the postsynaptic neurone (Spike-Timing-Dependent Plasticity, STDP) (Dan and Poo 2006). In the first method, the usual rule is that the high frequency stimulation produces facilitation at the synapse whereas low frequency depresses the synapse. In the second method, the usual rule is that if the input precedes the postsynaptic action potential then the synapse will strengthen whereas the opposite occurs if the presynaptic input follows the postsynaptic action potential. These rules, however, are only approximate guides, with many exceptions possible.

Repetitive TMS can reproduce these protocols in the human brain, and a large number of studies have revealed effects that are compatible with the idea that rTMS can produce short term effects on synaptic transmission in cerebral cortex that are consistent with them being equivalent to early stages of LTP and LTD (Ziemann et al. 2008). Most of the work has been performed on the motor cortex because a single TMS pulse produces a twitch of muscles on the contralateral side of the body that can be easily measured and used as an indicator of the excitability of the motor cortical output to spinal cord. Thus, a session of rTMS in which several hundred stimuli are applied in a short period leads to after-effects on excitability such that the response to a single TMS pulse is larger or smaller than at baseline for the next 30–60 min. Since these effects disappear if people are given a small dose of a drug that interferes with activity of the glutamatergic NMDA receptor, which is known to be an important contributor to LTP and LTD in animal experiments, they are termed "LTP/LTD-like" (Ziemann et al. 2008).

There are a large number of different types of rTMS protocol that differ in the frequency, intensity, and number of stimuli. Some of them tend to increase excitability and others to reduce it. Commonly used protocols are regular rTMS at 1–20 Hz; more complex patterned protocols are continuous theta burst stimulation (cTBS), which tends to reduce excitability and intermittent theta burst stimulation (iTBS) that increases excitability (Huang et al. 2005). In all cases, the effects are relatively short lasting compared with those reported in animal preparations. However, it is important to bear in mind that much of the basic animal work on LTP/LTD is performed in slice preparations from the hippocampus of dead animals. Far fewer studies have been performed on neocortex, and even fewer in awake animals (Werk and Chapman 2003; Werk et al. 2006). In the latter, it is difficult to induce reliable LTP/LTD. Such studies might be a much better model for plasticity studies in the human brain.

Work in humans has also replicated the paired presynaptic/postsynaptic approach to generating synaptic plasticity (STDP) (Stefan et al. 2000). A widely employed protocol is Paired Associative Stimulation (PAS) where a peripheral (median) nerve electrical stimulus is paired with TMS to motor cortex. If the stimuli are timed such that the peripheral stimulus reaches the cortex just prior the TMS pulse, a facilitatory effect is observed. With a slightly later peripheral nerve stimulus, calculated to reach cortex just after the TMS pulse, there is an inhibitory effect (Muller-Dahlhaus et al. 2010).

## Transcranial Direct Current Stimulation (TDCS)

TDCS conventionally involves applying a constant 1–2 mA electrical current for 10–20 min through two large (often 5 × 7 cm) electrodes on the scalp (Nitsche and Paulus 2000). Only a small proportion of the current actually flows into the brain; the majority travels in the lower resistance pathway of the skin and scalp between the two electrodes. Unlike TMS, TDCS does not activate axons and produce action potentials in cortical neurones. Calculations suggest that it might change the voltage across neuronal membranes (polarisation). This is typically about −70 mV (inside negative); TDCS might change this up to 0.5 mV (Woods et al. 2016).

With such a small stimulus, how do we know it will have any effect on the brain? There were many experiments in 1950s and 1960s in which effects of TDCS were investigated in animal brain (Bindman et al. 1962, 1964). These showed that during TDCS, ongoing activity of neurones near the anode (positive pole of the stimulator) increased whereas it decreased near the cathode. Furthermore, following 10 min TDCS, there were after-effects on discharge rates that persisted for minutes or even hours. However, it is important to note that in these experiments, the TDCS electrodes were placed directly on the surface of the brain and the polarising effect was many times larger than can ever be achieved in the human brain using electrodes on the scalp. In fact recent experiments in animal suggest that the minimal electric fields that have any detectable action on neural activity are of the order 1 V/m. Calculations show that this is about the maximum possible levels that might be achieved in human brain (Woods et al. 2016).

The conclusion is that the effects of TDCS are likely to be quite subtle. Nevertheless, since TDCS will affect a large number of neurones simultaneously, it may well have measurable effects on behaviour. Initial experiments showed that 10 min of 1 mA TDCS over human motor cortex changed its excitability for about 30 min afterwards. If the anode was over motor cortex and cathode over the contralateral orbit, the effect was excitatory (Nitsche and Paulus 2000). If the electrode polarities were reversed then the effect was inhibitory. As with rTMS, drug studies suggested that these after-effects were equivalent to early stages of LTP and LTD at synaptic connections in cortex (Liebetanz et al. 2002).

## Effects of TMS and TDCS on Motor Learning in Healthy Volunteers

Broadly speaking, we can distinguish two main types of motor learning: skill learning and adaptation learning (Krakauer and Mazzoni 2011). Skill learning, such as learning to putt a golf ball into a hole, involves incremental improvements in performance that accrue over an extended period of training. This is sometimes termed 'non-rule-based' learning since participants do not know in advance how to improve their performance. It can be distinguished from a more rapid form of learning in which a previously learnt movement is performed in the presence of a visuomotor or somatosensory perturbation—motor adaptation. In this case, participants are aware of the error and can use this information to update a previously learned skill. From a rehabilitation perspective, skill learning is perhaps more important since this is what we imagine patients have to do when re-learning how to use their affected limbs after a stroke.

There is good evidence that the motor cortex is involved in the early stages of skill learning. When volunteers learn a new motor skill, the improvement in performance during training (skill acquisition) is associated with a rapid increase in motor cortex excitability (Muellbacher et al. 2001; Nudo et al. 1996). After training has finished, a second process, known as consolidation occurs, in which the motor memories are strengthened further so that they persist for longer and become more resistant to interference by subsequent motor activity. This process may involve the motor cortex, at least for a short period of time after completing training. Thus, consolidation can be blocked by using TMS to interfere with activity in primary motor cortex immediately after training but not 6 h later (Muellbacher et al. 2002). There may also be offline performance gains so that performance is better on the day after training than it was at the end of training on the previous day (Brashers-Krug et al. 1996; Walker et al. 2002) Finally, it should be noted that an already consolidated motor memory may become susceptible to further alteration when it is reactivated (Nader and Hardt 2009) (reconsolidation) and modulation of motor cortex activity also influences this process, suggesting that this region may transiently interact with the stored memory trace during movement execution.

Several experiments have shown that rTMS and TDCS can influence both online and offline learning. Most studies using rTMS are designed to 'prime' the motor cortex for subsequent learning, such that training occurs during a period of increased excitability. For example, 10 Hz rTMS delivered to the motor cortex for 2 s immediately before each training block enhanced training in a sequential motor learning task (Kim et al. 2004). Similarly, Teo et al. tested the influence of intermittent TBS on subsequent training in a ballistic thumb movement task and found that iTBS enhanced the rate and extent of performance gains (Teo et al. 2011).

Reis and colleagues studied the effect of applying anodal tDCS to the active motor cortex during training in an isometric pinch task over 5 days, and found that learning was enhanced over sham and that the benefit was sustained at 3 months (Reis et al. 2009). Interestingly, tDCS in this study specifically enhanced offline gains between sessions, suggesting an effect on consolidation processes taking place after completion of each training session.

The conclusion is that, forms of skill learning in healthy volunteers can be enhanced by protocols of TDCS and rTMS.

## Application in Rehabilitation

Experiments in animals show that injury to the cortex is accompanied by changes in synaptic connections that can involve axon sprouting and formation of new synapses not only in the cortex itself but also at other subcortical sites and even in spinal cord [see review by (Nudo 2006)]. These changes are the natural reaction of neurones to the sudden absence of their usual interactions with damaged tissue. For example, neurones in rat premotor cortex that previously interacted with neurones in motor cortex sprout new axons following M1 damage that avoid the damaged motor area and aim for the sensory cortex. Similarly, the strength of reticulospinal connections with spinal motorneurones increases after damage to the corticospinal system (Zaaimi et al. 2012). It appears as if absence of corticospinal inputs promotes strengthening of other remaining inputs.

This change in anatomy means that movements are accompanied by different patterns of activity after stroke. Thus, fMRI shows that after stroke movements are made with different patterns of BOLD activation than those seen in an undamaged brain. Activation is more widespread and often involves increased activity of premotor areas bilaterally as well as other areas of cortex (Ward et al. 2004). There are also changes in motor output maps evoked at rest. In humans, not only is there reduced excitability of the output from the motor cortex of the stroke hemisphere, but there is also a shift in location of the main output zone, often in an anterior direction (Traversa et al. 1997). The changes are even clearer in primate experiments, in which detailed mapping of motor cortex can be performed before and after circumscribed lesions (Nudo 1997).

An important question is whether these naturally occurring changes are modified (improved) by rehabilitation therapy. That is, does therapy change the excitability or pattern of connections over and above what is seen in the absence of therapy? The answer seems to be a clear 'yes' in experiments in rodents, but things are less clear in primates. For example, in rat stroke models, task training seems to increase dendritic length and complexity in the motor cortex of the undamaged hemisphere (Biernaskie et al. 2004). In primates, work by Nudo's group has shown that training prevents the usual shrinkage in the area of the hand representation following focal damage to a small part of that area (sometimes termed 'use it or lose it'). However, the same group also writes 'It should be noted that in the primate studies to date, it has not been possible to demonstrate differences in motor abilities as a function of post-infarct experience'. Thus, the connection between map size in M1 and recovery of function may be less direct than might have been anticipated (Nudo et al. 2001). In humans, long-term training of arm use with constraint induced therapy in chronic stroke patients is accompanied with an increase in excitability of the corticospinal output (Liepert et al. 2000) although there is no direct evidence that this involves changes at the level of synaptic connections in cortex. It could equally well involve strengthening of connections within the spinal cord.

Nevertheless, given this background it seems a reasonable assumption that rehabilitation training may interact with the normal processes of 'rewiring' the motor system after stroke. Indeed, it is a common assumption that therapy can speed or optimise this 'rewiring' process. If this is correct it seems likely that rTMS and TDCS, both of which can transiently influence the strength of synaptic connections between neurones in cerebral cortex, would be able to influence the effectiveness of therapy.

## Does Non-invasive Brain Stimulation Enhance Motor Recovery from Stroke?

There have been a large number of relatively small scale trials of the effect of brain stimulation on motor recovery after stroke (Lefaucheur et al. 2014). The majority have studied effects in chronic stroke patients >6 months post stroke; only a few have examined effects in acute and sub-acute stages. Some trials have involved only a single session of stimulation and examined the immediate after-effects whereas others have employed multiple treatment sessions, usually once per day for 10 consecutive weekdays.

The general design has been similar to studies of brain stimulation on healthy learning: stimulation has targeted motor cortex with the aim of improving the response to standard physical therapy. rTMS is given immediately prior to therapy whereas TDCS is given either before or during therapy. Finally, stimulation is either excitatory to the stroke hemisphere, with the intention of improving its response to therapy; or inhibitory to the non-stroke hemisphere on the assumption that this may reduce competitive inhibition from the non-stroke to the stroke hemisphere (Lefaucheur et al. 2014).

There have been many recent summaries and meta-analyses of these trials, and in general most papers report a small positive effect; a few papers report a null effect; and no papers report negative effects. The conclusions are limited by the heterogeneity of the patients studied as well as the differences in stimulation protocols that have been used. However, the conclusion is that there may be a possible benefit but more work needs to be done to confirm this and to estimate whether it is clinically meaningful.

As an example of the variability of approaches and final outcomes, consider two very recently reported trials from the UK from Triccas et al. (TT trial) and Altman et al. (AL trial) (Allman et al. 2016; Triccas et al. 2015), both of which examine the effect of TDCS on the response to physical therapy for arm movement after stroke. The headline result emphasised in the two papers is that TDCS had no additional effect in a robotic therapy (TT trial) whereas TDCS significantly enhanced the response to a form of graded manual therapy (AL trial). The impression one gets is that one trial has a null result whereas the other is strikingly positive. However, a comparison of both trials shows that the results are in fact quite similar. It also highlights the difficulties in attempting to compare such trials, even when the TDCS protocol was the same in each trial.

The trials share some important similarities (see Table 1) such as the parameters of TDCS and the duration of each training session. They are also similar in terms of the baseline status of the patients, although the TT study contains lower functioning patients than the AL study. However, they differ in the type of therapy and (robotic versus graded manual therapy) and the total number of therapy sessions.

The TT was designed with a single outcome measure, which was the change in Fugl-Meyer (FM) score 3 months after treatment, whereas the AL trial had three different outcome measures, FM, ARAT (Action Research Arm Test) and WMFT (Wolf Motor Function Test). The FM scores were common to both trials and therefore can be directly compared. At baseline the scores are similar and the overall change at 3 months is around 7–8 points, which is above the usual clinically relevant threshold of 5 points. Thus, both trials were successful from the patient viewpoint.

In terms of the real versus sham comparisons, there was no difference between groups in the TT trial, whereas there was a (non-significant) tendency for the real group to benefit more in the AL trial (approx. 2.5 points more, see Table 1). This (FM) result was not the one highlighted in the paper; the emphasis was placed instead on the scores of the ARAT and WMFT which were significantly greater in the real TDCS group than sham, and it is on this basis that the AL trial authors conclude that TDCS improves the response to therapy after stroke.

**Table 1** Comparison of the variables and results from two trials

| | Triccas et al. (2015) | Allman et al. (2016) |
|---|---|---|
| Number of patients (real, sham) | 22 (11, 11) | 24 (12, 12) |
| Time post stroke (mean) | >2 months (19.6) | >6 months (33) |
| FMA baseline (range) | 32 (4–61) | 39 (Not available) |
| ARAT baseline | 20 (approximately calculation from data presented) | 23 (approximately calculation from data presented) |
| Therapy type | Robotic assisted arm reaching | Graded Repetitive Arm Supplementary Program (GRASP) |
| Number of sessions | 18 (2–3 per week) | 9 consecutive working days |
| Change in FM at 3 months | 7.2 (real); 7.1 (sham) | 9.3 (real); 6.7 (sham) |
| Change in ARAT | Mean of real and sham = 9 | 10 (real); 6 (sham) |

Thus a detailed comparison of the two studies shows that in fact they have very similar results, with the effect of therapy clearly dominating any additional effect of TDCS.

## Why Are the Results of Therapeutic Trials of Non-invasive Brain Stimulation Unclear?

The main reason for this is that, as in the examples above, the additional effects of stimulation, if they exist, are small. This does not mean that they may not be relevant, but that they are difficult to observe within the large variability in patients' individual responses to treatment. The result of this is that small trials may by chance detect a clear effect, and these trials are the ones most likely to reach publication.

One common way around this problem is to perform a meta-analysis of many small studies in order to obtain some estimate of the true effect size. However, as can be seen in the examples above, trials can differ in very many ways that could influence the results, and that makes it very difficult indeed to perform meaningful meta-analyses of published material. At present it seems as if the only solution is to perform a large scale and well-controlled trial.

In the meantime is there any possibility that brain stimulation methods can be made more effective so that therapeutic effects are improved? At the present time many groups are attempting to devise new protocols, for example, quadruple rTMS or multi-electrode focal TDCS to improve their effectiveness (Hamada and Ugawa 2010; Kuo et al. 2013). However, the protocols are usually evaluated in terms of their effect on excitability of motor cortex; experiments on behavioural effectiveness have yet to be performed. Other groups have pointed out that a problem with all non-invasive methods is that they cannot discriminate very well between the types of neurone that are activated: some may be inhibitory, some excitatory, some are interneurons, and some projection neurones. If all are activated simultaneously (and the evidence suggests that they are) then they may have very different overall effects. These groups are using modified types of stimulation in an attempt to target more clearly subsets of neurones at the site of stimulation (D'Ostilio et al. 2016). Finally, there is increasing interest in the prospect of "brain state controlled" stimulation. Data show that the effect of non-invasive methods depends on the state of the brain at the time the stimulation is applied (Gharabaghi et al. 2014). The new approach is to use EEG recording to identify brain states so that stimulation can be applied at specific times when it might prove more effective.

## Take Home Message

The responses to all present forms of non-invasive brain stimulation are highly variable between individuals as well as day-to-day within an individual. This is a stumbling block to using the methods as effective therapeutic treatments. Two new approaches are presently being trialled to achieve better targeted stimulation. One employs adjustable pulse parameters for stimulation while the other uses EEG monitoring to produce stimulation in optimal brain states. Both seem to provide a greater reliability than conventional methods, but further testing is required.

## References


Allman C, Amadi U, Winkler AM, Wilkins L, Filippini N et al (2016) Ipsilesional anodal tDCS enhances the functional benefits of rehabilitation in patients after stroke. Sci Transl Med 8: 330re1

Barker AT, Freeston IL, Jalinous R, Jarratt JA (1987) Magnetic stimulation of the human brain and peripheral nervous system: an introduction and the results of an initial clinical evaluation. Neurosurgery 20:100–109

Biernaskie J, Chernenko G, Corbett D (2004) Efficacy of rehabilitative experience declines with time after focal ischemic brain injury. J Neurosci 24:1245–1254

Bindman LJ, Lippold OC, Redfearn JW (1962) Long-lasting changes in the level of the electrical activity of the cerebral cortex produced bypolarizing currents. Nature 196:584–585

Bindman LJ, Lippold OC, Redfearn JW (1964) The action of brief polarizing currents on the cerebral cortex of the rat (1) during current flow and (2) in the production of long-lasting after-effects. J Physiol 172:369–382

Brashers-Krug T, Shadmehr R, Bizzi E (1996) Consolidation in human motor memory. Nature 382:252–255

D'Ostilio K, Goetz SM, Hannah R, Ciocca M, Chieffo R et al (2016) Effect of coil orientation on strength-duration time constant and I-wave activation with controllable pulse parameter transcranial magnetic stimulation. Clin Neurophysiol 127:675–683

Dan Y, Poo MM (2006) Spike timing-dependent plasticity: from synapse to perception. Physiol Rev 86:1033–1048

Gharabaghi A, Kraus D, Leao MT, Spuler M, Walter A et al (2014) Coupling brain-machine interfaces with cortical stimulation for brain-state dependent stimulation: enhancing motor cortex excitability for neurorehabilitation. Front Hum Neurosci 8:122

Hamada M, Ugawa Y (2010) Quadripulse stimulation–a new patterned rTMS. Restor Neurol Neurosci 28:419–424

Huang YZ, Edwards MJ, Rounis E, Bhatia KP, Rothwell JC (2005) Theta burst stimulation of the human motor cortex. Neuron 45:1–6

Kim YH, Park JW, Ko MH, Jang SH, Lee PK (2004) Facilitative effect of high frequency subthreshold repetitive transcranial magnetic stimulation on complex sequential motor learning in humans. Neurosci Lett 367:181–185

Krakauer JW, Mazzoni P (2011) Human sensorimotor learning: adaptation, skill, and beyond. Curr Opin Neurobiol 21:636–644

Kuo HI, Bikson M, Datta A, Minhas P, Paulus W et al (2013) Comparing cortical plasticity induced by conventional and high-definition 4 × 1 ring tDCS: a neurophysiological study. Brain Stimul 6:644–648

Lefaucheur JP, Andre-Obadia N, Antal A, Ayache SS, Baeken C et al (2014) Evidence-based guidelines on the therapeutic use of repetitive transcranial magnetic stimulation (rTMS). Clin Neurophysiol 125:2150–2206

Liebetanz D, Nitsche MA, Tergau F, Paulus W (2002) Pharmacological approach to the mechanisms of transcranial DC-stimulation-induced after-effects of human motor cortex excitability. Brain 125:2238–2247

Liepert J, Bauder H, Wolfgang HR, Miltner WH, Taub E, Weiller C (2000) Treatment-induced cortical reorganization after stroke in humans. Stroke 31:1210–1216

Malenka RC, Bear MF (2004) LTP and LTD: an embarrassment of riches. Neuron 44:5–21

Muellbacher W, Ziemann U, Boroojerdi B, Cohen L, Hallett M (2001) Role of the human motor cortex in rapid motor learning. Exp Brain Res 136:431–438

Muellbacher W, Ziemann U, Wissel J, Dang N, Kofler M et al (2002) Early consolidation in human primary motor cortex. Nature 415:640–644

Muller-Dahlhaus F, Ziemann U, Classen J (2010) Plasticity resembling spike-timing dependent synaptic plasticity: the evidence in human cortex. Front Synaptic Neurosci 2:34

Nader K, Hardt O (2009) A single standard for memory: the case for reconsolidation. Nat Rev Neurosci 10:224–234

Nitsche MA, Paulus W (2000) Excitability changes induced in the human motor cortex by weak transcranial direct current stimulation. J Physiol 527(Pt 3):633–639

Nudo RJ (1997) Remodeling of cortical motor representations after stroke: implications for recovery from brain damage. Mol Psychiatry 2:188–191

Nudo RJ (2006) Mechanisms for recovery of motor function following cortical damage. Curr Opin Neurobiol 16:638–644

Nudo RJ, Milliken GW, Jenkins WM, Merzenich MM (1996) Use-dependent alterations of movement representations in primary motor cortex of adult squirrel monkeys. J Neurosci 16:785–807

Nudo RJ, Plautz EJ, Frost SB (2001) Role of adaptive plasticity in recovery of function after damage to motor cortex. Muscle Nerve 24:1000–1019

Reis J, Schambra HM, Cohen LG, Buch ER, Fritsch B et al (2009) Noninvasive cortical stimulation enhances motor skill acquisition over multiple days through an effect on consolidation. Proc Natl Acad Sci USA 106:1590–1595

Rothwell JC, Thompson PD, Day BL, Boyd S, Marsden CD (1991) Stimulation of the human motor cortex through the scalp, pp 159–200

Stefan K, Kunesch E, Cohen LG, Benecke R, Classen J (2000) Induction of plasticity in the human motor cortex by paired associative stimulation. Brain 123(Pt 3):572–584

Teo JT, Swayne OB, Cheeran B, Greenwood RJ, Rothwell JC (2011) Human theta burst stimulation enhances subsequent motor learning and increases performance variability. Cereb Cortex 21:1627–1638

Traversa R, Cicinelli P, Bassi A, Rossini PM, Bernardi G (1997) Mapping of motor cortical reorganization after stroke—a brain stimulation study with focal magnetic pulses. Stroke 28:110–117

Triccas LT, Burridge JH, Hughes A, Verheyden G, Desikan M, Rothwell J (2015) A double-blinded randomised controlled trial exploring the effect of anodal transcranial direct current stimulation and uni-lateral robot therapy for the impaired upper limb in sub-acute and chronic stroke. NeuroRehabilitation 37:181–191

Walker MP, Brakefield T, Morgan A, Hobson JA, Stickgold R (2002) Practice with sleep makes perfect: sleep-dependent motor skill learning. Neuron 35:205–211

Ward NS, Brown MM, Thompson AJ, Frackowiak RS (2004) The influence of time after stroke on brain activations during a motor task. Ann Neurol 55:829–834

Werk CM, Chapman CA (2003) Long-term potentiation of polysynaptic responses in layer V of the sensorimotor cortex induced by theta-patterned tetanization in the awake rat. Cereb Cortex 13:500–507

Werk CM, Klein HS, Nesbitt CE, Chapman CA (2006) Long-term depression in the sensorimotor cortex induced by repeated delivery of 10 Hz trains in vivo. Neuroscience 140:13–20
Woods AJ, Antal A, Bikson M, Boggio PS, Brunoni AR et al (2016) A technical guide to tDCS, and related non-invasive brain stimulation tools. Clin Neurophysiol 127:1031–1048
Zaaimi B, Edgley SA, Soteropoulos DS, Baker SN (2012) Changes in descending motor pathway connectivity after corticospinal tract lesion in macaque monkey. Brain 135:2277–2289
Ziemann U, Paulus W, Nitsche M, Pascual-Leone A, Byblow WD et al (2008) Consensus: motor cortex plasticity protocols. Brain Stimulat 164–82

# Part VI
# Human-Machine Interface

# Organizing and Reorganizing Coordination Patterns

**Rajiv Ranganathan and Robert A. Scheidt**

**Abstract** Understanding how the nervous system learns to coordinate the large number of degrees of freedom in the body to produce goal-directed movement is not only one of the central questions in theoretical movement neuroscience, but also has direct relevance for movement rehabilitation. In spite of the centrality of this issue, the literature on how a new coordination pattern is acquired and refined when first learning a novel task remains surprisingly small relative to studies that focus on modifications of already well-learned coordination patterns. In this chapter, we outline some of the reasons behind why the study of coordination continues to pose a serious challenge for movement neuroscience, particularly when it comes to systematically studying and testing hypotheses on how new coordination patterns are organized and reorganized with practice. We then describe a novel experimental paradigm—the body–machine interface (BoMI)—that has been developed and used over the last decade to examine this issue. The paradigm combines the control of a large number of degrees of freedom along with a linear mapping, which makes it appealing to examine how coordination of these high degrees of freedom is organized in a systematic fashion. Finally, we outline some of the new insights that this paradigm has provided into classic issues of motor learning such as the learning of high-dimensional spaces, generalization, and transfer.

**Keywords** Body–machine interface (BoMI) · Motor learning · Principal component (PC) · Task space · Null space · Glove-cursor task · Redundant movements · Muscle synergies

R. Ranganathan
Department of Kinesiology, Michigan State, 203 IM Circle, East Lansing, MI, 48824, USA

R.A. Scheidt (✉)
Department of Biomedical Engineering, Marquette University, Olin Engineering Center, Room 206, P.O. Box 1881, Milwaukee, WI 53201, USA
e-mail: scheidt@ieee.org



J. Laczko and M.L. Latash (eds.), *Progress in Motor Control*,
Advances in Experimental Medicine and Biology 957,
DOI 10.1007/978-3-319-47313-0_18

## The Problem of Movement Coordination

Whether it be the performance of an Olympic gymnast or a violin virtuoso, the ability of the nervous system to precisely coordinate, both spatially and temporally, the movement of numerous degrees of freedom (muscles and joints) is simply staggering. It is fitting then that Bernstein (Bernstein 1967) posed the problem of movement coordination as the problem of how the nervous system masters these "redundant degrees of freedom." Understanding how the nervous system organizes patterns of coordination through experience is not only critical for a theoretical understanding of movement control, but also is important for sensorimotor rehabilitation, which seeks to encourage the nervous system to form and/or reorganize functionally useful patterns of coordination after neurological injury, such as stroke.

In spite of the central role of movement coordination in motor learning and rehabilitation, a majority of research in motor learning has largely avoided the problem of how patterns of coordination are learned and organized, with the notable exception of research in motor development (e.g., Gessell 1946; Thelen et al. 1996; Adolph et al. 2012). In an early review of motor learning research in the 1970s and 1980s, Newell (1991) commented that "the laws of learning… have been tested on tasks that require subjects only to learn the scaling of movement amplitude, movement time or force output in an already established coordination mode" (p. 218). Despite tremendous technological advances in the way we can measure and analyze multi-joint movement in the 25 years since that review, current theories on motor learning are still largely driven by experimental paradigms (e.g., reaching, locomotion, key pressing, and isometric force production) in which the underlying coordination pattern is already well acquired, and where there is no or limited redundancy. In fact, there has been a resurgence of the same sentiment from several investigators pointing out that these paradigms may not fully capture the entire spectrum of motor learning—c.f. "learning versus adaptation" (Bastian 2008) and "adaptation versus skill learning" (Krakauer and Mazzoni 2011).

In this chapter, we outline two main challenges in studying the learning (and relearning) of movement coordination. We describe our experimental approach using the body–machine interface (BoMI) paradigm and describe how it allows us to address these challenges. We then review how the BoMI has provided new insights into key issues in motor learning and conclude with future directions on how these insights may be used to address challenges in learning and movement rehabilitation.

### *Challenge #1: Coordination Is Messy*

The acquisition of coordination patterns occurs in an early stage of motor learning, meaning that the study of coordination is an experimentalists' nightmare. Performance is characterized by large within- and between-subject variability and discrete transitions in the qualitative nature of coordination occur commonly during learning. Consider for example the development of overarm throwing: changes with practice are not only characterized by improvement in throwing velocity, but performance moves through several qualitative stages involving changes in coordination between the legs, the trunk, and the arm. In addition, progression through these stages is discontinuous (nonlinear) and shows significant intersubject variability (Roberton et al. 1979; Langendorfer and Roberton 2002). Even when coordination tasks are simplified in the laboratory—for example, when learning to move a rollerball (Liu et al. 2006)—learning curves are extremely variable and do not obey the typical power law (Newell and Rosenbloom 1981) or exponential functions (Heathcote et al. 2000; Smith et al. 2006) that are found in tasks where the coordination pattern had already been acquired.

In addition to high variability in the data, another experimental challenge is that many novel tasks require a relatively long learning time (days or weeks, depending on the task to be learned) before participants achieve relatively stable performance. This makes it difficult to conduct experiments to observe the effect of an independent variable. Thus, in spite of the rich descriptive accounts of how coordination patterns are acquired through learning and development in several activities (such as reaching, locomotion and throwing), it is difficult to directly test specific hypotheses on the how the nervous system learns novel coordination patterns, and identify factors that govern this stage of learning.

## *Challenge #2: Coordination Is Hard to Quantify and Harder to Explain*

The second major challenge in the study of movement coordination is the nature of the experimental data that is typically recorded. Motor learning studies, borrowing heavily from their roots in experimental psychology (for a review, see Adams 1987), have traditionally focused on the analysis of "responses" or movement outcomes (e.g., movement time, performance errors, etc.) with very little emphasis on the how the movement itself unfolds in time (i.e., in terms of movement kinematics, forces, or muscle activities). Given that movement data can be very high dimensional (depending on the number of degrees of freedom involved), how do we both measure and compress massive amounts of data to test specific hypotheses? What tools are available to measure the acquisition of a "coordination pattern," and whether coordination has changed dramatically (i.e., qualitatively) with learning? Here we argue that there are two related issues with this challenge—the first issue deals with describing and quantifying coordination, and a second issue is to explain the extent to which coordination depends on specific features of the motor task such as the spatial geometry of movements and forces to be produced.

To address the issue of quantifying coordination patterns, early work on coordination (specifically locomotion) focused on using angle–angle plots (e.g., for a review see Wheat and Glazier 2006). A separate, large body of literature offered a "dynamical systems" perspective of motor coordination, based on the idea that even though coordination itself is high dimensional, it can be captured in terms of low-dimensional collective variables (Kelso 1995). Perhaps the most well known of these collective variables was the use of relative phase between two effectors in rhythmic movements to capture transitions in bimanual coordination (Kelso 1984). However, in spite of its immense impact, the experimental strategy of trying to identify collective variables has been much harder to generalize to complex multi-joint movement, especially in nonrhythmic activities.

More recently, there have been advances in using dimensionality reduction methods to understanding high-dimensional data obtained during experimental study of motor coordination. One hypothesis that has received considerable attention is the idea that the nervous system coordinates complex, multidegree of freedom movement by combining smaller, temporally stable, building blocks or modules (sometimes referred to as "motor primitives" or "synergies"). Principal component analysis (PCA) has been extensively used in the identification of kinematic synergies (Santello et al. 1998), whereas nonnegative matrix factorization (NNMF) and factor analysis have been preferred for understanding muscle synergies in activities like posture (Ting and McKay 2007), reaching (Tresch et al. 1999), and locomotion (Ivanenko et al. 2004). Although there persists vigorous debate as to what gives rise to these synergies (Kutch and Valero-Cuevas 2012; de Rugy et al. 2013), these studies demonstrate that regularities in the performance of goal-directed behaviors can be captured by describing coordination as combinations of a small number of kinematic, kinetic, and/or electromyographic patterns of behavior.

The second major challenge in the study of movement coordination is that even if coordination patterns can be fully described and quantified by such motor primitives, it is critical to have a framework in which to explain how the experimentally identified motor primitives relate to—and depend on—specific features of the motor task (e.g., the spatial geometry of movements and forces). In other words, how does one determine whether observed changes in movement coordination are potentially relevant or irrelevant to task performance? Such definitions are usually difficult to define in real-world tasks like throwing a baseball or playing the violin due to a lack of sufficient understanding of the coordination patterns that differentiate skilled and unskilled performance. To address this challenge, several related approaches (Scholz and Schoner 1999; Müller and Sternad 2004; Cusumano and Cesari 2006; Torres et al. 2011) have provided a way to unambiguously define a priori "task-space" and "null-space" performances, that is, patterns of coordination that do versus do not contribute to task performance. We advocate for the idea that advancing understanding of how task- and null-space performances change with learning will provide new insights into in how patterns of multi-joint coordination evolve during learning and adaptation. More specifically, using this framework will help move the study of coordination from an account that simply describes changes in coordination into an account of being to explain why these coordination changes occur.

In summary, considering both the experimental and methodological challenges, there remains a significant knowledge gap in understanding of how coordination patterns are learned during practice. In particular, there is a critical need for an experimental paradigm where (i) the coordination pattern is novel but can still be learned reasonably quickly, and (ii) there is an a priori way to understand how coordination patterns learned during practice actually relate to specific features of the task.

## The Body–Machine Interface Paradigm

The experimental paradigm we currently use to address this critical need is called a body–machine interface. Similar to its more well-known counterpart, the brain–machine interface, a body–machine interface takes a large number of input signals (in this case, motions of various body segments) and maps them onto a task space (cf. Mosier et al. 2005; Liu et al. 2011; Casadio et al. 2012). Take for example the task of moving a mouse to control the motion of a cursor on a computer display; people readily and immediately respond to instructions such as "move the cursor rapidly 5 cm to the left," despite the fact that this action requires very different patterns of muscle activations and interjoint coordination depending on the hand's initial position relative to the torso. This simple ability demonstrates that the central nervous system is competent to capture geometric properties of the space in which actions take place (e.g., measures of "distance" and "direction") and to use that information to plan and execute arbitrary movements. How then, does the spatial geometry of a task influence the patterns of multi-joint coordination used to learn and adapt that task? We describe below one form of a BoMI that is particularly well suited to address this question: an instrumented data glove that maps finger motions into cursor motions on a computer display.

We asked participants to don an instrumented glove (CyberGlove, Immersion Technologies, Inc.) and to then practice using finger motions to move a computer cursor between several different target locations on a computer screen (Mosier et al. 2005; Liu and Scheidt 2008; Liu et al. 2011; Ranganathan et al. 2013) (cf., Fig. 1a). This is a novel and challenging task that most people can learn to perform with some proficiency within the span of 1 to 2 hours. Glove signals, $\boldsymbol{H}$, encoding finger joint angles are mapped onto two-dimensional cursor motions, $\boldsymbol{P}$, using an affine transformation matrix $\boldsymbol{A}$:

$$\boldsymbol{P} = \begin{bmatrix} x \\ y \end{bmatrix} = \boldsymbol{A} \cdot \boldsymbol{H}^{\mathrm{T}} = \begin{bmatrix} a_{1,1} & a_{1,2} & \cdots & a_{1,19} & a_{1,20} \\ a_{1,1} & a_{2,2} & \cdots & a_{2,19} & a_{2,20} \end{bmatrix} \cdot \begin{bmatrix} h_1 & h_2 & \ldots & h_{19} & 1 \end{bmatrix}^{\mathrm{T}},$$

where $a_{1,20}$ and $a_{2,20}$ are translation parameters used to center the projection of the hand's resting posture onto the center of the display screen. The remaining $a_{i,j}$ coefficients determine how changes in each finger joint sensor value contributes to

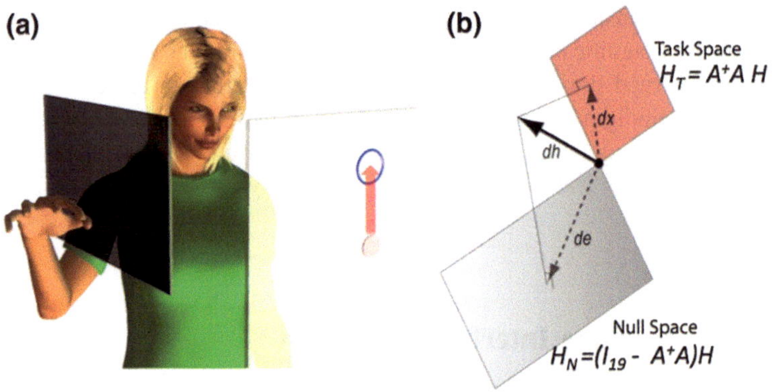


**Fig. 1** **a** Schematic of the glove-cursor interface. Hand postures are measured and mapped to position of a cursor on the screen. **b** Representation of task space and null space in the glove-cursor task—the hand posture $H$ can be separated into a task-space component $H_T$ that influences the position of the cursor, and a null-space component $H_N$ which has no influence on the cursor position

motion along the $x$ and $y$ axes of the screen. It is important to note that this task is redundant in the sense that capturing a single target location on the screen can be achieved with multiple finger joint configurations. So whereas the screen has a well-defined Euclidean metric (i.e., the distance between any two points $\boldsymbol{P_1}$ and $\boldsymbol{P_2}$ is the length of the straight segment that joins them), there is no obvious or natural definition of distance within the "signal space" of finger articulations because $\boldsymbol{A}$ partitions articulation space (the 19-dimensional vector of glove signals) into a task space within which the cursor moves, and a complementary null space that includes all combinations of finger motions that do not contribute to cursor motion (Fig. 1b).

## *Defining the Hand-to-Screen Mapping*

The $\boldsymbol{A}$ matrix provides a many-to-one mapping of hand postures onto cursor positions on the screen, thereby defining the spatial geometry of the redundant glove-cursor task. Of the many possible ways to define the $\boldsymbol{A}$ matrix, one particular approach facilitates the interpretation of results in glove-cursor studies of learning, adaptation, facilitation, and interference. In this approach, participants are coached into forming hand gestures corresponding to the 24 static fingerspelling characters of the American Manual Alphabet (AMA), which span a very large range of possible hand configurations (cf., Fig. 2a). The data are decomposed using Principal Component Analysis (PCA) to identify those combinations of joint motions that capture the greatest amount of variability within the data set. As shown in Fig. 2b, the vast majority of finger joint variability within the static characters of

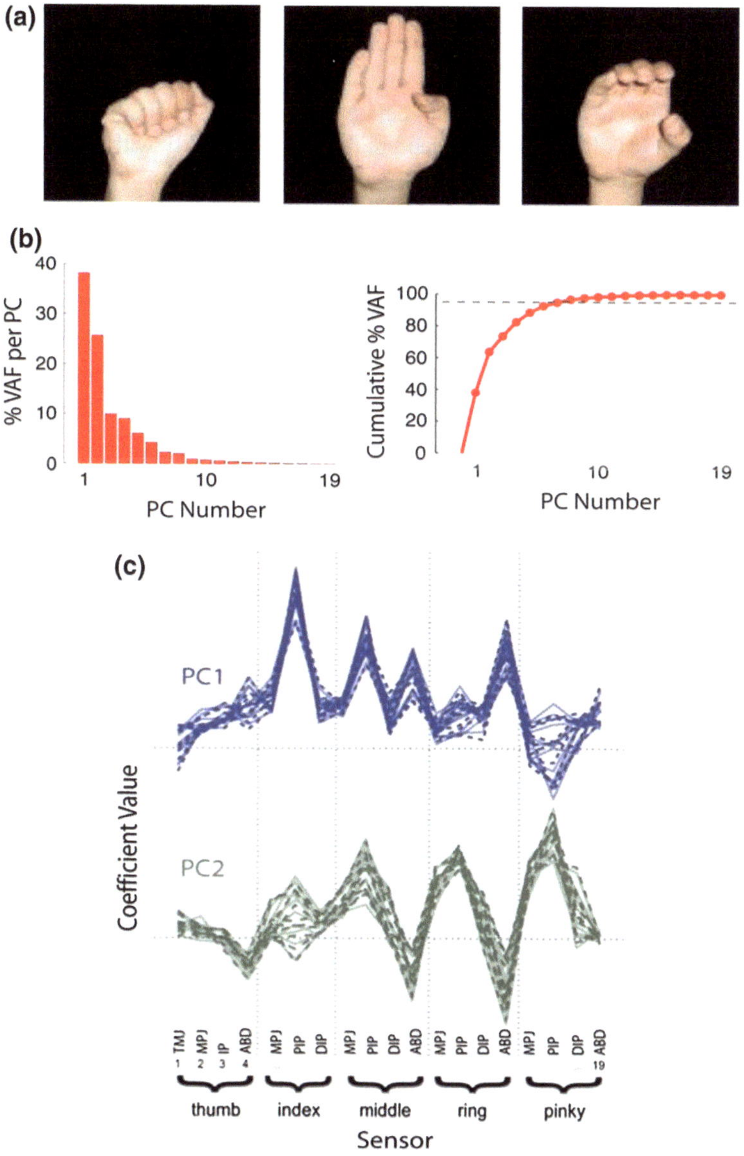


**Fig. 2** Construction of the map between the hand posture and the cursor. **a** Participants use letters from the American Manual Alphabet (AMA) when wearing the cyberglove. **b** PCA reveals that a majority of the variance can be explained by a small number of (4–7) eigenvectors. **c** Repeatability of the coefficients of PC1 and PC2 when the glove is taken off and on multiple times

the AMA is accounted for by the first 8–10 Principal Components (PCs). This suggests that forming hand gestures as in fingerspelling requires at minimum 8–10 control dimensions, depending on one's choice of threshold differentiating signal from noise. The ***A*** matrix is then composed by stacking one or more of the first several PCs on top of each other as row vectors. Using just one PC yields a one-dimensional task, using two PCs yields a two-dimensional task, and so on. The ***A*** matrix we have used most often to define a 2D task is the one that uses the first 2 fingerspelling PCs (i.e., those that capture the most variability within the AMA data set).

Figure 2c plots the PC1-PC2 ***A*** matrix coefficients corresponding to each sensor for 10 participants. Solid lines correspond to the average values for each subject; dashed lines show the values computed for a single subject in each of 10 don/doff cycles. The ***A*** matrix coefficients were similar across participants and across donning and doffing cycles. Consequently, all participants using the PC1-PC2 ***A*** matrix perform virtually identical tasks, even on different days.

## Learning in High-Dimensional Spaces: Performance Changes in "Task" and "Null" Spaces

Mosier and colleagues (Mosier et al. 2005) developed the glove-cursor BoMI to determine specifically the extent to which the central nervous system acquires patterns of coordination that capture the Euclidean properties of task space (the space in which actions take place), and to explore how that representation constrains subsequent planning and execution of movements in high-dimensional spaces.

In a first set of experiments, two groups of participants trained for 4 days to move the cursor between pairs of screen targets on a 2 × 3 grid (cf., Fig. 3a). Here, the ***A*** matrix was composed first by associating arbitrary hand postures with the four corners of the display screen and then using linear least squares regression to fit the matrix coefficients to glove signals recorded with the hand in those postures. Participants could practice the task under two different feedback conditions. In No Vision trials (NV), participants received no visual feedback of cursor motion during movement. After each target capture attempt however, the cursor reappeared and participants were required to bring the cursor to the desired target before the next movement could start. In Vision trials (V), participants did receive visual feedback of cursor motion during movement on training trials 2, 3, 5, 6, 8, and 9. Testing trials 1, 4, 7, and 10 were identical to NV trials, and were used to provide a fair comparison of performance under the two different training conditions.

At the beginning of practice on Day 1, all participants exhibited tremendous performance variability. Subjects were slow to acquire the targets because cursor paths were long, convoluted, and as likely to be directed orthogonal to the ideal, straight line path to the target as they were to be directed along that path (Fig. 3a). After about an hour's practice in both subject groups, the time between target presentation and target capture decreased, final position errors decreased, and cursor paths straightened (i.e., they were more likely to be directed straight to the target

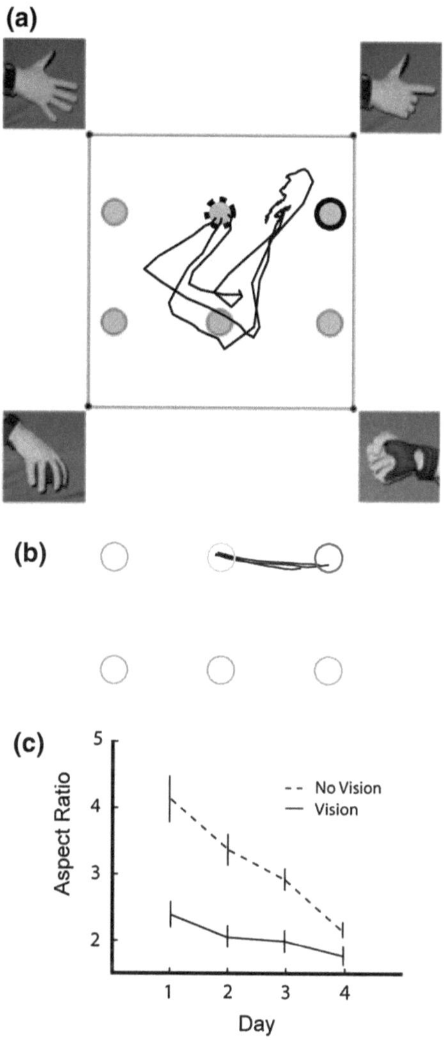


**Fig. 3** Improvements in cursor control with practice. **a** Cursor paths initially in learning (the *corner* pictures indicate calibration postures for the corners of the workspace). **b** Cursor paths after learning. **c** Aspect ratio (a measure of linearity of cursor paths) as a function of practice and visual feedback

than not). Further straightening of cursor paths and improvements in target capture accuracy were seen on subsequent days of training (cf., Fig. 3b), suggesting that consolidation of motor memories enhances these aspects of performance (Fig. 3c). Observation of practice-related performance changes—decreased movement times, increased target capture accuracy, increased cursor path rectilinearity—was confirmed in subsequent glove-cursor studies performed with distinct, independently sampled participant groups and with $\boldsymbol{A}$ matrices defined as described in Fig. 1 (Liu and Scheidt 2008; Liu et al. 2011; Ranganathan et al. 2013; Farshchiansadegh et al. 2015).

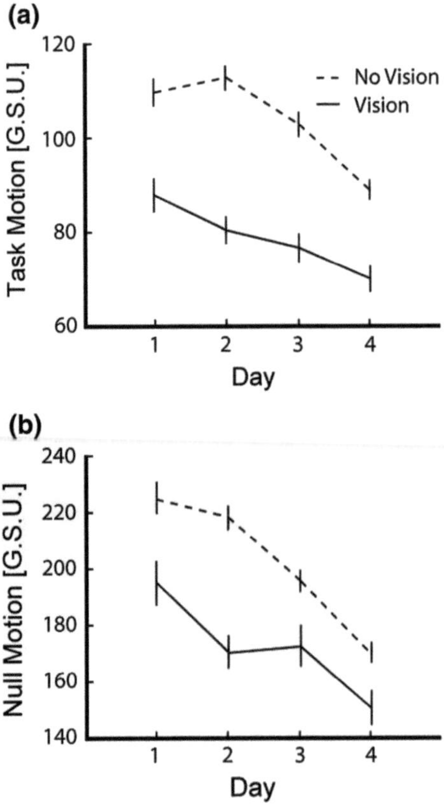


**Fig. 4** Change in the **a** task-space motion and **b** null-space motion as a function of practice and visual feedback. Note how both decrease with practice even though decreasing null-space motion has no effect on task performance

Mosier and colleagues analyzed how behavior evolved within the task and null spaces by first projecting changes in finger posture from the start to the end of each movement into components residing in the task and null subspaces (Mosier et al. 2005), and then comparing the amount of motion (i.e., path length) within each subspace as practice progressed. Consistent with an increase in cursor path straightness, Mosier observed that task-space motion decreased significantly across days, with the presence of visual feedback leading to a smaller amount of task-space motion overall (Fig. 4a). Mosier also observed a similar reduction in null-space motion across days in both feedback conditions (Fig. 4b). This observation is consistent with the idea that when confronted with a sensorimotor task with redundant degrees of freedom, the nervous system learns the geometric structure of the task (i.e., a representation of the Euclidean space on which hand movements were remapped) and improves performance by selectively operating within that space. Counter to the predictions of optimal feedback control and the "minimum intervention principle" (Todorov and Jordan 2002), which proposes that deviations from the average trajectory are only corrected when they interfere with task performance, the results of Mosier and colleagues show that when learning a novel coordination task, null-space motions also decrease with practice, even though they are irrelevant to task performance. This suggests the formation of an inverse map (i.e., one that maps desired cursor position onto appropriate hand posture), a construct that is discussed in greater detail below in the section on coordination in articulation space.

### *Generalization of Newly Acquired Patterns of Coordination*

In a second set of experiments, Mosier and colleagues required participants to train over a restricted region of the task space and tested target capture performance on two unpracticed sets of targets: Three of the targets, the interpolation set, were internal to the trained region of task space, whereas the other three targets, the extrapolation set, were outside the trained region (Mosier et al. 2005) (Fig. 5a). Participants performed 60 interpolation and extrapolation movements before (baseline) and after (test) 500 movements between the training targets. Participants also performed an additional 60 interpolation and extrapolation movements after a 6-h pause (retention).

Immediately after training, target capture errors were significantly reduced relative to baseline performance for both the interpolation and extrapolation target sets. The performance benefits of this learning persisted across a 6-h pause (Fig. 5b). Virtually the same pattern of error reduction was seen for both sets of generalization targets, further supporting the hypothesis that participants learn the geometric structure of the task. These generalization results were confirmed in a subsequent study by Liu and Scheidt (2008), who controlled the quality, quantity, and timing of visual feedback of performance, thus clarifying that visual feedback of endpoint motion is indeed necessary in this task for learning a new spatial relationship between endpoint and articulation spaces.

## Coordination in Articulation Space—Learning and Adapting Task-Space Geometry

In order to capture unpracticed targets in the glove-cursor task, participants had to learn how to generate hand gestures that would project, under matrix $\boldsymbol{A}$, onto those novel targets. In order to generalize beyond the training targets, participants effectively had to learn how to invert matrix $\boldsymbol{A}$, which maps the highly redundant set of finger joint signals onto cursor motions. Because $\boldsymbol{A}$ has more columns than rows, it does not have a unique inverse. This is a direct consequence of the fact that in this redundant task, there exist an infinite number of different finger configurations that project onto any target location on the display screen. Because participants readily generalize beyond the trained target set, they undeniably solve this "ill-posed problem." While this problem has been investigated in well-learned tasks like

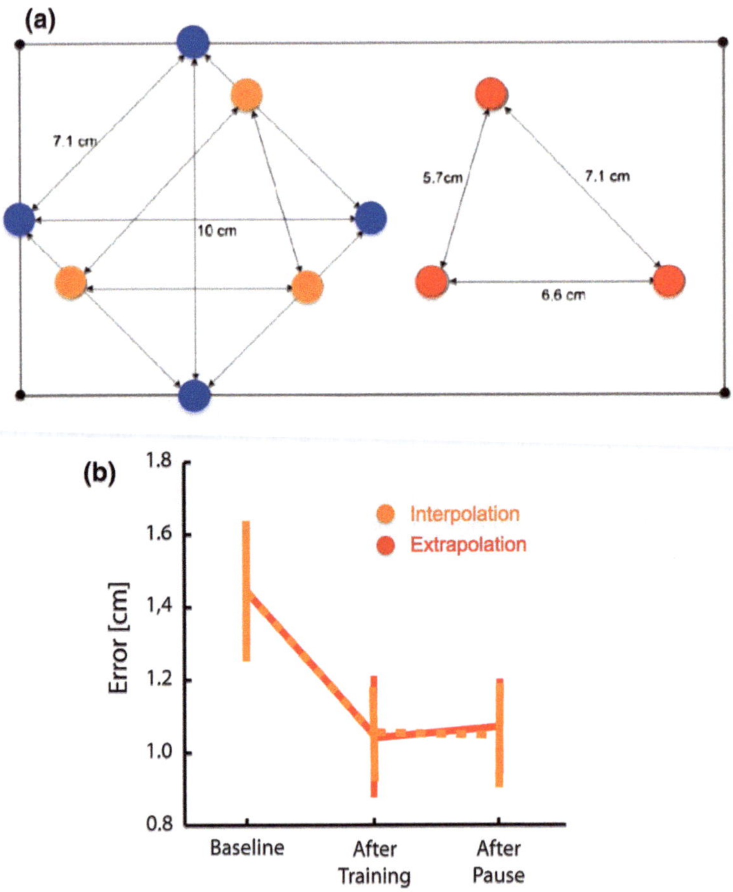


**Fig. 5** Generalization of learning **a** Participants practiced on the shaded targets (in the cardinal directions), and generalization of learning was tested using both an interpolation set and an extrapolation set. **b** Euclidean error at the beginning, end of training, and a delayed period showed that generalization occurred for both interpolation and extrapolation sets. *Blue* training target set; *Orange* interpolation set; *Red* extrapolation set

reaching (Torres and Zipser 2002; Biess et al. 2007; Zipser and Torres 2007), the structure of the glove-cursor task permits us to determine how participants do so in a novel task.

In a very real sense, each participant's solution to the problem of how to invert matrix $\boldsymbol{A}$ reflects how he or she has imported the geometry of task space (as defined by matrix $\boldsymbol{A}$) into articulation space. Essentially, we seek to estimate from the experimental data a matrix $\boldsymbol{B}$, which maps desired displacements of cursor positions onto appropriate changes in finger joint configurations:

$$\Delta H = B \cdot \Delta P$$

We can estimate the ***B*** matrix using a least squares fit to the data (Liu et al. 2011; see also Farshchiansadegh et al. 2015):

$$B_{\text{est}} = \Delta H \cdot \Delta P^{\text{T}} \cdot (\Delta P \cdot \Delta P^{\text{T}})^{-1},$$

where superscript T indicates the matrix transpose operation and superscript −1 indicates the matrix inverse operation. As participants learn to generalize beyond the trained target set, each person's ***B*** matrix comes to describe the motor representation of task-space geometry that they have acquired through practice. That is, $\boldsymbol{B}_{\textbf{est}}$ reflects how the participant adjusts each articulatory degree of freedom when challenged to produce a desired change in cursor location. Although each person is likely to form a unique solution to the redundancy problem, we have found this formulation for $\boldsymbol{B}_{\textbf{est}}$ to be useful for studying how patterns of coordination can change in response to changing task demands.

In the study of (Liu et al. 2011) for example, two groups of young adults participated in a multisession experiment where they practiced capturing screen targets under a fixed ***A*** matrix for about hour on Day 1. Both groups returned to the lab on Day 2 and after reacquainting with the baseline map, one group performed three blocks of 36 movements each wherein they practiced capturing screen targets while experiencing a stepwise rotational distortion of visual feedback:

$$\boldsymbol{P}_{\text{T}} = \begin{bmatrix} x_{\text{rotated}} \\ y_{\text{rotated}} \end{bmatrix} = \boldsymbol{T}_{\text{R}}\boldsymbol{P} = \begin{bmatrix} \cos(45°) & -\sin(45°) \\ \sin(45°) & \cos(45°) \end{bmatrix} \begin{bmatrix} x \\ y \end{bmatrix}$$

The other group performed three blocks of movements wherein they experienced a stepwise, scaling distortion of visual performance feedback:

$$\boldsymbol{P}_{\text{T}} = \begin{bmatrix} x_{\text{scaled}} \\ y_{\text{scaled}} \end{bmatrix} = \boldsymbol{T}_{\text{S}}\boldsymbol{P} = \begin{bmatrix} 1.8 & 0 \\ 0 & 1.8 \end{bmatrix} \begin{bmatrix} x \\ y \end{bmatrix}$$

In both groups, training with the distorted visual feedback progressed for about one-half hour, after which time the baseline mapping was reinstated for two more blocks of trials.

During Day 1 training, initial target capture attempts were inefficient and dominated by exploratory behavior whereas cursor trajectories late in training were generally straight and terminated close to the goal (Fig. 6a). This ability was readily recalled during baseline practice on Day 2 in both groups (Fig. 6b).

Both groups of participants exhibited an adaptive feedforward strategy compensating for the rotation or scaling distortions. For the rotation group, imposing the stepwise counterclockwise rotation caused cursor trajectories to initially deviate counterclockwise, but later curve back to the desired final position (Fig. 6b, top). With practice under the altered mapping, trajectories regained their original rectilinearity with a learning rate constant of 33 ± 5 trials. When the baseline map was

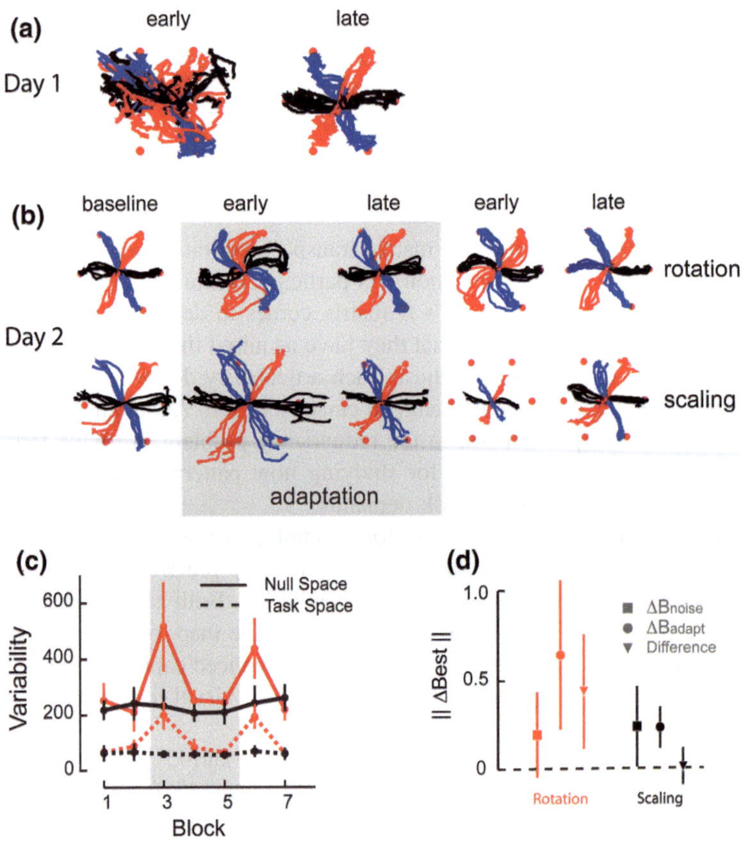


**Fig. 6** Adapting to task geometry. **a** Cursor paths at the beginning and end of Day 1 showing learning of the task. **b** Cursor paths on Day 2 during baseline, adaptation (with either rotation or scaling), and aftereffects. **c** Changes in null-space and task-space variability for both scaling (*black*) and rotation (*red*). **d** Change in the estimated inverse map showing reorganization in the rotation, but not the scaling perturbation

restored, the initial portion of the cursor trajectories deviated clockwise relative to trajectories made during baseline practice. These aftereffects of training with rotated cursor feedback were transient. For the scaling group, initial exposure to the visual distortion resulted in cursor trajectories that far overshot their goal (Fig. 6b, bottom). Further practice under the altered mapping reduced these extent errors with a learning rate time constant of 12 ± 3 trials. Restoring the baseline map resulted in initial cursor movements that undershot their goal. As for the rotation group, aftereffects of training were largely eliminated by the end of the washout period. These task-space adaptations are similar to those observed during studies of adaptation to rotations and scalings during traditional motor adaptation tasks such as reaching and pointing (cf., Krakauer et al. 2000).

Unlike reaching and pointing, however, the glove-cursor task permits analysis within the complete space of highly redundant control variables that impact task performance. This capability allowed (Liu et al. 2011) to ask how patterns of coordination within articulation space reorganize when a geometrical transformation $T_R$ or $T_S$ is imposed on the geometry of cursor motion. Do participants preserve their original (baseline) inverse map or do they interpret the transformed environment as a new task geometry to be learned? Both are viable options. In the first case, participants would maintain their baseline model of the map (i.e., matrix $\boldsymbol{B}$) and apply an inverse transformation $\mathrm{T}^{-1}$ to the targets. In the second case, participants would abandon their baseline model and create a whole new model of the map. The first option requires participants to interpret the new map as a transformation of the old one (as was actually the case). The second option requires participants to learn the new map de novo.

Liu and colleagues (Liu et al. 2011) found that performance variability in both the task and null subspaces increased transiently after imposing and removing the rotation (but not the scaling) distortion (Fig. 6c). This finding suggests that the rotation and scaling distortions induced categorically different reorganizations of coordination. In a follow-on analysis, Liu et al. (2011) examined the extent to which $\boldsymbol{B}_{\mathbf{est}}$ changed during adaptation to the rotation and scaling distortions. To do so, they computed $\boldsymbol{B}_{\mathbf{est}}$ using values of $\Delta H$ and $\Delta P$ from trials at the end of the adaptation trial block (ADAPT) as well as at the beginning (BL1) and end (BL2) of baseline trials. They evaluated how well the $\boldsymbol{B}_{\mathbf{est}}$ obtained after adaptation ($\boldsymbol{B}_{\mathbf{ADAPT}}$) was predicted by rotation ($\boldsymbol{T}_{\mathbf{R}}$) or scaling ($\boldsymbol{T}_{\mathbf{S}}$) of the $\boldsymbol{B}_{\mathbf{est}}$ obtained at the end of baseline practice ($\boldsymbol{B}_{\mathbf{BL2}}$) by computing the difference magnitude $\Delta\boldsymbol{B}_{\mathbf{ADAPT}}$:

$$\Delta\boldsymbol{B}_{\mathbf{ADAPT}} = \left|\left|\boldsymbol{B}_{\mathbf{ADAPT}} - \boldsymbol{B}_{\mathbf{BL2}}\mathrm{T}^{-1}\right|\right|$$

To test the significance of this change, they compared $\Delta\boldsymbol{B}_{\mathbf{ADAPT}}$ to the difference magnitude estimated using movements recorded at the beginning ($\boldsymbol{B}_{\mathbf{BL1}}$) and end ($\boldsymbol{B}_{\mathbf{BL2}}$) of baseline training:

$$\Delta\boldsymbol{B}_{\mathrm{NOISE}} = \left|\left|\boldsymbol{B}_{\mathbf{BL1}} - \boldsymbol{B}_{\mathbf{BL2}}\right|\right|$$

For participants exposed to the rotation distortion, $\boldsymbol{B}_{\mathbf{est}}$ after adaptation could not reasonably be characterized as a rotated version of the baseline mapping because $\Delta\boldsymbol{B}_{\mathbf{ADAPT}}$ far exceeded $\Delta\boldsymbol{B}_{\mathbf{NOISE}}$ for these participants (Fig. 6d, red). The within-subject difference between $\Delta\boldsymbol{B}_{\mathbf{ADAPT}}$ and $\Delta\boldsymbol{B}_{\mathbf{NOISE}}$ was significantly greater than zero (Fig. 6d, red solid bar). Thus the rotational distortion induced these participants to reorganized patterns of coordination within articulation space to form new inverse hand-to-screen mappings that differed substantially from the original mapping. By contrast, $\Delta\boldsymbol{B}_{\mathbf{ADAPT}}$ did not exceed $\Delta\boldsymbol{B}_{\mathbf{NOISE}}$ for scaling participants; this suggests that scaling participants simply contracted their baseline inverse map to compensate for the imposed scaling distortion. Taken together, the results reveal that people invoke distinct compensatory strategies when confronted with rotation and scaling distortions of visual feedback, thereby demonstrating power of the BoMI paradigm to provide insight into how the human brain organizes and reorganizes patterns of coordination in the production of goal-directed actions.

## Facilitation and Interference in the Learning of Related, Redundant Motor Tasks

Due to redundancy of control in the BoMI paradigm, there are multiple options or solutions that can be used to achieve task goals. Ranganathan and colleagues (Ranganathan et al. 2014) sought to characterize the extent to which it is possible to influence the participant's solution in articulation space by exploiting the phenomenon of "transfer" (i.e., how practice on a prior task influences the coordination pattern used when learning a novel task). There exist common cases of positive transfer (e.g., learning to roller skate facilitates subsequent learning of ice skating) as well as negative transfer (e.g., learning to play tennis interferes with subsequent learning of badminton). Transfer (positive or negative) can be explained by invoking concepts of "similarity" or "dissimilarity" between the tasks involved (Osgood 1949). However, experimental testing of this explanation in multi-joint movements has been challenging because heretofore there has been no obvious way to either quantify similarity between coordination patterns or to design motor tasks that require the use of similar or dissimilar coordination patterns.

The BoMI paradigm provided a novel way to examine this question. Ranganathan and colleagues (Ranganathan et al. 2014) used the glove-cursor task to construct several tasks that had different degrees of overlap or similarity with a specific "criterion" task. By requiring participants to train on pairs of tasks that either do or do not share task-space dimensions, the authors aimed to address questions about how the brain represents extrinsic task space within the sensorimotor system, and how such representations determine whether prior learning will facilitate or interfere with subsequent task learning. While the phenomenon of transfer has been explored in adaptation studies under the idea of structural learning (Braun et al., 2009, 2010), similarity between tasks in those studies was based on the type of visuomotor perturbations experienced, not in terms of movement coordination patterns themselves.

Ranganathan and colleagues selected a two-dimensional criterion task where the A matrix was based on PC1 and PC2 from the fingerspelling task. All study participants had to learn the criterion task. The authors then constructed several "prior" tasks that had different degrees of overlap with the criterion task. Given that principal components are orthogonal to each other, the prior tasks were designed to either share or not share the criterion task dimensions. Three of the prior tasks had significant overlap with the criterion task as they shared PC projection vectors with the criterion task's task space; the A matrix for the "1D task" was based on PC1, the $\boldsymbol{A}$ matrix for the "2D task" was identical to the criterion task (the experimental control condition), whereas the $\boldsymbol{A}$ matrix for the "3D task" was based on PC1, PC2, and PC3. Two additional prior tasks had no overlap with the criterion task's task space; the ***A*** matrix for the "1D-Null task" was based on PC3, whereas the ***A*** matrix for the "2D-Null task" was based on PC3 and PC4. Participants were split into five different groups: the 2D group did not have a "prior" task and trained only on the criterion task (serving as the control group), whereas the remaining four groups trained on a different prior task on Day 1 (1D, 3D, 1D-Null, or 2D-Null) before training on the criterion task during the next 2 days (Fig. 7a).

Transfer to the criterion task depended on the amount of overlap between prior task and the criterion task hand-to-screen mappings (Fig. 7b). Participants who initially learned a mapping that overlapped with the mapping in the criterion task (i.e., the 1D and the 3D tasks) showed improved initial task performance (i.e., positive transfer) on Day-2 training in the criterion task compared to participants who learned the task from scratch. By contrast, participants who learned a prior mapping in the null space of the criterion task (the 1D-Null and 2D-Null groups) had longer movement times (i.e., negative transfer). Moreover, and unlike adaptation tasks where interference is washed out in a few trials, interference effects in

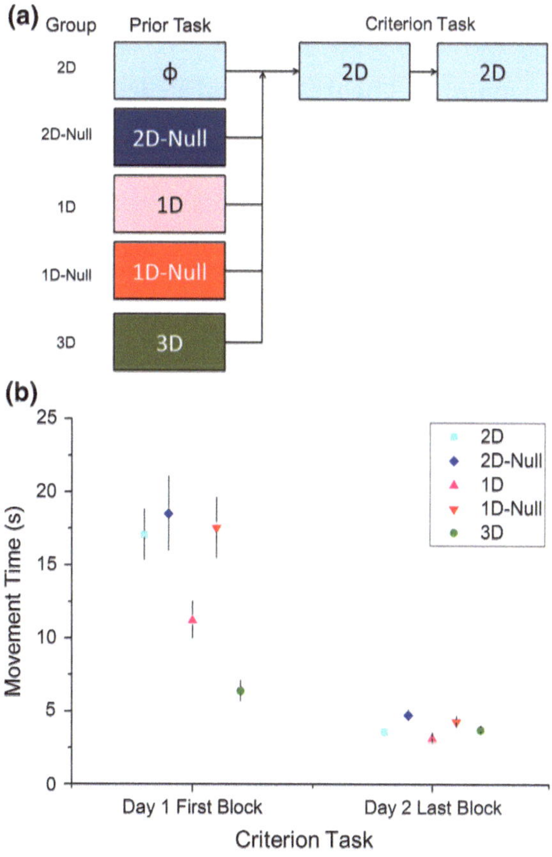


**Fig. 7** Facilitation and interference when learning multiple maps. **a** Practice schedule of groups. The 2D group did not have a prior task as it was always tested only on the criterion task **b** Movement time early and late in the criterion task showing significant facilitation for 1D and 3D groups, and prolonged interference for the 1D-Null and 2D-Null groups

the 2D-Null group were sustained over 2 days of practice. This interference mimics real-life phenomena where learning can result in atypical movement habits that can sometimes be maladaptive (e.g., the "chickenwing" backswing in golf, or compensatory trunk movements during reaching after a stroke).

The extent to which the task space of the prior and criterion tasks did or did not overlap resulted in positive or negative transfer, respectively. Participants exhibited a strong tendency to explore preferentially along the task space of the prior task even when performing the criterion task; they did not learn the criterion task entirely de novo, but instead appeared to adapt their prior mapping to accommodate the new criterion mapping. Persistence of null-space coordination patterns acquired through prior practice of tasks that did not enforce the geometry of the criterion task impeded future improvements in criterion task performance, and this interference appears to be difficult to unlearn. We suggest that the ability of the glove-cursor task to quantify and manipulate the degree of similarity between tasks may enable future studies to gain additional insight into the phenomenon of transfer and how it may be possible to modify movement coordination through specific training schedules and programs, an idea that has relevance for movement rehabilitation science.

## Examining Theoretical and Practical Issues in Sensorimotor Control Using BoMIs

In addition to the issues raised so far, BoMIs have also been used to provide theoretical insight into other motor control issues. For example, an unresolved question in the field of motor control is whether control of the large number of muscles in the body is simplified by grouping muscles into a smaller number of "synergies" (also referred to as "motor modules"). Approaches range from detailed musculoskeletal modeling (Valero-Cuevas et al. 2015) to intraspinal and intracortical stimulation (Giszter et al. 2007; Overduin et al. 2015). A specific point of contention is whether the low dimensionality observed in goal-directed behavior is indicative of a control strategy adopted by the nervous system (Tresch et al. 1999; d'Avella et al. 2003), or whether they simply arise from the biomechanical constraints of the body when performing the task (Kutch and Valero-Cuevas 2012). Answering this question is methodologically challenging because: (i) it is difficult to measure muscle activity from all of the independently controlled muscle elements surrounding multiple joints, and (ii) even if this could be measured, the exact mapping of how muscle activity determines (and is determined by) kinematic and kinetic performance is complex and difficult to disentangle empirically.

BoMIs provide a unique opportunity to answer this question, as it is possible to create "virtual bodies" where the exact mapping between muscle activity and behavior can be determined by the experimenter (Nazarpour et al. 2012; Berger et al. 2013; de Rugy et al. 2013). For example, using surface EMG recordings from the superficial muscles in the arm, it is possible to drive a virtual arm whose behavior is determined only by those muscle recordings. Moreover, because the interface is virtual, the biomechanics of the virtual body can also be altered by performing "virtual surgeries" (Berger et al. 2013). For example, agonistic muscles generating similar mechanical actions about a joint (such as the biceps and brachioradialis) normally are activated simultaneously. However, to examine if these muscles are controlled as a "single unit" (i.e., a muscle synergy), it is possible to create a virtual arm where these two muscles are functional antagonists (e.g., the biceps could drive the virtual arm into elbow flexion, whereas the brachioradialis could drive the elbow into extension). A strong case for fixed muscle synergies can be made if the muscles continue to be activated together, even when their function in the virtual world is made to conflict. In one study, Berger and colleagues found that it was possible to learn novel patterns of muscle coordination when learning to control a virtual object, but learning was faster when that control required muscle activity consistent with preexisting patterns of intermuscular coordination (Berger et al. 2013). In this way, BoMIs can give unique insight into the neural and biomechanical bases of muscle synergies.

In addition to the theoretical impact, BoMIs also have practical implications in the field of rehabilitation. One of the challenges in developing assistive technologies for people with limited movement abilities is to provide a single interface that can help them interact with their environment in multiple ways—such as controlling a cursor, a robot arm or even a wheelchair. Although this has traditionally been the domain of brain–machine interfaces (Serruya et al. 2002; Wolpaw et al. 2002; Lebedev and Nicolelis 2006)—where neural signals are mapped to the control of an external device—it is worth noting that brain–machine interfaces are ideally suited only for a small number of cases where there is almost a complete absence of movement. On the other hand, a significant portion of people with movement impairments still possess residual motor function. BoMIs can be designed to exploit these residual movements to enable these individuals to control external devices. In addition to being fast and noninvasive, a key advantage of BoMIs over brain–machine interfaces is that they can be dynamically adapted to the changing movement repertoire of the individual. This makes this interface especially appealing in cases where movement abilities may change with time—either due to growth and physical development (Lee et al. 2016) or due to rehabilitation and partial functional recovery.

Recently, Mussa-Ivaldi and colleagues have developed a BoMI for participants with high-level spinal cord injury (SCI) (Casadio et al. 2010; Thorp et al. 2016). The BoMI consisted of a wearable sensor system, which consisted of four, wireless inertial measurement units (IMUs) that were attached to the upper body. Similar to the glove-cursor task, a high-dimensional vector of signals from the IMUs were mapped to the low-dimensional position of a cursor on a computer screen, which allowed participants to communicate (type text) and play computer games using movements of the upper body. The interface also allowed the IMU signals to directly control the velocity of a powered wheelchair using an IMU-to-wheelchair mapping. Despite variations in the level of injury and movement abilities, participants with SCI were able to learn to control the cursor and the wheelchair within a few sessions of practice. More importantly, they retained this ability over time. The ability to use the BoMI for real-time control over virtual and real-world objects provides a more natural, continuous mode of control that exceeds the capabilities of discrete methods of control like head switches or sip/puff controllers. Moreover, using upper body movements instead of other modalities that overlap with communication such as gaze control or mouth sticks, participants using the BoMI retain the freedom to communicate while operating the interface.

## Take Home Message

The problem of motor coordination remains a challenging issue for movement neuroscientists. The BoMI—a simple, noninvasive experimental paradigm—has already provided substantial insight into how the nervous system organizes and reorganizes coordination patterns. We anticipate that the richness of this experimental paradigm will stimulate interest into reexamining classic theoretical issues in motor learning such as transfer and the role of variability in the acquisition and adaptation of novel patterns of coordination.

The BoMI also has great potential to advance the current state of the art of movement rehabilitation and assistive devices. Recent studies using a BoMI to facilitate movement coordination in individuals who have had a stroke (Wright et al. 2013) or SCI (Pierella et al. 2015) show promise, and it remains to be seen whether this type of practice could serve as a useful adjuvant to more traditional types of therapy. Moreover, the BoMI provides a high-throughput and noninvasive way to provide multidimensional control (e.g., to control multiple external devices) using a single interface. A challenge for future research is to develop BoMI interfaces yielding dexterous control of prostheses including a relatively high number of independent degrees of freedom, such as those that would allow independent control of the arm, wrist, and fingers.

**Acknowledgments** The authors thank Prof. Ferdinando Mussa-Ivaldi for his insightful and formative discussions, which were integral to the conception and execution of each of the glove BoMI experiments described in this chapter.

## References

Adams JA (1987) Historical review and appraisal of research on the learning, retention, and transfer of human motor skills. Psychol Bull 101:41–74

Adolph KE, Cole WG, Komati M, Garciaguirre JS, Badaly D, Lingeman JM, Chan GL, Sotsky RB (2012) How do you learn to walk? Thousands of steps and dozens of falls per day. Psychol Sci 23:1387–1394

Bastian AJ (2008) Understanding sensorimotor adaptation and learning for rehabilitation. Curr Opin Neurol 21:628–633
Berger DJ, Gentner R, Edmunds T, Pai DK, d'Avella A (2013) Differences in adaptation rates after virtual surgeries provide direct evidence for modularity. J Neurosci 33:12384–12394
Bernstein N (1967) The coordination and regulation of movement. Pergamon Press, Oxford
Biess A, Liebermann DG, Flash T (2007) A computational model for redundant human three-dimensional pointing movements: integration of independent spatial and temporal motor plans simplifies movement dynamics. J Neurosci 27:13045–13064
Braun DA, Aertsen A, Wolpert DM, Mehring C (2009) Motor task variation induces structural learning. Curr Biol 19:352–357
Braun D, Mehring C, Wolpert D (2010) Structure learning in action. Behav Brain Res 206: 157–165
Casadio M, Pressman A, Fishbach A, Danziger Z, Acosta S, Chen D, Tseng HY, Mussa-Ivaldi FA (2010) Functional reorganization of upper-body movement after spinal cord injury. Exp Brain Res 207:233–247
Casadio M, Ranganathan R, Mussa-Ivaldi FA (2012) The body-machine interface: a new perspective on an old theme. J Mot Behav 44:419–433
Cusumano JP, Cesari P (2006) Body-goal variability mapping in an aiming task. Biol Cybern 94:367–379
d'Avella A, Saltiel P, Bizzi E (2003) Combinations of muscle synergies in the construction of a natural motor behavior. Nat Neurosci 6:300–308
de Rugy A, Loeb GE, Carroll TJ (2013) Are muscle synergies useful for neural control? Front Comput Neurosci 7:19
Farshchiansadegh A, Ranganathan R, Casadio M, Mussa-Ivaldi FA (2015) Adaptation to visual feedback delay in a redundant motor task. J Neurophysiol 113:426–433
Gessell A (1946) The ontogenesis of infant behavior. In: Carmichael L (ed) Manual of child psychology. John Wiley, New York, pp 295–331
Giszter S, Patil V, Hart C (2007) Primitives, premotor drives, and pattern generation: a combined computational and neuroethological perspective. Prog Brain Res 165:323–346
Heathcote A, Brown S, Mewhort DJ (2000) The power law repealed: the case for an exponential law of practice. Psychon Bull Rev 7:185–207
Ivanenko YP, Poppele RE, Lacquaniti F (2004) Five basic muscle activation patterns account for muscle activity during human locomotion. J Physiol (Lond) 556:267–282
Kelso JAS (1984) Phase transitions and critical behavior in human bimanual coordination. Am J Physiol 246:R1000–R1004
Kelso JAS (1995) Dynamic patterns: the self-organization of brain and behavior. MIT Press, Cambridge
Krakauer JW, Mazzoni P (2011) Human sensorimotor learning: adaptation, skill, and beyond. Curr Opin Neurobiol 21:636–644
Krakauer JW, Pine ZM, Ghilardi MF, Ghez C (2000) Learning of visuomotor transformations for vectorial planning of reaching trajectories. J Neurosci 20:8916–8924
Kutch JJ, Valero-Cuevas FJ (2012) Challenges and new approaches to proving the existence of muscle synergies of neural origin. PLoS Comput Biol 8:e1002434
Langendorfer SJ, Roberton MA (2002) Individual pathways in the development of forceful throwing. Res Q Exerc Sport 73:245–256
Lebedev MA, Nicolelis MA (2006) Brain-machine interfaces: past, present and future. Trends Neurosci 29:536–546
Lee MH, Ranganathan R, Kagerer FA, Mukherjee R (2016) Body-machine interface for control of a screen cursor for a child with congenital absence of upper and lower limbs: a case report. J Neuroeng Rehabil 13:34
Liu X, Scheidt RA (2008) Contributions of online visual feedback to the learning and generalization of novel finger coordination patterns. J Neurophysiol 99:2546–2557
Liu YT, Mayer-Kress G, Newell KM (2006) Qualitative and quantitative change in the dynamics of motor learning. J Exp Psychol Hum Percept Perform 32:380–393

Liu X, Mosier KM, Mussa-Ivaldi FA, Casadio M, Scheidt RA (2011) Reorganization of finger coordination patterns during adaptation to rotation and scaling of a newly learned sensorimotor transformation. J Neurophysiol 105:454–473
Mosier KM, Scheidt RA, Acosta S, Mussa-Ivaldi FA (2005) Remapping hand movements in a novel geometrical environment. J Neurophysiol 94:4362–4372
Müller H, Sternad D (2004) Decomposition of variability in the execution of goal-oriented tasks: three components of skill improvement. J Exp Psychol Hum Percept Perform 30:212
Nazarpour K, Barnard A, Jackson A (2012) Flexible cortical control of task-specific muscle synergies. J Neurosci 32:12349–12360
Newell KM (1991) Motor skill acquisition. Annu Rev Psychol 42:213–237
Newell A, Rosenbloom PS (1981) Mechanisms of skill acquisition and the law of practice. In: Anderson JR (ed) Cognitive skills and their acquisition. Erlbaum, Hillsdale, NJ, pp 1–55
Osgood CE (1949) The similarity paradox in human learning; a resolution. Psychol Rev 56: 132–143
Overduin SA, d'Avella A, Roh J, Carmena JM, Bizzi E (2015) Representation of muscle synergies in the primate brain. J Neurosci 35:12615–12624
Pierella C, Abdollahi F, Farshchiansadegh A, Pedersen J, Thorp EB, Mussa-Ivaldi FA, Casadio M (2015) Remapping residual coordination for controlling assistive devices and recovering motor functions. Neuropsychologia 79:364–376
Ranganathan R, Adewuyi A, Mussa-Ivaldi FA (2013) Learning to be lazy: exploiting redundancy in a novel task to minimize movement-related effort. J Neurosci 33:2754–2760
Ranganathan R, Wieser J, Mosier KM, Mussa-Ivaldi FA, Scheidt RA (2014) Learning redundant motor tasks with and without overlapping dimensions: facilitation and interference effects. J Neurosci 34:8289–8299
Roberton MA, Halverson LE, Langendorfer S, Williams K (1979) Longitudinal changes in children's overarm throw ball velocities. Res Q 50:256–264
Santello M, Flanders M, Soechting JF (1998) Postural hand synergies for tool use. J Neurosci 18:10105–10115
Scholz JP, Schoner G (1999) The uncontrolled manifold concept: identifying control variables for a functional task. Exp Brain Res 126:289–306
Serruya MD, Hatsopoulos NG, Paninski L, Fellows MR, Donoghue JP (2002) Instant neural control of a movement signal. Nature 416:141–142
Smith MA, Ghazizadeh A, Shadmehr R (2006) Interacting adaptive processes with different timescales underlie short-term motor learning. PLoS Biol 4:e179
Thelen E, Corbetta D, Spencer JP (1996) Development of reaching during the first year: role of movement speed. J Exp Psychol Hum Percept Perform 22:1059–1076
Thorp EB, Abdollahi F, Chen D, Farshchiansadegh A, Lee MH, Pedersen JP, Pierella C, Roth EJ, Seanez Gonzalez I, Mussa-Ivaldi FA (2016) Upper body-based power wheelchair control interface for individuals with tetraplegia. IEEE Trans Neural Syst Rehabil Eng 24:249–260
Ting LH, McKay JL (2007) Neuromechanics of muscle synergies for posture and movement. Curr Opin Neurobiol 17:622–628
Todorov E, Jordan MI (2002) Optimal feedback control as a theory of motor coordination. Nat Neurosci 5:1226–1235
Torres EB, Zipser D (2002) Reaching to grasp with a multi-jointed arm. I. Computational model. J Neurophysiol 88:2355–2367
Torres EB, Heilman KM, Poizner H (2011) Impaired endogenously evoked automated reaching in Parkinson's disease. J Neurosci 31:17848–17863
Tresch MC, Saltiel P, Bizzi E (1999) The construction of movement by the spinal cord. Nat Neurosci 2:162–167
Valero-Cuevas F, Cohn B, Yngvason H, Lawrence E (2015) Exploring the high-dimensional structure of muscle redundancy via subject-specific and generic musculoskeletal models. J Biomech 48:2887–2896

Wheat JS, Glazier PS (2006) Measuring coordination and variability in coordination. In: Davids K, Bennett SJ, Newell KM (eds) Movement system variability. Human Kinetics, Champaign, IL, pp 167–181

Wolpaw JR, Birbaumer N, McFarland DJ, Pfurtscheller G, Vaughan TM (2002) Brain-computer interfaces for communication and control. Clin Neurophysiol 113:767–791

Wright ZA, Rymer WZ, Slutzky MW (2013) Reducing abnormal muscle coactivation after stroke using a myoelectric-computer interface: a pilot study. Neurorehabil Neural Repair 28:443–451

Zipser D, Torres E (2007) Computing movement geometry: a step in sensory-motor transformations. Prog Brain Res 165:411–424

# A Computational Index to Describe Slacking During Robot Therapy

**Davide Piovesan**

**Abstract** Movement facilitation has a fundamental role in the rehabilitation treatment of stroke survivors. However, its action mechanisms are still poorly understood. An open question is to identify the effect of the level of assistance on the recovery process. To address this topic, new insight on voluntary control and movement strategies during rehabilitation must be gained. Robot-assisted arm movements were examined in a task where subjects had to reach distal targets in the presence of assistive forces. As the training proceeded, subjects improved their performance and exercised with both the initial force level of the first session and with progressively decreasing levels of assistive force. We found that when stroke survivors became to execute voluntary movements with lower forces levels they decreased their voluntary control in the presence of higher forces, following a minimum effort trajectory. These findings provide a new important insight for the rehabilitation of stroke survivors, suggesting that passive mobilization and exercise with constant force, although useful for muscular reinforcement, may have a detrimental effect on voluntary control and movements planning.

**Keywords** Manipulandum · Assistive · Rehabilitation · Adaptive controllers · Challenge-based · Dystonia · Rigidity · Viscosity · Damping · Degrees of freedom (DOF) · Reference trajectory · Proprioception · Neuro-rehabilitation

## Introduction

When we move our arm in space from one point to another, a force applied to our hand may change how we perform the movement and therefore how our brain controls the motion of the arm. This well-known concept has been extensively used in motor control studies to investigate how the brain learns to move in unknown, through predictable environments (Lackner and Dizio 1994; Shadmehr and

---

D. Piovesan (✉)
Department of Mechanical Engineering, Gannon University, Erie, PA, 16501, USA
e-mail: piovesan001@gannon.edu



J. Laczko and M.L. Latash (eds.), *Progress in Motor Control*,
Advances in Experimental Medicine and Biology 957,
DOI 10.1007/978-3-319-47313-0_19

Mussa-Ivaldi 1994; Thoroughman and Shadmehr 2000; Scheidt et al. 2001). In these studies when subjects perform reaching movements in the presence of a deterministic force field, they build a model of the external dynamics and learn to apply a force that counteracts the effect of the perturbing force field.

The ability to change the way we plan and control movements by using force fields has a far reach potential in rehabilitation. However the role of external forces in the recovery process is still poorly known. To this end, several research groups proposed different strategies to enhance the rehabilitation process for individuals with neuromotor disabilities.

Abdollahi et al. (2013) suggested that the introduction of force field augmenting the endpoint errors of subjects have better effects on movement planning. Huang et al. (Huang et al. 2010; Huang and Patton 2011) tested on stroke survivors the efficacy of destabilizing force fields based on negative viscosity, while several researches found resistive training to be beneficial in individuals affected by multiple sclerosis (Taylor et al. 2006; Dalgas et al. 2007; Motl and Pilutti 2012). These methods are just few examples of a broader category of robot-mediated rehabilitative exercises that have been recently defined as "challenge based" approaches (Marchal-Crespo and Reinkensmeyer 2009).

Most of the recently proposed protocols in the robotics rehabilitation field are built according to this principle. We are now aware, that repetitive exercises based on high levels of assistance and passive mobilization of the limbs, although beneficial for muscle reinforcement, have limited effects on the recovery process (Kaelin-Lang et al. 2005; Hogan et al. 2006).

To elicit plasticity (i.e., a permanent change in control strategy), movements have to be associated with a task and a volitional effort (Nudo 2006), stimulating a reorganization process that is similar to the learning of a motor skill (Krakauer et al. 2012). Different studies suggested to keep the assistance of the impaired movement at the minimum that evokes a functional response and proposed different assistive paradigms such as "assistance as needed" (Wolbrecht et al. 2008; Squeri et al. 2011) or "minimally assistive" approaches (Casadio et al. 2009b). In this context, there is the need to adapt the assistance to the residual ability of each individual given the progress of the disease and the improvement due to the ongoing therapy. Several approaches were proposed to automatically adjust the assistance and the exercise difficulty based on performance (Krebs et al. 2003) or on models that describe individual subject' characteristics (Blaya and Herr 2004; Riener et al. 2005). These protocols led to promising results (Vergaro et al. 2010) and a great margin of improving can be anticipated when modifying the therapy as function of the subjects' performance computed in real time during the robotic exercise thus building adaptive controllers based on subjects' characteristics.

Emken et al. (2007) suggested that the use of adaptive controllers is essential because humans are "greedy optimizers," thus tending to exploit the assistive forces decreasing the amount of voluntary control. On the other hand, the role of the assistive forces on the recovery process and more specifically on the voluntary control of stroke survivors has not been directly investigated, yet.

The aim of this study is to investigate the control strategy of stroke survivors with a training protocol in which the level of assistive forces is always decreased down to the minimum force that can evoke a voluntary response. As the training proceeded and the recovery occurred, volunteers were able to perform the same exercise with lower levels of forces. We develop a metric that could clearly differentiate the amount of voluntary movement for each trial correlating the velocity profile of the subjects' hand with those of an ideal trajectory in which the assistive force would interact only with passive arm mechanics. We investigated the effect of the initial (maximum) assistive force level when subjects were able to move in the presence of significantly lower forces. We hypothesized that subjects could either (i) use the same amount of voluntary control they used at the beginning, (ii) relay less on the assistive force and more on the voluntary control since their movement abilities were improved, or (iii) be "greedy optimizers" and decrease their voluntary control exploiting the forces.

## Materials and Methods

We tested the influence of a constant level of assistive force provided by a planar manipulandum on the arm trajectories of stroke survivors during different phases of robot-assisted training. Subjects may make use of the force provided by the robot in different ways. On one hand, subjects could exploit the assistance by complying with the force profile. This would generate non-rectilinear trajectories; nevertheless, the voluntary control would be reduced at the minimum and reaching of the target would always be achieved. On the other hand, subjects could maximize their voluntary control, using the assistive force only to facilitate the release of their motor plan. These two conditions are the extreme of a broad spectrum of strategies that can be utilized by the subject, where switching between mechanisms can occur in either direction. Subjects might rely more on force at the beginning of the therapy and starting to engage their voluntary control as the recovery proceeds with the rehabilitation treatment. Conversely, the voluntary control could start at the beginning of the therapy and then be abandoned to exploit more the assistance provide by the force as the exercise became easier due to the recovery.

If subjects were to fully exploit the assistive force, the trajectory to reach the target would arise from the interaction of the force with the subject's passive mechanics. We performed a set of numerical simulations to identify such passive trajectories and we correlated the prediction of the simulation with the performance of nine stroke survivors, who underwent a robot-mediated training of 10 sessions. Switching between strategies would be characterized by a change in correlation between the simulated trajectories and the experimental data. The correlation would decrease switching from a passive to an active strategy and vice versa. If a change of strategy does occur, we will expect a change in correlation between the measured and the predicted performance

(a) between trials at the beginning and at the end of the training with the same force level
(b) between trials in the last session with decreasing force levels, in such way that the performance with lower level of force would be similar to the behavior in the previous sessions with higher force.

A control experiment was run to test the effect of assistive force on voluntary control in five able bodied individuals with no history of previous neurological disorder and with normal level of movement skills.

## *Experimental Task*

The task was to hit a set of seven equally spaced targets arranged at distal positions almost to the workspace limit (C layer). The exercise was composed of blocks where subjects starting from the three different positions (A Layer) had to reach each of the seven targets, presented in random order. Each block consisted of 21 outward movements of amplitude about 20 cm. The workspace was centered with respect to the shoulder joint contralateral to the lesion.

The planar robotic manipulandum provided assistive forces helping stroke survivors to accomplish the task. In the blocks where visual feedback was provided, subjects saw their arm, the robot, and a computer monitor (1 m away at eye level) that displayed the end-effector position and the target to reach (Fig. 1). The target and the cursor corresponding to the end-effector position were represented as round disks of different colors and 1 cm radius. In the blocks where visual feedback was not provided, subjects were blindfolded and they have to reach the targets only relying on the haptic cue of the assistive force.

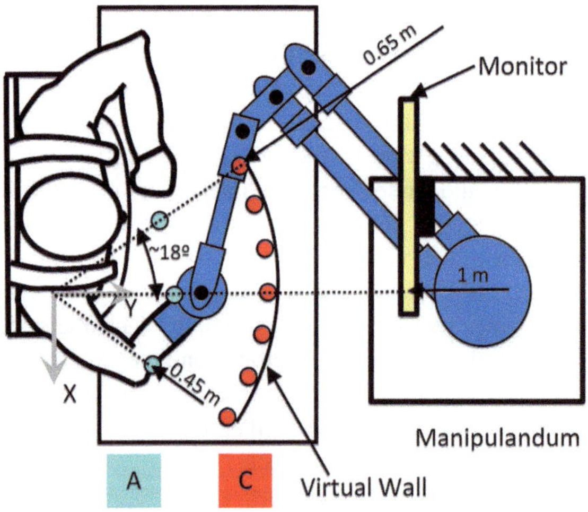


**Fig. 1** Experimental setup. *A* represent the starting point layer, *C* represents the targets layer. All combinations of targets was performed for a total of 3 × 7 = 21 reaches per session

The haptic rendering of the environment was generated according to the equation:

$$\mathbf{F}_e = \left\{ G(F_a, t)\frac{(\mathbf{x}_T - \mathbf{x}_e)}{|\mathbf{x}_T - \mathbf{x}_e|} - \begin{bmatrix} B_e & 0 \\ 0 & B_e \end{bmatrix} \dot{\mathbf{x}}_e + \mathbf{F}_w(\mathbf{x}_e, K_w) \right\} \quad (1)$$

where terms in bold (e.g., $\mathbf{u}$) represent vectors and terms in italic (e.g., $u$) represent scalars. In Eq. (1) $\mathbf{F}_e$ is the force provided by the manipulandum, $\mathbf{x}_T$ is the target position, $\mathbf{x}_e$ is the position of the end effector, and $F_a$ is the selected level of the assistive force in the trial (see Table 1 for maximum level). The term $G(F_a, t)$ starts from 0, increasing linearly to $F_a$ with a rise time of 1 s; therefore, enabling a smooth activation of the force field. The two additional terms represent a viscous field to stabilize the arm posture and a rigid wall. While the viscous field is always active, the rigid wall engages only beyond the targets' level, which provided a representation of the boundary of the workspace. The coefficient $B_e$ was empirically determined to be 12 Ns/m as a trade-off between stability and dissipated energy. The stiff virtual wall was rendered with a 1000 N/m elastic coefficient $K_w$(Casadio et al. 2009a) and used only as safety boundary. When the movement was done properly, the virtual wall was never hit. The training consisted of 10 sessions of a duration that ranged from 45 to 75 min. Each session started with the same initial force, selected by the therapist as the minimal force allowing the subject to initiate the movement. After the first two blocks, the therapist could decide to extend the exercise with additional blocks. In these blocks, the levels of force were lowered, in accordance with the subject's residual ability and fatigue. For each subsequent session, while starting always with the first imposed force of the first session, subjects experienced a further decrease of assistive force, where the ultimate goal would be (when possible) to reach the target with no assistive force. When subjects

**Table 1** Subject characteristics

| | Age | FM before | FM after | Ash | G | E | WT | HT | $F_a$ |
|---|---|---|---|---|---|---|---|---|---|
| S1 | 72 | 6 | 8 | 3 | M | I | 90 | 165 | L |
| S2 | 69 | 12 | 18 | 1+ | F | I | 60 | 155 | R |
| S3 | 57 | 17 | 21 | 3 | M | I | 90 | 170 | L |
| S4 | 34 | 13 | 23 | 1+ | F | I | 67 | 178 | R |
| S5 | 30 | 6 | 9 | 2 | F | I | 57 | 168 | L |
| S6 | 46 | 6 | 13 | 2 | F | H | 50 | 155 | L |
| S7 | 55 | 36 | 41 | 1 | F | H | 55 | 165 | L |
| S8 | 59 | 5 | 8 | 3 | F | I | 65 | 170 | R |
| S9 | 53 | 41 | 45 | 1 | F | H | 58 | 168 | R |

Subjects data. *Age* years. *FM* upper arm Fugl-Meyer score, max 66/66; before, after, and after three months with respect to the robot therapy sessions, *Ash* ashworth score, gender: *M* male, *F* female; *E* etiology: *I* ischemic, *H* hemorrhagic; *WT* weight [kg]: *HT* height [m]; $f_a$ level of assistive force at which we estimated the stiffness [n] reached each target, the assistive force and visual feedback were switched off for 1 s before the following target appeared on the screen. The kinematic response to this sudden drop in assistive force was used to estimate the arm impedance using the time frequency technique described in Piovesan et al. (2009, 2011a, b, 2012).

Nine chronic stroke survivors with different levels of impairment participated in the experiment and their clinical and anthropometric data are listed in Table 1. Subjects held with their right impaired hand the end effector of a planar manipulandum (Casadio et al. 2006). Their shoulders and wrist movements were restricted by using custom made holders.

## Simulation

We simulated the experiment in silico to predict the expected movement performance during training assuming that subjects would use a reference trajectory generated by the assistive force acting on the passive properties of their arm.

### *Model of the Coupling Between Robot and Human Arm*

A simple schematic of the coupling between robot and human arm mechanics in the Cartesian space for a single degree of freedom is depicted in Fig. 2. The system is governed by the following equation:

$$M_x^i \ddot{x}_i + C_x^i\left(\dot{x}_i - \dot{x}_{ref_i}\right) + K_x^i\left(x_i - x_{ref_i}\right) + D_x^i \dot{x}_i + R_x^i x_i = F_{ei} \quad (2)$$

where subscript $x$ refers to the properties of the arm in the Cartesian space. The variable $x_i$ is the $i$th coordinate of the point of contact between the hand and robot's

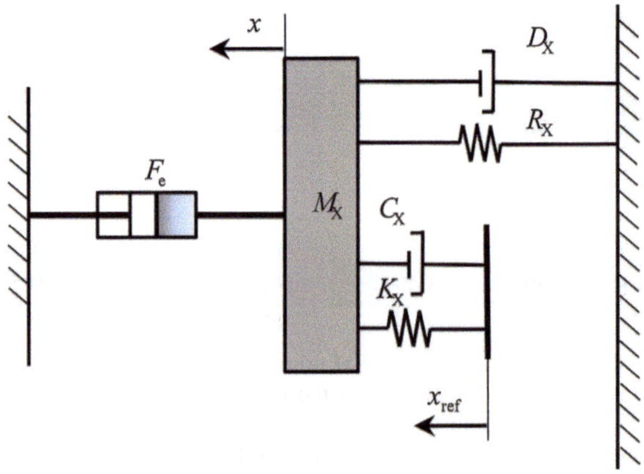


**Fig. 2** Mechanical schematics of the interaction between the human arm and the robot: $[M_x, C_x, K_x]$ are the inertia, damping, and stiffness of the arm. $F_e$ is the assistive force, $x_{ref}$ is the reference trajectory for the subject motor plan. $[R_x, D_x]$ are the rigidity and viscosity of the arm

end effector, calculated in the subjects' reference frame. $F_{ei}$ is the force provided by the manipulandum from Eq. (1), $x_{ref_i}$ is the reference trajectory for the subjects' motor plan. $M$, $K$, $C$, $R$, and $D$, are the matrices that represent inertia, stiffness, damping, rigidity, and viscosity proprieties of the subjects arm. According to this formalism stiffness and rigidity, as well as damping and viscosity refers to different arm properties that we want to clarify in the following paragraph.

One of the consequences of stroke is the development of an intrinsic arm dystonia that might result in a dominant flexion pattern. Dystonia generates a position-dependent force that can vary with the degree of impairment. Dystonic forces are relative to a specific equilibrium position and the starting point of the trajectory was located in its vicinity. With the hand in such position, applying a displacement to it in the direction away from the body gives rise to a roughly linear increase in force opposing the movement. Approximating the position-dependent force with a linear spring is common in the literature, and the angular coefficient of rigidity for elbow movement has been reported to reach values up to 8 Nm/rad (Schmit et al. 1999). In this work, we will refer to the feature characterizing dystonia as a linear coefficient generating force with respect to the trajectory starting point, using the term "coefficient of rigidity" or simply "rigidity."

Conversely, we will adopt the term "coefficient of stiffness" or simply "stiffness" to the parameter that generates a position-dependent reaction force with respect to a moving point on a reference trajectory. Hence, rigidity and stiffness have two different reference points. During a point-to-point movement, rigidity generates a force field that hinders the movement. This field is always present even in unimpaired individuals but tend to increase due to atrophy and muscle contractions. Stiffness generates a position-dependent force that maintains the hand on the intended trajectory if a disturbance occurs. The stiffness reference point changes in time, moving synchronously with the intended trajectory. In general, stiffness is one order of magnitude larger than rigidity. However, the force at the end effector associated to them can be comparable since the rigidity is multiplied by the amplitude of the movement, while stiffness is multiplied by the deviation of the movement from the intended trajectory.

Similar considerations can be made for viscosity and damping. The former acts in parallel to rigidity and generates a force field proportional to the instantaneous velocity of the movement. The latter acts in parallel to stiffness and generates a force field proportional to the rate of change of the real trajectory with respect to the intended one.

## *Identification of the Reference Trajectory Based on External Force*

The reference trajectory based on external force (EF) can be calculated solving the Eq. (2) for all degrees of freedom (DOF) imposing, $\mathbf{R}_x$, $\mathbf{D}_x$, $\mathbf{F}_e$ and assuming that the reference trajectory coincide with the actual trajectory. Notice that $B_e$ in Eq. (1) is embedded in $\mathbf{F}_e$ and it depends on the velocity of the end effector. Since the arm is modeled as a double pendulum and not as a point mass, the inertial matrix $\mathbf{M}_x$ is not diagonal. Hence, the EF trajectory is not a straight line due to the effect of centripetal and Coriolis force about the joints. The trajectory is curved but it is the most advantageous in order to reach the target in terms of minimum modulation of joint torque (i.e., the external force is doing the work for you). The solution of Eq. (2) given the aforementioned constraints will be used as reference trajectory so that $x_{EF} = x_{\text{ref}}$.

From the experimental data, we obtained the reaching time $T$ of each assisted movement. The time was estimated as the period comprised between the instant in which subjects increase the hand absolute reaching velocity above 10 % of the absolute maximum to the instant in which they decrease the velocity below such threshold, permanently. To estimate the EF trajectory requires some assumption of limb mechanics. The arm was modeled as a two DOF system where the shoulder has a fixed center of rotation. Arm inertial parameters for the estimation of $\mathbf{M_X}$ were estimated based on a subject's weight and height (Zatsiorsky and Seluyanov 1983). The efficacy of this method with respect to others is discussed in (Piovesan et al. 2011c). The endpoint stiffness $\mathbf{K_X}$, and damping $\mathbf{C_X}$, were measured using a newly developed technique, which can estimate the parameters from the values of the arm residual vibration after the target is reached and the assistive force is suddenly switched off. The description of the technique can be found in Piovesan et al. (2012, 2013b), while experimental results used in the present paper are reported in Piovesan et al. (2011a, b, 2013a).

The implemented rigidity and viscosity were estimated using an algorithm described in the next section. Hence, the trajectory of the hand in the Cartesian space was computed using an inverse dynamic routine implemented in Simulink®. The assistive force was implemented as described in Eq. (1).

**Rigidity estimation**

Given the desired reaching time of the movement and the external force applied in the experiment, we implemented a first simulation to find the maximum allowed rigidity. Setting the reference position for the rigidity at the starting point, the farther from such position the hand is displaced, the larger is the force generated by the rigidity.

Setting the external force $\mathbf{F}_e$, we will iteratively change the rigidity and viscosity so that the hand would reach the target at the same time $T$ of the real trajectory. Since the assistive force is switched off when the target is reached, if the set rigidity is too low, then the movement will be too fast, and a recoil will be present. Conversely, if the resistive force generated by the rigidity is too high, the end-effector would not reach the target in the desired time.

We set the starting rigidity and viscosity matrices at the joints to

$$\mathbf{R}_\theta = \begin{bmatrix} R_{11} & R_{12} \\ R_{21} & R_{22} \end{bmatrix} = \begin{bmatrix} 3.5 & 0.5 \\ 0.5 & 2.0 \end{bmatrix} \frac{\text{Nm}}{\text{rad}}$$
$$\mathbf{D}_\theta = 0.1 \cdot \mathbf{R}_\theta \frac{\text{Nms}}{\text{rad}} \tag{3}$$

which is approximately the normal joint "rigidity" of an unimpaired individual, with no load applied (Wiegner and Watts 1986). The index 1 refers to the shoulder, 2 refers to the elbow. Since rigidity and viscosity of stroke survivors are presumably going to be higher than such values, we considered a multiplicative coefficient $\rho > 1$ to account for the impairment and modulate the resistive torque as follows:

$$\rho\left(\mathbf{R}_\theta \cdot \boldsymbol{\theta} + \mathbf{D}_\theta \cdot \dot{\boldsymbol{\theta}}\right) = \boldsymbol{\tau}_r$$

Given the assistive force $\mathbf{F}_e$ and reaching time $T$, the parameter $\rho$ changes the stiffness in our simulations isotropically so to obtain a trajectory that reaches the target with no residual recoil. In this work, the maximum value of the coefficient of "rigidity" was 2, producing a maximum rigidity of 7 Nm/rad, well within the physiological range reported in the literature (Schmit et al. 1999).

## Data Analysis

We compared the resulting task trajectories using the cross-correlation function between the velocity time profile of the data and the velocity signals of the simulations.

$$\Re_{\text{data,sim}}(\tau) = \frac{\text{xcorr}(\|v_{\text{data}}(t)\| \cdot \|v_{\text{sim}}(t-\tau)\|)}{\text{autocorr}(\|v_{\text{data}}(t)\|)} \tag{4}$$

where $\|v_i(t-\tau)\| = \sqrt{\dot{x}_i^2(t) + \dot{y}_i^2(t-\tau)}$.

The correlation coefficients are the values of the cross-correlation function when the lag $\tau = 0$.

We used this method for comparing the movement strategies adopted in the same condition and with the same level of force in different training sessions.

The coefficients were statistically analyzed using repeated measure analysis of variance (ANOVA) with subject as a random factor taking into account two fixed factors such as, either: target location and sessions with same assistive force level, or the target location and force level within the same session

Furthermore, while the experiment consisted of 10 sessions, the initial sessions are quite difficult to investigate using the proposed approach, since subjects' movements were quite segmented and at very low speed. Hence, we performed the proposed comparison starting from the 4th session, where speed and timing were compatible with single movement trajectories.

If the correlation between the simulated and the real performance will change with training, we plan a further analysis that will verify the effect of force. We expect that by decreasing the forces in the last session the correlation between the measured and the predicted trajectories will change.

Specifically, we hypothesize that

(a) the trajectories at the end of the training will be more correlated with the simulated ones than the trajectories with a same force level at beginning of the training
(b) the performance with lower level of force in the last session would be similar with the behavior (lower correlation) in previous sessions with higher assistive force.

## Results

The first two histogram bars from the left in Fig. 3 show the correlation coefficients between the experimentally acquired velocity profiles and those calculated in simulation when following a minimum effort trajectory at the maximum level of force. The distribution of coefficients is calculated at the fourth and last session (mean and standard deviation) for our sample population of stroke survivors. We can observe a statistically significant increase in the correlation coefficients at the last session (see Table 2), supporting the hypothesis that when a substantial improvement in voluntary motion is achieved, a level of assistance beyond subjects need induces a decrease in voluntary control generally referred as "slaking" (Reinkensmeyer et al. 2009; Reschechtko et al. 2015). The third and fourth histogram bars from the left show a similar behavior when the exercise is executed in absence of visual feedback. Hence, we can infer that executing the task blindfolded—i.e., without any visual feedback and only relying on the haptic cue of the assistive forces—does not introduce any statistically significant changes in the "slaking" behavior.

To verify if in the last session the voluntary control not maintained at high force level is still present at lower levels, we performed an analysis to assess the change

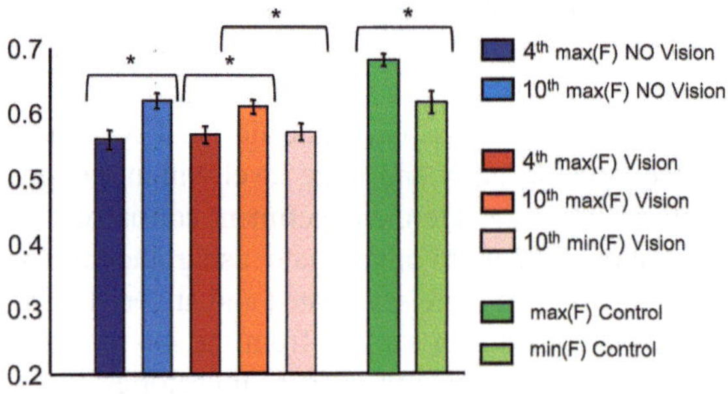


**Fig. 3** Correlation coefficients between experimental data and simulations (mean and standard error)

**Table 2** Two-way anova with subject as a random factor among sessions and directions

| Source | *df* | $\Re_{\text{data},EF}(0)$ 4th versus 10th at $\max(F_a)$ | |
|---|---|---|---|
| | | $F(1, df)$ | $p$ |
| Session | 1 | 19.7 | <0.0001 |
| Direction | 6 | 0.91 | 0.4841 |
| Vision | 1 | <0.001 | 0.9546 |
| Session * Direction | 6 | 0.68 | 0.6673 |
| Session * Vision | 1 | 0.5 | 0.478 |
| Vision * Direction | 6 | 0.47 | 0.8296 |

**Table 3** Two-way anova with subject as a random factor among force intensity and directions

| Source | *df* | $\Re_{\text{data},EF}(0)$ control 25 N versus 5 N | | $\Re_{\text{data},EF}(0)$ stroke $\max(F_a)$ versus $\min(F_a)$ at 10th | |
|---|---|---|---|---|---|
| | | $F(1, df)$ | $p$ | $F(1, df)$ | $p$ |
| Force | 1 | 16.12 | 0.0001 | 6.83 | 0.0093 |
| Direction | 6 | 7.55 | <0.001 | 0.65 | 0.692 |
| Force * Direction | 6 | 4.2 | 0.0005 | 0.49 | 0.819 |

in correlation coefficients between simulation and experimental data when the assistive forces reach their minima at the 10th session. In the fifth histogram bar, we can observe a marked reduction of the correlation coefficients with respect to those calculated in the same session at higher assistive levels (see also Table 3). This is an indication that for lower assistance, subjects are not likely to use the minimum effort trajectory as their reference thus implying that a different, not passive, control is used.

The phenomenon described above is also reproducible in control subjects, where the correlation between minimum effort trajectory and experimental data is higher for high assistive force levels, and tend to decrease for minimal level of assistance. Statistical significance for the dependency of the aforementioned changes to the assistive force levels is reported in Table 3.

## Discussion

In this work, we tested the hypothesis that during robot-mediated therapy stroke survivors alter their motor plan as they increase movement proficiency. The hypothesis encompassed the possibility for the subject to use the trajectory generated by the assistive force on the passive mechanics (inertia, rigidity, and viscosity) as reference.

We developed a method to account for the limb rigidity and viscosity, based on a series of inverse dynamics simulations. While the simulations returned a result that was physiologically plausible, we acknowledge that for multi-degrees of freedom model the rigidity matrix $\mathbf{R}_\theta$ might not be symmetric. The asymmetry could have an effect on the final trajectory. We also observed some biomechanical constrains that can cause the subject to use different strategies when interacting with an assistive force in Eq. (1). The endpoint velocity of a right handed subject is higher when moving from the contralateral to the ipsilateral side (left to right), even though the external force and the movement distance are the same. This observation points out the possibility to chose different strategies when moving in different directions. One of the reasons for a subject to use a force generated trajectory lies on the out coming velocity profiles, which depends on the value of rigidity and viscosity. With a coefficient of rigidity so that the hand can reach the target, the peak velocity is larger than the experimental data. The introduction of stiffness and damping also increases the velocity profile. This aspect might induce the subject to favor the EF trajectory.

Finally, we confirmed that knowing the real stiffness of the hand is important to properly model the movements of an impaired individual. While the motor planning seems to remain intact, the outcome result is strongly influenced by altered stiffness and damping values.

## Take Home Message

Robot-assisted therapy was applied to a group of stroke survivors and control individuals.

A set of simulations was used to compute the minimum effort trajectory generated by the interaction between assistive force acting on the subject hand when the subject was reaching for a target.

When the difference between the reference trajectory and the hand trajectory is negligible, the arm is equivalent to a passive system where the assistive force acts upon viscosity, rigidity, and inertia of the arm. In this case the endpoint path is characteristically curved. When considering the path of a right arm moving outward, the convex side of the path points laterally. The main contributors to such skewness are the Coriolis and centrifugal forces arising from the asymmetry of the inertial matrix.

After 10 sessions of robotic treatment all starting with an initial force chosen by the therapist we had hypothesized the following three scenarios:

i. At the end of the treatment the subjects could use the same amount of voluntary control they used at the beginning. This would result in no statistically significant change of the $\Re_{\mathrm{data},EF}(0)$ index for the initial level of force. This would not necessarily indicate that the therapy was not effective as the subjects might be able to perform movements with a lower level of assistive force at which they were incapable to move prior the treatment.

ii. The subjects would rely less on the assistive force and more on the voluntary control since their movement abilities were improved. This would decrease the $\Re_{\text{data},EF}(0)$ correlation index in favor of a more straight trajectory with minimum jerk velocity.

iii. The subjects would be “greedy optimizers” and decrease their voluntary control exploiting the forces, thus resulting in an increased $\Re_{\text{data},EF}(0)$ index.

We verified hypothesis (iii) for both control subjects and impaired individuals. When the assistive force becomes more than necessary for only initiating the movement, subjects relinquished their voluntary control

## References

Abdollahi F, Lazarro EDC, Listenberger M et al (2013) Error augmentation enhancing arm recovery in individuals with chronic stroke a randomized crossover design. Neurorehabilitation Neural Repair, 1545968313498649

Blaya JA, Herr H (2004) Adaptive control of a variable-impedance ankle-foot orthosis to assist drop-foot gait. IEEE Trans Neural Syst Rehabil Eng 12:24–31. doi:10.1109/TNSRE.2003.823266

Casadio M, Morasso PG, Sanguineti V, Arrichiello V (2006) Braccio di Ferro: a new haptic workstation for neuromotor rehabilitation. Technol Health Care 13:1–20

Casadio M, Morasso P, Sanguineti V, Giannoni P (2009a) Minimally assistive robot training for proprioception enhancement. Exp Brain Res 194:219–231

Casadio M, Morasso P, Sanguineti V, Giannoni P (2009b) Minimally assistive robot training for proprioception enhancement. Exp Brain Res 194:219–231

Dalgas U, Stenager E, Ingemann-Hansen T (2007) Multiple sclerosis and physical exercise: recommendations for the application of resistance-, endurance-and combined training. Multiple sclerosis

Emken JL, Benitez R, Sideris A, Bobrow JE, Reinkensmeyer DJ (2007) Motor adaptation as a greedy optimization of error and effort. J Neurophysiol 97:3997–4006. doi:10.1152/jn.01095.2006

Hogan N, Krebs HI, Rohrer B et al (2006) Motions or muscles? Some behavioral factors underlying robotic assistance of motor recovery. J Rehabil Res Dev 43:605

Huang FC, Patton JL (2011) Evaluation of negative viscosity as upper extremity training for stroke survivors. In: 2011 IEEE international conference on rehabilitation robotics (ICORR), pp 1–6

Huang FC, Patton JL, Mussa-Ivaldi FA (2010) Manual skill generalization enhanced by negative viscosity. J Neurophysiol 104:2008–2019. doi:10.1152/jn.00433.2009

Kaelin-Lang A, Sawaki L, Cohen LG (2005) Role of voluntary drive in encoding an elementary motor memory. J Neurophysiol 93:1099–1103. doi:10.1152/jn.00143.2004

Krakauer JW, Carmichael ST, Corbett D, Wittenberg GF (2012) Getting neurorehabilitation right what can be learned from animal models? Neurorehabilitation Neural Repair 26:923–931

Krebs HI, Palazzolo JJ, Dipietro L et al (2003) Rehabilitation robotics: performance-based progressive robot-assisted therapy. Auton Robots 15:7–20

Lackner JR, Dizio P (1994) Rapid adaptation to Coriolis force perturbations of arm trajectory. J Neurophysiol 72:299–313

Marchal-Crespo L, Reinkensmeyer DJ (2009) Review of control strategies for robotic movement training after neurologic injury. J Neuroengineering Rehabil 6:1–15. doi:10.1186/1743-0003-6-20

Motl RW, Pilutti LA (2012) The benefits of exercise training in multiple sclerosis. Nat Rev Neurol 8:487–497

Nudo RJ (2006) Plasticity. NeuroRx 3:420–427. doi:10.1016/j.nurx.2006.07.006

Piovesan D, Dizio P, Lackner JR (2009) A new time-frequency approach to estimate single joint upper limb impedance. In: Engineering in medicine and biology society (EMBS): annual international conference of the IEEE, vol 1, pp 1282–1285

Piovesan D, Casadio M, Morasso P, Giannoni P (2011a) Influence of visual feedback in the regulation of arm stiffness following stroke. In: Engineering in medicine and biology society (EMBS): annual international conference of the IEEE, pp 8239–8242

Piovesan D, Casadio M, Mussa-Ivaldi FA, Morasso PG (2011b) Multijoint arm stiffness during movements following stroke: implications for robot therapy. In: 2011 IEEE international conference on rehabilitation robotics (ICORR), pp 1–7

Piovesan D, Pierobon A, DiZio P, Lackner JR (2011c) Comparative analysis of methods for estimating arm segment parameters and joint torques from inverse dynamics. J Biomech Eng 133:031003

Piovesan D, Pierobon A, DiZio P, Lackner JR (2012) Measuring multi-joint stiffness during single movements: numerical validation of a novel time-frequency approach. PLoS ONE 7:e33086. doi:10.1371/journal.pone.0033086

Piovesan D, Morasso P, Giannoni P, Casadio M (2013a) Arm stiffness during assisted movement after stroke: the influence of visual feedback and training. IEEE Trans Neural Syst Rehabil Eng 21:454–465. doi:10.1109/TNSRE.2012.2226915

Piovesan D, Pierobon A, DiZio P, Lackner JR (2013b) Experimental measure of arm stiffness during single reaching movements with a time-frequency analysis. J Neurophysiol 110:2484–2496. doi:10.1152/jn.01013.2012

Reinkensmeyer DJ, Akoner OM, Ferris DP, Gordon KE (2009) Slacking by the human motor system: computational models and implications for robotic orthoses. In: Engineering in medicine and biology society, 2009. EMBC 2009. Annual international conference of the IEEE, pp 2129–2132

Reschechtko S, Zatsiorsky VM, Latash ML (2015) Task-specific stability of multi-finger steady-state action. J Mot Behav 47:365–377. doi:10.1080/00222895.2014.996281

Riener R, Lunenburger L, Jezernik S, Anderschitz M, Colombo G, Dietz V (2005) Patient-cooperative strategies for robot-aided treadmill training: first experimental results. IEEE Trans Neural Syst Rehabil Eng 13:380–394. doi:10.1109/TNSRE.2005.848628

Scheidt RA, Dingwell JB, Mussa-Ivaldi FA (2001) Learning to move amid uncertainty. J Neurophysiol 86:971–985

Schmit BD, Dhaher Y, Dewald JP, Rymer WZ (1999) Reflex torque response to movement of the spastic elbow: theoretical analyses and implications for quantification of spasticity. Ann Biomed Eng 27:815–829

Shadmehr R, Mussa-Ivaldi F (1994) Adaptive representation of dynamics during learning of a motor task. J Neurosci 14:3208–3224

Squeri V, Basteris A, Sanguineti V (2011) Adaptive regulation of assistance 'as needed' in robot-assisted motor skill learning and neuro-rehabilitation. In: 2011 IEEE international conference on rehabilitation robotics (ICORR), IEEE, pp 1–6

Taylor NF, Dodd KJ, Prasad D, Denisenko S (2006) Progressive resistance exercise for people with multiple sclerosis. Disabil Rehabil 28:1119–1126. doi:10.1080/09638280500531834

Thoroughman KA, Shadmehr R (2000) Learning of action through adaptive combination of motor primitives. Nature 407:742–747

Vergaro E, Casadio M, Squeri V, Giannoni P, Morasso P, Sanguineti V (2010) Self-adaptive robot training of stroke survivors for continuous tracking movements. J Neuroengineering Rehabil 7:13. doi:10.1186/1743-0003-7-13

Wiegner AW, Watts RL (1986) Elastic properties of muscles measured at the elbow in man: I. Normal controls. J Neurol Neurosurg Psychiatry 49:1171–1176

Wolbrecht ET, Chan V, Reinkensmeyer DJ, Bobrow JE (2008) Optimizing compliant, model-based robotic assistance to promote neurorehabilitation. IEEE Trans Neural Syst Rehabil Eng 16:286–297. doi:10.1109/TNSRE.2008.918389

Zatsiorsky V, Seluyanov V (1983) The mass and inertia characteristics of the main segments of the human body 30. In: HaKK Matsui (ed) International congress of biomechanics: biomechanics VIII-B. Human Kinetics, Champaign, pp 1152–1159

# Toward a Proprioceptive Neural Interface That Mimics Natural Cortical Activity

**Tucker Tomlinson and Lee E. Miller**

**Abstract** The dramatic advances in efferent neural interfaces over the past decade are remarkable, with cortical signals used to allow paralyzed patients to control the movement of a prosthetic limb or even their own hand. However, this success has thrown into relief, the relative lack of progress in our ability to restore somatosensation to these same patients. Somatosensation, including proprioception, the sense of limb position and movement, plays a crucial role in even basic motor tasks like reaching and walking. Its loss results in crippling deficits. Historical work dating back decades and even centuries has demonstrated that modality-specific sensations can be elicited by activating the central nervous system electrically. Recent work has focused on the challenge of refining these sensations by stimulating the somatosensory cortex (S1) directly. Animals are able to detect particular patterns of stimulation and even associate those patterns with particular sensory cues. Most of this work has involved areas of the somatosensory cortex that mediate the sense of touch. Very little corresponding work has been done for proprioception. Here we describe the effort to develop afferent neural interfaces through spatiotemporally precise intracortical microstimulation (ICMS). We review what is known of the cortical representation of proprioception, and describe recent work in our lab that demonstrates for the first time, that sensations like those of natural proprioception may be evoked by ICMS in S1. These preliminary findings are an important first step to the development of an afferent cortical interface to restore proprioception.

T. Tomlinson · L.E. Miller (✉)
Department of Physiology, Feinberg School of Medicine, Northwestern University, 303 E. Chicago Avenue, Chicago, IL 60611, USA
e-mail: lm@northwestern.edu

L.E. Miller
Department of Physical Medicine and Rehabilitation, Northwestern University, 710 North Lake Shore Drive, Chicago, IL, USA

L.E. Miller
Department of Biomedical Engineering, Northwestern University, 2145 Sheridan Road, Evanston, IL 60208, USA



J. Laczko and M.L. Latash (eds.), *Progress in Motor Control*,
Advances in Experimental Medicine and Biology 957,
DOI 10.1007/978-3-319-47313-0_20

**Keywords** Somatosensation · Intracortical microstimulation (ICMS) · Somatosensory cortex · Prosthesis

The 15 years since brain–computer interfaces (BCIs) became a reality have seen a host of technical improvements that have moved the field from two- or three-dimensional control of a cursor to multidimensional control of a robotic limb (Hochberg et al. 2006; Collinger et al. 2013), to the controlled activation of multiple paralyzed muscles (Ethier et al. 2012). However, in nearly all BCIs, the feedback guiding these movements is exclusively visual. The only exceptions we are aware of is a study that used passive movements of a monkeys limb to supply proprioceptive feedback (Suminski et al. 2010), and a recent study in which stimulation in S1 of monkeys encoded the vector from hand to target, allowing the monkey's to guide reaching in the absence of other cues (Dadarlat et al. 2015). Researchers are increasingly recognizing that supplying somatosensory input directly *to* the user's brain is as important, and arguably more difficult a problem, than is extracting motor command information *from* the brain (Weber et al. 2012; Bensmaia and Miller 2014). Most of the experimental work within this area has focused on the sense of touch in monkeys (Romo et al. 1998; Fitzsimmons et al. 2007; Tabot et al. 2013) or whisking in rodents (Venkatraman and Carmena 2011). In contrast very little progress has been made in reproducing proprioception to provide kinesthetic feedback.

There are multiple points at which a somatosensory afferent neural interface might be implemented. For patients with intact CNS, such as amputees, the nerve stump and dorsal roots are possible implant sites. One advantage of a peripheral nerve interface is the further processing of these signals as they propagate centrally. The relative simplicity of the information carried by peripheral afferent fibers might be considered another advantage. Since both muscle spindles and Golgi tendon organs (GTOs) have been accurately modeled (Mileusnic et al. 2006; Mileusnic and Loeb 2006), in principle, their firing rates can be mimicked. However, mimicking these inputs is complicated by the unknown state of the descending gamma system that normally serves to alter both the static sensitivity of spindles to muscle length, and their dynamic sensitivity to length changes (Eldred et al. 1953; Lennerstrand 1968). Inferring muscle lengths and forces based on measured limb state would also be a substantial challenge.

Central interfaces on the other hand, present the obvious advantage of applying not only to persons with limb amputation, but also those with spinal cord injury, for whom peripheral interface sites are of no use. Furthermore, neurons in S1 integrate multimodal afferent signals to form a higher-level representation of limb state. Neurons within area 2 of the primary somatosensory cortex (S1) appear to signal the direction of hand movement, whether those movements are active reaching movements, or perturbations of the hand (Soso and Fetz 1980; Prud'homme and Kalaska 1994; London and Miller 2013). Cortical interfaces have the potential to tap into this higher-level kinematic representation. However, these same neurons are influenced by movement-related forces as well as kinematics (Jennings et al. 1983; Prud'homme and Kalaska 1994; London et al. 2011), making independent control of the two modalities an additional challenge.

In the rest of this chapter we will focus on the implementation of cortical, rather than peripheral interfaces and we direct the reader to the following additional sources for information on the latter approaches (Navarro et al. 2005; Raspopovic et al. 2014; Tan et al. 2014; Saal and Bensmaia 2015). We will survey the underlying neurophysiology and anatomy that gives rise to the sense of proprioception, followed by a discussion of prior and current attempts using electrical stimulation to recruit these areas in the restoration of sensation.

## Proprioception Is an Essential Part of Normal of Motor Control

Sherrington first used the term proprioception to define the sense of body position (Sherrington 1906). Proprioception is now commonly defined as the sensory information that contributes to a sense of joint position and motion. It arises primarily from muscle spindles and GTOs that respond to muscle length and force, respectively. However, joint receptors and cutaneous afferents responding to skin stretch also likely contribute (Burke et al. 1988; Macefield et al. 1990; Weerakkody et al. 2009). Sir Charles Bell called proprioception the "sixth sense" (Bell and Shaw 1865), as it is often overlooked compared to the five main senses of which we are more consciously aware.

Beyond its somewhat limited role in our conscious sense of limb position and motion, proprioception plays a critical subconscious role in the planning and control of limb movement. Chronic loss of proprioception produces cortical remapping (Sanes et al. 1988) and profound changes in motor ability (Sanes et al. 1984; Sainburg et al. 1993; Gordon et al. 1995). Patients suffering chronic loss of proprioception due to large fiber neuropathy make looping, discoordinated reaching movements (Fig. 1). While these patients can compensate to a certain extent through vision, this compensation is effortful, and normal ease of motion is never restored. Claude Ghez and his colleagues showed that these deficits arise from a loss of coordination between musculature at the elbow and shoulder that prevents these patients from compensating for the intersegmental inertial dynamics of the limb (Sainburg et al. 1993). They hypothesized that this compensation relies on a model of limb dynamics, the accuracy of which depends on proprioceptive feedback. Interestingly, in some respects the movements of robotic limbs controlled by current BCIs look similar to those of these deafferented patients. This similarity suggests that some of the remaining functional inadequacy of BCIs may result from the lack of somatosensory feedback, including both touch and proprioception.

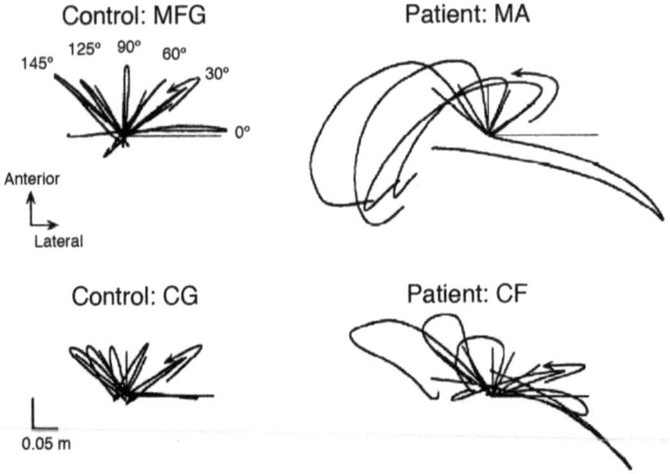


**Fig. 1** Reaching trajectories in normal subjects and subjects with peripheral deafferentation. Plots show the trajectories of a series of rapid out and back reaches to targets at several angles. Control subjects (*left*) were able to direct their movements accurately and make rapid reversals of direction, while deafferented subjects made errors in the initial reach direction and large looping motions (Adapted from Sainburg et al. 1995, Fig. 3)

## Cortical Representation of Proprioception

Primary somatosensory cortex (S1) is composed of four distinct regions, two of which (areas 3a and 2) receive thalamic inputs originating from muscle receptors and are considered to be proprioceptive in nature (Pons et al. 1985; Krubitzer et al. 2004). Area 3a is a functionally earlier cortical processing stage for proprioceptive signals than area 2, in that its thalamic input arises primarily from muscles, with a small number of cutaneous responses perhaps due to the indeterminate boundary between 3a and 3b (Friedman and Jones 1981). Additionally, 3a has only weak corticocortical connections wtih cutaneous areas 3b and 1 (Huffman and Krubitzer 2001). In contrast, area 2 receives a combination of muscle and cutaneous inputs from the thalamus, as well as strong corticocortical inputs from cortical areas 3a and 3b (Huffman and Krubitzer 2001). Both areas have a weak somatotopy with the proximal limb represented medially, and the hand more lateral, but the spatial resolution of these maps is nothing like that of the detailed maps in the cutaneous areas 3b and 1. In some sense this is inevitable given inputs that arise from muscles, including those that span more than one joint. Despite the lack of large-scale structure, neighboring neurons do tend to share similar responses to limb motion (Soso and Fetz 1980; Weber et al. 2011).

Despite these anatomical differences between the two areas, there has been relatively little attention paid to the potential functional differences between them. Neurons located in the proximal limb parts of both 3a and 2 generally fire in bursts during arm movements, typically with firing rates that vary approximately sinusoidally with the direction of hand motion and increase with the speed of motion, much like the similar behavior of neurons in M1 (Prud'homme and Kalaska 1994; London and Miller 2013). It is thus useful to represent a given neuron with a "preferred direction" (PD). These neurons typically discharge during both active and passive movement, the relative magnitude of the two components varying from neuron to neuron (Soso and Fetz 1980; Prud'homme and Kalaska 1994; London and Miller 2013). Neurons that respond strongly during both active and passive motions typically have PDs for the two types of movement that are well aligned (London and Miller 2013), thus simplifying the representation of limb movement, and the requisite mapping from limb state to patterns of ICMS.

On the other hand, in addition to the phasic, movement-related discharge, many proprioceptive neurons also have a tonic component that varies approximately with the instantaneous position of the hand. This combined movement and postural tuning also occurs in varying proportions across neurons (Gardner and Costanzo 1981; Prud'homme and Kalaska 1994; London and Miller 2013). The directionality of the postural tuning appears to be similar to the movement-related PDs (Prud'homme and Kalaska 1994). Restoring a full sense of proprioception will presumably require supplying both these components.

Beyond the single neuron studies described above, we have recently used chronically implanted arrays to combine information from many simultaneously recorded neurons in order to reconstruct detailed information about kinematic limb state. Figure 2a is a scatter plot of the mean firing rate for two of 41 area 2 neurons measured in a 300 ms window following limb perturbations in four different directions (indicated by the symbol color). The structure in this plot is caused by the differing directional sensitivity of these two neurons. Although it is impossible to depict the corresponding 41-dimensional plot for the entire 41-neuron sample population directly, there are methods to reduce the dimensionality of this "neural space" to a much smaller number. Figure 2b shows the results of a factor analysis, which can be thought of as the particular projection of these 41-dimensional points onto a plane that retains as much of the original information as possible. Not only were points corresponding to a particular target tightly clustered in this 2-D neural space, the geometry of the two spaces was approximately isomorphic. This suggests that the neurons in area 2 carry information that conveys, approximately linearly, the position of the hand following perturbations in different directions.

The dynamics of limb movements can also be captured accurately using the "decoder" methods that have much more commonly been applied to M1 neurons (Chapin et al. 1999; Serruya et al. 2002; Vargas-Irwin et al. 2010; Wodlinger et al. 2015). Trajectories during planar, random-target reaching movements were predicted from area 2 neurons with mean $R^2$ between measured and predicted signals of 0.59 for position and 0.66 for velocity (Weber et al. 2011). Acceleration was predicted more poorly with $R^2 = 0.44$. In that study, a neuron dropping analysis revealed that these predictions were as accurate for neurons that had primarily cutaneous receptive fields as those with more muscle-like responses. This counterintuitive result is nonetheless consistent with earlier observations of the reliable modulation of area 1 and 3b neurons during center-out reaching movements (Cohen

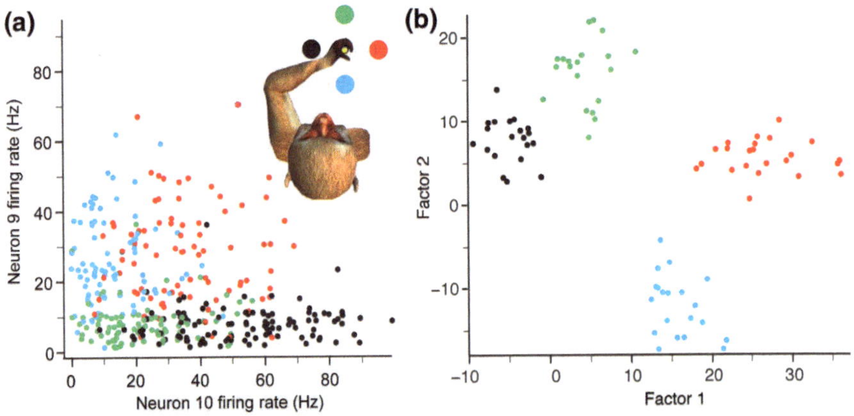


**Fig. 2** Neural discharge related to limb movements imposed by force perturbations on the monkey's hand in different directions. **a** Each *circle* represents the mean firing rate of two (of 41) neurons during a 300 ms window following a limb perturbation. Perturbation directions are coded by color, as shown in the *inset* diagram. Some separation of the neuronal response by reaching target is evident, but there is considerable overlap for the four reach directions. **b** Factor analysis applied to the full 41-dimensional space of concurrently measured firing rates. The perturbation directions are clearly separated in this 2D factor space, which has a geometry very much like that of the perturbation directions

et al. 1994), the discharge of cutaneous peripheral afferents during rotations of the knee (Edin 2001), and of the perception of externally imposed finger movements in the presence of local anesthesia (Edin and Johansson 1995).

In addition to purely kinematic responses, S1 neurons also modulate during isometric force production with directionally tuned firing rates (Jennings et al. 1983; Prud'homme and Kalaska 1994; London et al. 2011). During reaching while compensating for loads in different directions, firing rates in S1 are well fit by sinusoidal functions of force direction as well as movement (Fig. 3). Under these conditions, the force PDs ("load axis" in that study) tend to be roughly anti-aligned with the kinematic PDs. Although the depth of modulation during movement and isometric tasks co-varies significantly across neurons (Jennings et al. 1983; London et al. 2011), neurons with strong tonic kinematic tuning to position tend to have stronger force representation (Prud'homme and Kalaska 1994). This differential sensitivity to force might account for some of the differing responses of S1 neurons during kinematically similar active and passive movements. (Soso and Fetz 1980; Prud'homme and Kalaska 1994; London and Miller 2013) These observations raise important questions about how the CNS resolves these differences to evoke perceptually similar kinesthetic sensations during active and passive movements. It also raises questions of the impact this sensitivity to muscle activity might have on attempts to elicit kinesthetic sensations by intracortical microstimulation (ICMS).

While there have been quite a number of studies that have attempted to distinguish between different kinematic coordinate representations within M1 (Evarts 1969; Caminiti et al. 1991; Scott and Kalaska 1995; Kakei et al. 1999; Morrow

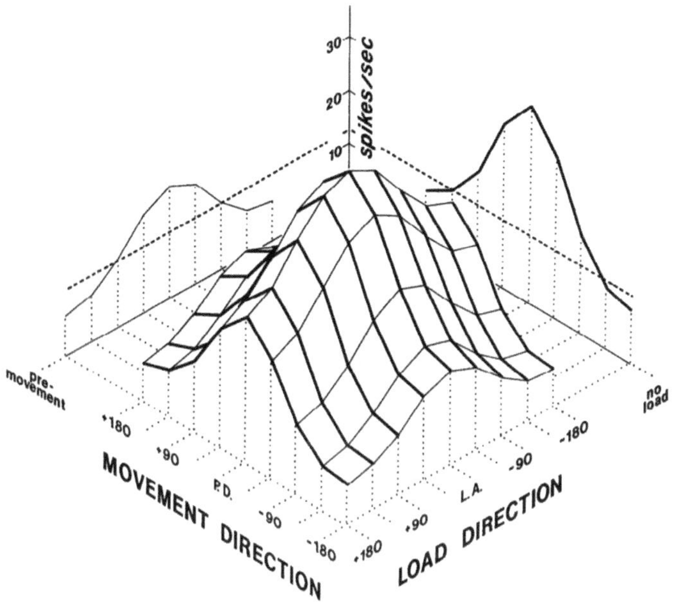


**Fig. 3** Representation of the combined effect of reach direction and external loads on the activity of proprioceptive neurons in S1. The plotted surface represents the response of 93 neurons, averaged after alignment to the movement and force PDs. The *marginal curves* represent the neural responses to unloaded reaching and to isometric force (Adapted from Prud'homme and Kalaska 1994, Fig. 11A)

et al. 2007; Oby et al. 2013), the analogous studies have not been done in S1. Existing studies have largely assumed a hand-centered, Cartesian representation and simply described the resulting mappings. It remains possible that neurons in 3a or 2 might actually be better described in terms of muscle lengths. Alternately, given the multimodal nature of the receptive fields in Area 2, it may be that a transformation occurs between the representations of kinematics in these two areas. Only with an accurate model of the relation between limb state and the cortical activity in S1, can we hope to recreate naturalistic patterns of cortical activity through electrical stimulation.

Similar uncertainty exists for the force representation in S1. Although contact forces applied to the hand and digits are represented in cutaneous areas 3b and 1 (Sinclair and Burton 1991; Tremblay et al. 1996), it seems very unlikely that they would also be represented in the proximal limb regions of areas 3a and 2. Instead, the endpoint force-related tuning in these areas seems more likely to represent muscle forces. Under conditions in which endpoint force is well correlated with muscle force (for example, isometric contractions) it can be predicted quite accurately from the activity of multiple area 2 neurons (London et al. 2011). Even during reaching movements against purely inertial loads, predictions remain reasonably accurate. However, when additional random forces are added to dissociate the direction of endpoint motion and endpoint force, the predictions become much worse, suggesting the modulation of area 2 neurons is more directly related to muscle forces, rather than the measured endpoint force.

## Use of Intracortical Microstimulation to Restore Sensation

Anyone who has placed their tongue on the poles of a 9-V battery knows something of the odd sensation that can be evoked when the nervous system is activated electrically. In the late eighteenth century, Alessandro Volta is credited with the first electrical activation of the auditory system when he inserted two metal rods into his ears and connected them with a battery. He described a very unpleasant "boom" within his head. Substantially refined two centuries later, the cochlear implant has been used to restore hearing to 300,000 deaf patients worldwide as of 2012 (Yawn et al. 2015).

ICMS has been shown to evoke detectable percepts from a variety of cortical areas, including human visual cortex (Bak et al. 1990; Schmidt et al. 1996), rat (Otto et al. 2005a; Koivuniemi and Otto 2012) and cat (Wang et al. 2012) auditory cortex, rat barrel fields (Talwar et al. 2002; Houweling and Brecht 2008; Venkatraman and Carmena 2011; Bari et al. 2013; Thomson et al. 2013), and monkey somatosensory cortex (O'Doherty et al. 2009; Zaaimi et al. 2013; Dadarlat et al. 2015; Kim et al. 2015b). There is currently tremendous interest in using this stimulation to replace natural sensation when it has been lost.

The earliest application of direct cortical stimulation to induce sensation was probably in the primary visual cortex, resulting in the perception of small dots of light dubbed "phosphenes." Phosphenes were first produced experimentally in the late 1920s by neurosurgeons in Germany, Lowenstein and Borchardt (Löwenstein and Borchardt 1918), Foerster (1929) and Fedor Krause and Heinrich Schum (Krause and Schum 1932), who described localized sensations of light, the position of which depended on the location of the stimulus near the occipital pole (Lewis and Rosenfeld 2016). The possibility that this electrically elicited sensation might be exploited to provide a form of artificial vision was pursued independently 40 years later in England by Giles Brindley at Cambridge University (Brindley and Lewin 1968) and in the Untied States at the University of Utah by William Dobelle (Dobelle and Mladejovsky 1974). These experiments used approximately 1 mm-sized electrodes placed on the surface of the visual cortex, and required currents as large as 10 mA to evoke the visual effects (Fig. 4a, b).

Artificial vision was pursued further through the neuroprosthesis program at the NIH, using penetrating intracortical electrodes. Experiments included intraoperative stimulation of three normally sighted patients undergoing surgical treatment of epilepsy (Bak et al. 1990) and chronic stimulation over a period of four months in a patient who had been blind for 22 years due to glaucoma (Schmidt et al. 1996). Phosphenes were produced in all these patients with threshold currents ranging from 1 to 10 μA, much lower than those necessary for the surface stimulation of

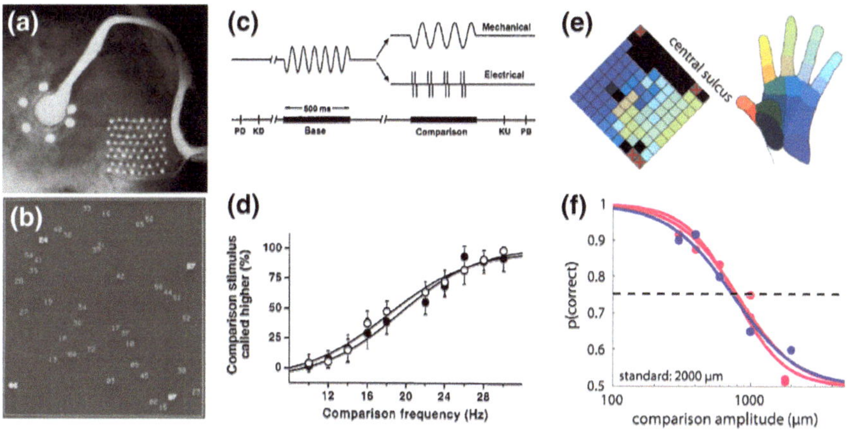


**Fig. 4** Cortical interfaces used to restore sensation. **a** Radiograph of electrodes implanted on the surface of V1 in a blind human patient (Figure **a** adapted from Dobelle et al. 1976, Fig. 1). **b** Map of the location of phosphenes generated by the electrodes in (**a**). (Figure **b** adapted from Dobelle et al. 1974, Fig. 3). **c, d** Stimulus trains delivered by penetrating electrodes in S1 allow monkeys to discriminate frequency as they would with vibratory stimulation of the fingers. **c** Frequency discrimination task using either mechanical vibration of the fingertip or ICMS in tactile area 3b. The monkey was presented a baseline mechanical stimulus of fixed frequency and a subsequent comparison stimulus and asked to report whether the latter was higher or lower frequency. **d** Psychophysical responses for mechanical (*filled*) and ICMS (*open*) comparisons, which were not significantly different (Figures **c** and **d** adapted from Romo et al. 1998, Figs. 1 and 2). **e** Color coded receptive field locations of recordings made by a Utah array implanted in area 2. **f** Psychometric performance comparing the relative intensity of two sequential indentations of the skin on the finger (*blue curves*) or two sequential ICMS trains applied to the corresponding electrode (*red curves*). As in the frequency discrimination task, these curves were statistically indistinguishable. The ICMS current was computed from indentation amplitude using an empirically determined mapping function between the natural and artificial percepts (Figures **e** and **f** adapted from Tabot et al. 2013, Figs. 1 and 3)

Brindley and Dobelle. Further work led to development of two different types of high-density electrode arrays made from silicon, one effort led by Ken Wise at the University of Michigan (Hoogerwerf and Wise 1994) and another by Richard Normann at the University of Utah (Jones et al. 1992). The resultant "Michigan" and "Utah" electrode arrays, currently available from NeuroNexus and Blackrock microsystems, respectively, have undergone considerable further development, and continue to be used in a wide range of recording and stimulation studies.

Stimulus intensity can be graded by frequency, pulse width, or current and used to modulate the reliability with which the subject can detect the stimulus (Fridman et al. 2010). The increase in detection rate with current intensity (pulse amplitude or duration) results from the activation of a larger tissue volume, whereas increases in frequency presumably lead to higher firing rates in recruited neurons. Studies report detection thresholds between 2 and 40 μA in sensory areas ranging from visual areas in humans (Bak et al. 1990; Schmidt et al. 1996) auditory areas in rat and cat (Koivuniemi and Otto 2012; Wang et al. 2012) and somatosensory areas in rats

(Houweling and Brecht 2008; Bari et al. 2013) and monkeys (Tabot et al. 2013; Zaaimi et al. 2013; Kim et al. 2015a). There is typically a gradual increase in the detection rate with increased current. This monotonic relation between stimulus and perceptual intensity provides a basis for modulating the stimulation to achieve percepts of different magnitudes.

Kevin Otto and colleagues trained rats to discriminate between ICMS delivered on different electrodes in auditory cortex. With 68 μA stimuli, rats were unable to distinguish electrodes with less than 750 μm separation (Otto et al. 2005b). The effective electrode density is thus coupled to the maximum current, potentially limiting the variety of distinct sensations that may be delivered. Further, some researchers have reported loss of efficacy at high stimulus currents. In V1 phosphenes lost color and shrank in size as currents were increased (Schmidt et al. 1996). These investigators also noted that high current on a single electrode occasionally produced multiple phosphenes. Similarly, in experiments stimulating area MT, effects on perceived visual motion were lost above 80 μA (Murasugi et al. 1993). These effects may reflect a saturation in the recruitment of neurons near the electrode combined with the recruitment of a more distant, and likely less functionally homogeneous populations. Combined, these findings set a limit on the maximum currents that may be used for increasing the strength of artificial percepts.

To add to the complexity, electrical stimulation activates not only the soma of nearby neurons but also axons passing near the electrode. As a consequence, the evoked activity is not focused exclusively near the electrode, but instead includes sparse activation of neurons within a several millimeter radius (Histed et al. 2009). This suggests that the problem of activating a functionally homogeneous population of neurons may be more severe than that suggested by a simple model of current spread.

An adjunct method for increasing the robustness of the evoked sensation may be to stimulate multiple electrodes. In stimulus detection studies, detection threshold current at each electrode within a group decreased modestly with the use of two or four electrodes in tactile areas of S1 (Kim et al. 2015b), and more dramatically in a separate study of area 2, (Zaaimi et al. 2013). In the latter study the effects were particularly large, even supralinear, with five and seven electrodes. Distributing stimulation across many electrodes and reducing the current at each one has the twin advantages of potentially increasing the functional homogeneity of the activated neurons within a smaller volume (Weber et al. 2011), and reducing the potential for damage to the surrounding tissue or electrodes. A recent study found no histologic damage or functional deficits following six months of stimulation on multiple electrodes with currents up to 100 μA (Rajan et al. 2015). These results suggest that easily detected currents are well below those expected to cause damage, but it is difficult to extrapolate to the cumulative effect of decades of stimulation that would be required from a useful prosthesis.

Beyond demonstrating the ability to detect stimulation, a number of animal studies have shown that ICMS can also be used to provide salient information. Temporally patterned stimulation in rat auditory cortex-enabled rats to perform a frequency discrimination task in which ICMS replaced an auditory cue (Otto et al.

2005b). In somatosensory cortex, ICMS has been used to provide salient information in a number of different tasks, allowing animals to distinguish between patterns of stimulation in cortex and to compare the percepts due to ICMS and natural stimuli. In an early study, cortical barrel field stimulation was used to guide a rat's open field locomotion, stimulation in right or left S1 signaling the corresponding rewarded turn (Talwar et al. 2002). Another study showed that rats performing an object detection task were able to locate virtual objects on the basis of the timing of ICMS delivered to barrel as the whisker passed through the virtual object (Venkatraman and Carmena 2011). The rats explored the virtual object by whisking back and forth, much as they would against an actual object, suggesting that the sensation produced by ICMS was qualitatively similar to that driven by deflection of the whiskers.

In experiments with monkeys, stimulus frequency within proprioceptive area 3a provided cues that allowed monkeys to distinguish between rewarded and unrewarded targets in a simple cued reaching task (London et al. 2008). The group of Miguel Nicolelis has performed more elaborate experiments in which monkeys used brain control to move an avatar hand among multiple virtual targets (O'Doherty et al. 2009). Target contact triggered ICMS within area 1. Two targets were associated with different temporal patterns of stimulation referred to as "virtual textures." Monkeys were able to explore the targets and learn the pattern associated with the rewarded target.

In contrast to the discrimination experiments described above, in at least two other studies, animals learned to interpret ICMS that provided continuous feedback during motion. In one case, ICMS was delivered to barrel cortex of rats and modulated in intensity according to the rat's distance from an infrared light above a water port. ICMS intensity increased as rats drew closer to the port, allowing them to locate it through this artificial somatosensory percept (Thomson et al. 2013). Having learned the basic task with a visible light cue, the rats learned to use the IR-driven ICMS cue in about a month. In another experiment, monkeys learned to reach to invisible targets guided by feedback that represented an error vector from the current hand position to the unseen target (Dadarlat et al. 2015). Feedback was provided either by a random moving-dot flow field or the strength of somatosensory ICMS delivered across eight electrodes. The frequency of simulation was determined by the projection of the error vector onto eight evenly spaced unit vectors assigned arbitrarily to the eight electrodes. After 3–4 months of paired visual and ICMS feedback, the monkeys learned to use ICMS to guide reaches in the absence of the visual information.

## Biomimicry in Afferent Cortical Interfaces

Work in patients with limb amputation has shown that physiologically appropriate cutaneous feedback improves the patient's embodiment of the prosthesis (Cincotti et al. 2007; Marasco et al. 2011). Likewise, it is reasonable to suspect that learning to use an ICMS interface that mimics natural sensations would be faster, and ultimately perhaps more effective than learning arbitrary associations with unnatural sensations or arbitrarily modulated ICMS (Bensmaia and Miller 2014). However, the extent to which artificially evoked activity must mimic that of the natural afferent inputs in order to be useful remains a critical question.

Ranulfo Romo performed a now classic series of experiments in the somatosensory system, in which monkeys identified which of two sequential mechanical stimuli applied to the fingertip had a higher frequency (Romo et al. 1998). The monkeys were able to complete the task when one or both of the mechanical stimuli was replaced with an ICMS train of the same frequency in the tactile area 3b (Fig. 4c, d). This experiment was the first to demonstrate that ICMS in somatosensory areas can evoke percepts with some naturalistic components. Similar experiments have been done more recently using punctate pressure stimuli (Tabot et al. 2013). The monkeys in that experiment were able to report both the location and magnitude of effects evoked by area 1 or 3b stimulation, and to compare them to their mechanical analogs (Fig. 4e, f). ICMS with higher currents led to a sensation of higher pressure, allowing construction of mapping functions between the desired perceptual effects and ICMS. These maps, and the receptive field locations characteristic of each electrode, were ultimately used to generate discriminable sensations through ICMS based on forces applied to sensors on a prosthetic limb, demonstrating a plausible method to provide cutaneous feedback to a patient. These effects rely on the somatotopy of S1, and the fact that neighboring neurons in both areas tend to have largely overlapping receptive fields.

Extrapolaind from these studies, a reasonable approach to reproducing a sense of limb movement in a particular direction would be to electrically activate the neurons that naturally signal movement in that direction. Figure 5 suggests the design of such an interface, using an array of electrodes that activate groups of neurons with identified PDs. By activating neurons in an appropriate spatiotemporal sequence, it may be possible to reproduce a sensation of the natural movement that elicits the same sequence of neural activity.

A potential challenge to using the natural behavior of neurons to design biomimetic interfaces is that the sensations resulting from ICMS do not always correspond to those normally signaled by the activated neurons. Stimulation of V1, for instance, does not produce the perceptual experience that might be anticipated based on the physiological characteristics of single neurons recorded in V1. Rather than the oriented bars that might be anticipated based on the response properties of complex cells (Hubel and Wiesel 1962), ICMS gives rise primarily to the sensation of fixed spots of colored light. Perhaps this is the consequence of the combined activation of many cells with different orientation selectivity. On the other hand, stimulation of the higher visual areas MT and MST can, in some conditions, elicit a sensation of visual motion that is consistent with the visual motion sensitivity of the neighboring neurons (Salzman et al. 1992; Celebrini and Newsome 1995). These experiments combined ICMS with the moving-dot stimuli in manner that biased the monkey's judgment of the actual dot motion toward the directional preference of the electrically activated neurons. This effect was present only with currents well

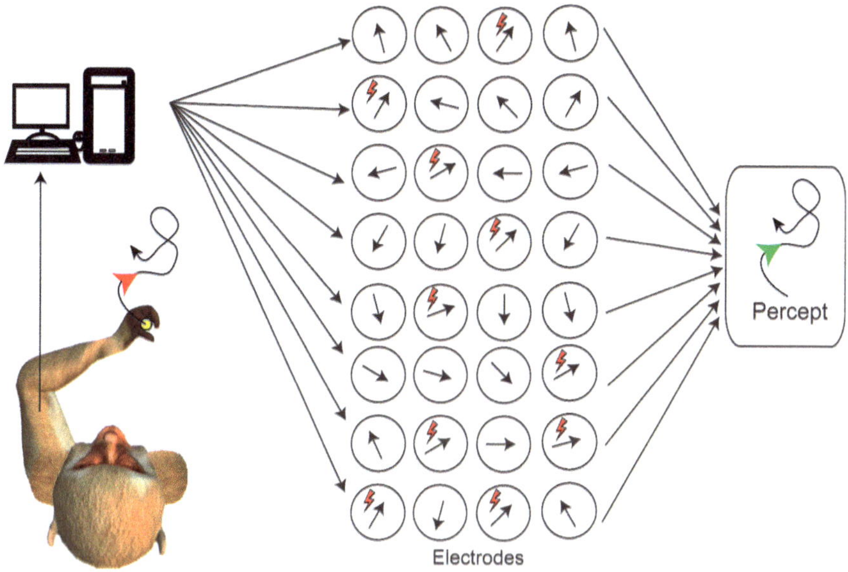


**Fig. 5** Schematic representation of a biomimetic neural prosthesis for proprioception. Each electrode, indicated by a *black circle*, activates a volume of neural tissue with an experimentally-determined, multi-unit PD, indicated by the *black arrow*. We speculate that stimulating an appropriate set of electrodes (*red sparks*) with PDs similar to the instantaneous direction of limb movement (*red arrow* on hand path) will evoke a corresponding sense of limb motion (*green arrow*). A complex spatiotemporal pattern of stimulation might then be used to convey feedback of the entire movement trajectory. Our experiments, described later, suggest that stimulating a set of electrodes with a single, well-defined PD causes a sensation of hand movement in that direction. The effect of stimulating electrodes with dissimilar PDs has yet to be determined

below 80 μA (Salzman et al. 1992) and required the electrode to be positioned at a depth within a column in which neurons encountered over 150 μm of electrode travel had similar preferred directions (Salzman et al. 1992).

## Generation of Artificial Proprioception Through ICMS

The strong representation of hand velocity by neurons in area 2 and the similarity between these neurons and the motion sensitivity of neurons in MT/MST suggests a useful experimental paradigm that we have begun to explore, examining the hypothesis that stimulating in area 2 will alter the monkey's perception of hand motion. More specifically, just as ICMS in MT and MST biases perception of motion, ICMS in S1 should increase the activity in the neurons near the stimulated electrodes, biasing the monkey's perceived hand motion toward the PD of the stimulated neurons.

We trained a monkey to perform two reaching tasks, while grasping the handle of a planar robotic manipulandum. Movement of the handle controlled the position of a cursor on a screen and servomotors could be used to apply forces to the handle. The first task required the monkey to reach to a sequence of randomly positioned targets in order to generate movements with different speeds and in many directions. We recorded multi-unit activity and used these data to compute preferred directions for each electrode. We used multi-unit, rather than discriminated single units, to better represent the set of neurons that would be activated by ICMS.

The monkey also learned a two alternative forced choice task designed to test his ability to discriminate the direction of force pulses applied to the handle (Fig. 6). After moving the cursor to a central target, an additional pair of targets appeared, located symmetrically on a line passing through the central target. After a random hold period, the robot applied a 500 ms force pulse that displaced the monkey's hand a few centimeters. The monkey then reached to the outer target that was closer to the direction of the force pulse. The monkey was rewarded for a correct choice and given an audible error signal if the reach was to the wrong target.

To evaluate the monkey's performance in the direction identification task, we computed the proportion of trials in which he chose one of the targets and fit these responses as a sigmoidal function of the perturbation direction (black curve in Fig. 7a). The monkey could judge perturbations near either target quite accurately. Between these extremes, the psychometric function passed through the point of subjective equality (PSE), the angle for which the monkey perceived the perturbation to be intermediate between the two targets. In this example, PSE fell very close to 90° in the coordinate system aligned to the target axis.

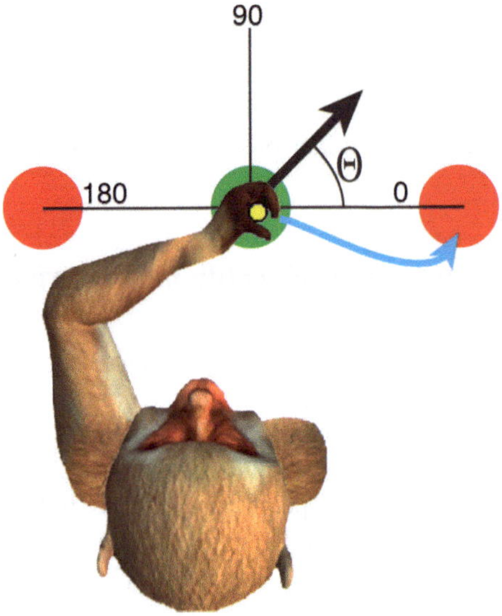


**Fig. 6** Direction identification task. The monkey held its hand in a central location (*green circle*) while a physical perturbation (*black arrow*) was applied to the hand. The monkey was required to reach to the *red* target that was nearer to the pertubation direction (*blue arrow*). If the monkey reached to the correct target he received a liquid reward. This task increased in difficulty as the angle of the perturbation (θ) approached 90°, midway between the two targets

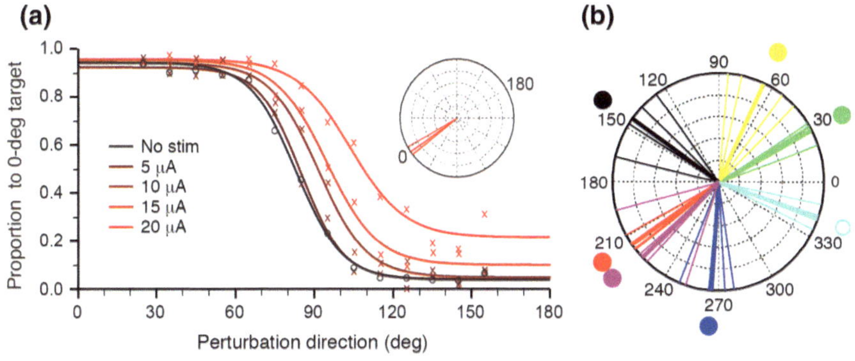


**Fig. 7** Biased perception of perturbation direction resulting from concurrent ICMS in area 2. **a** Psychometric curves showing the monkey's performance on the direction identification task of Fig. 6. Perturbation direction is relative to the target axis, where 0° corresponds to the PD of the stimulated electrodes. In this example, data were collected over seven days, with the PD target at 211° in absolute coordinates. Black circles and curve indicate the proportion of the monkey's choice of the PD (0°) target as a function of the direction of the mechanical perturbation. Red curves indicate the monkey's responses with progressively increasing stimulus currents. Stimulation caused more frequent reaches to the PD target, shifting the psychometric curve up and to the right. **b** Summary of the PD groupings of the seven electrode sets tested during these experiments. Filled circles around the perimeter indicate sets for which stimulation biased the monkey's movements toward the corresponding PD target. A single set of electrodes (*open cyan circle*) failed to produce the predicted result

On any given ICMS session, we selected four electrodes that had similar PDs and aligned the target axis to their mean PD. Thus one target was in the direction of the PD and the second in the anti-PD direction. We delivered ICMS to all four electrodes concurrently with the perturbation, using currents of 5, 10, 15, or 20 μA. The sensation caused by the ICMS appeared to combine with the natural sensation of the force pulse, producing a biased perception of the force direction (red curves and symbols in Fig. 6a). The psychometric curves shifted up and to the right, introducing a large bias toward the zero-degree PD target for 90° perturbations, and shifting the monkey's PSE toward the PD. Furthermore, the bias was graded with current. 5 μA stimulation had no significant effect, while 20 μA was strong enough that the monkey was never fully confident in selecting the anti-PD target, even when the perturbation was aimed directly toward it.

In a series of experiments over a period of 18 weeks we tested seven different groups of electrodes (Fig. 7b). Six of these resulted in an effect essentially like that shown in Fig. 7a, with the bias direction congruent with the PD of the stimulated electrodes. The seventh set of electrodes produced no detectable effect. Subsequent analysis showed that these electrodes had larger PD variability over time than the other sets, suggesting that the failure to induce a bias may have been the result of instability across the several days of testing.

It is important to note that the ICMS-induced bias was not the result of training, as was thye case in several other studies. We rewarded the monkey based on accurate discrimination of the physical perturbation, rather than any feature of the stimulation. Since the monkey's performance was roughly optimal prior to stimulation, the bias had the effect of increasing the monkey's error rate, thereby lowering the amount of reward. This is the opposite of studies that train monkeys to associate arbitrary ICMS patterns with specific actions in order to receive rewards. The fact that reward rate decreased, suggests that the monkey was unable to learn to ignore the sensation due to ICMS, as doing so would have allowed him to return to the baseline level of reward. These observations suggest that the ICMS generated sensations that were sufficiently naturalistic that they combined predictably with the mechanical stimuli.

## Future Directions

Forty years ago, Dobelle and colleagues implanted an 8 × 8 grid of electrodes in a patient blinded 10 years earlier by a gunshot wound (Dobelle et al. 1976). They identified a set of six (among 60 some) phosphenes that formed a braille cell, which allowed the patient to read at a rate of 30 characters per second, "much" faster than he could read tactile braille. However, Dobelle emphasized that "cortical braille [should be viewed] primarily as a technique to begin investigation of dynamic pattern presentation, rather than as a basis for clinically useful prostheses." Dobelle's statement reflected the recognition that his prosthesis could transmit only the simplest of spatial patterns. The prospect of scaling it up to reproduce an approximation of a visual scene is daunting, to say the least. Not only is the density of existing electrode arrays insufficient to pixelate an entire scene, but even existing spacing can cause interactions between adjacent electrodes (Dobelle et al. 1976; Otto et al. 2005b).

The cortical representation of touch and proprioception suggest that each may have an advantage over the cortical visual prosthesis. Unlike the highly artificial visual phosphenes, small punctate stimuli that activate only a small number of neurons make up a good portion of natural tactile stimuli. It is plausible that an array or arrays with hundreds of electrodes might be used to deliver clinically useful feedback about object contact timing, location, and pressure to the user of a prosthetic limb (Tabot et al. 2013). On the other hand, the broad directional tuning of proprioception means that a large proportion of these neurons are activated to greater or lesser extent for most arm movements. The sense of limb movement results from their overall integrated activity. Here the considerable challenge will be in selecting and activating many electrodes with an appropriate time-varying intensity (see Fig. 5).

The biases shown in Fig. 7 suggest that ICMS can cause activation that combines meaningfully with naturally-evoked activity. In order to move beyond these initial experiments to larger groups of electrodes, we intend to test a vector summation model, in which the predicted sensation is the sum of the individual sensations from each electrode. Alternatively, ICMS is known to entrain neurons to the stimulus (Griffin et al. 2011), which may preclude simple linear integration of the effect across electrodes. In that case, stimulation in multiple areas with different functional properties could lead to a winner take all scenario in which the strongest signal dominates. The eventual desire to provide not just a sense of limb motion, but also limb position and movement-related forces presents challenges well beyond our current experimental or modeling efforts.

Another important difference between touch and proprioception is that the former is very much a part of our conscious perception, while the latter is less so. Perhaps the most important function of a proprioceptive prosthesis will be in the control of movement, a function that was not addressed at all by our preliminary experiments. We elected to stimulate in area 2 rather than area 3a, in large part because of its accessibility on the cortical surface. However, its multimodal inputs are an additional reason that area 2 is an appealing target, as the sense proprioception relies on all these inputs (Collins et al. 2005). It is interesting to speculate whether area 3a, which receives inputs primarily from the deep muscle and joint receptors, might make a more appropriate site for a prosthesis that specifically addresses the subconscious control of limb movement.

A potential confound of ICMS used for artificial sensation is the observation that the standard symmetric biphasic current pulse tends to recruit axons at lower threshold currents than cell bodies. This effect causes distributed recruitment of neurons and complicates biomimetic approaches based on recording the activity of neurons near the electrode. A potential way to mitigate this effect may be to use pulses with reduced cathodal amplitude, which instead biases recruitment toward the soma (McIntyre and Grill 2000, 2002). A recent stimulus detection study in rat barrel field found that this asymmetric waveform yielded lower detection thresholds than did symmetric pulses, despite the smaller cortical area it recruited (Bari et al. 2013). By limiting recruitment to smaller clusters of neurons, this approach might recruit populations with more uniform functional properties.

While ICMS represents the current state of the art for afferent somatosensory interfaces in large animals and humans, optogenetic techniques may eventually provide a more selective method (Fenno et al. 2011; Yizhar et al. 2011). Optogenetics uses viral transfection to alter the DNA of neurons, causing them to produce photosensitive ion channels, allowing neurons to be activated by pulses of light. While current methods for delivering focal sources of light do not match the spatial resolution of ICMS, this technique allows cells with particular phenotypes to be targeted, providing the potential to engage only inhibitory neurons or those projecting only to or from particular areas. It also eliminates the challenges imposed on concurrent recordings by electrical stimulus artifacts and can eliminate the problem of the activation of axons of passage. While these techniques have been used primarily in rodents, there is considerable interest in translating them to primate models where they show great promise for future application to afferent interfaces (Diester et al. 2011).

## Take Home Message

Only recently, the focus of neural interfaces turned to the critical need to restore somatosensation, which is lost along with the ability to execute movement, as a consequence of SCI or limb amputation. Most efforts to restore somatosensation at the cortical level have relied on learned associations between arbitrary stimulation patterns and either reward (O'Doherty et al. 2011; Venkatraman and Carmena 2011; Thomson et al. 2013) or other feedback modalities (London et al. 2008; Dadarlat et al. 2015). In a few studies attempting to restore cutaneous sensation, ICMS has been shown to reproduce naturalistic sensations (Romo et al. 1998; Tabot et al. 2013). However, no such evidence has previously been presented for proprioception. By delivering ICMS coincidently with force pulses applied to a monkey's hand, we biased the monkey's perception of the direction of the resulting movement. We suggest that the combination of ICMS-driven cortical activity and that due to the actual perturbation, altered the monkey's perception in a predictable manner. The biases shown in these experiments required no learning, and appear to reflect naturalistic sensations of arm movement. These preliminary findings are a first step toward the development of an afferent cortical interface to restore proprioception.

## References


Bak M, Girvin JP, Hambrecht FT, Kufta CV, Loeb GE, Schmidt EM (1990) Visual sensations produced by intracortical microstimulation of the human occipital cortex. Med Biol Eng Comput 28:257–259

Bari BA, Ollerenshaw DR, Millard DC, Wang Q, Stanley GB (2013) Behavioral and electrophysiological effects of cortical microstimulation parameters. PLoS ONE 8:e82170. doi:10.1371/journal.pone.0082170

Bell C, Shaw A (1865) The hand: its mechanism and vital endowments, as evincing design. Bell & Daldy, London

Bensmaia SJ, Miller LE (2014) Restoring sensorimotor function through intracortical interfaces: progress and looming challenges. Nat Rev Neurosci 15:313–325. doi:10.1038/nrn3724

Brindley GS, Lewin WS (1968) The sensations produced by electrical stimulation of the visual cortex. J Physiol 196:479–493

Burke D, Gandevia S, Macefield G (1988) Responses to passive movement of receptors in joint, skin and muscle of the human hand. J Physiol 402:347–361

Caminiti R, Johnson PB, Galli C, Ferraina S, Burnod Y (1991) Making arm movements within different parts of space: the premotor and motor cortical representation of a coordinate system for reaching to visual targets. J Neurosci 11:1182–1197

Celebrini S, Newsome W (1995) Microstimulation of extrastriate area MST influences performance on a direction discrimination task. J Neurophysiol 73:437–448

Chapin JK, Moxon KA, Markowitz RS, Nicolelis MA (1999) Real-time control of a robot arm using simultaneously recorded neurons in the motor cortex. Nat Neurosci 2:664–670. doi:10.1038/10223

Cincotti F, Kauhanen L, Aloise F et al (2007) Vibrotactile feedback for brain-computer interface operation. Comput Intell Neurosci

Cohen DAD, Prud'homme MJL, Kalaska JF (1994) Tactile activity in primate primary somatosensory cortex during active arm movements: correlation with receptive field properties. J Neurophysiol 71:161–172

Collinger JL, Wodlinger B, Downey JE et al (2013) High-performance neuroprosthetic control by an individual with tetraplegia. Lancet 381:557–564. doi:10.1016/S0140-6736(12)61816-9

Collins DF, Refshauge KM, Todd G, Gandevia SC (2005) Cutaneous receptors contribute to kinesthesia at the index finger, elbow, and knee. J Neurophysiol 94:1699–1706. doi:10.1152/jn.00191.2005

Dadarlat MC, O'Doherty JE, Sabes PN (2015) A learning-based approach to artificial sensory feedback leads to optimal integration. Nat Neurosci 18:138–144. doi: 10.1038/nn.3883. http://www.nature.com/neuro/journal/v18/n1/abs/nn.3883.html (supplementary-information)

Diester I, Kaufman MT, Mogri M et al (2011) An optogenetic toolbox designed for primates. Nat Neurosci. doi:10.1038/nn.2749 nn.2749 [pii]

Dobelle WH, Mladejovsky MG (1974) Phosphenes produced by electrical stimulation of human occipital cortex, and their application to the development of a prosthesis for the blind. J Physiol 243:553–576

Dobelle WH, Mladejovsky MG, Girvin JP (1974) Artificial vision for the blind: electrical stimulation of visual cortex offers hope for a functional prosthesis. Science 183:440–444

Dobelle WH, Mladejovsky MG, Evans JR, Roberts TS, Girvin JP (1976) "Braille" reading by a blind volunteer by visual cortex stimulation. Nature 259:111–112

Edin BB (2001) Cutaneous afferents provide information about knee joint movements in humans. J Physiol 531:289–297. doi:10.1111/j.1469-7793.2001.0289j.x

Edin BB, Johansson N (1995) Skin strain patterns provide kinaesthetic information to the human central nervous system. J Physiol 487:243–251. doi:10.1113/jphysiol.1995.sp020875

Eldred E, Granit R, Merton PA (1953) Supraspinal control of the muscle spindles and its significance. J Physiol 122:498–523

Ethier C, Oby ER, Bauman MJ, Miller LE (2012) Restoration of grasp following paralysis through brain-controlled stimulation of muscles. Nature 485:368–371. http://www.nature.com/nature/journal/v485/n7398/abs/nature10987.html (supplementary-information)

Evarts EV (1969) Activity of pyramidal tract neurons during postural fixation. J Neurophysiol 32:375–385

Fenno L, Yizhar O, Deisseroth K (2011) The development and application of optogenetics. Annu Rev Neurosci 34:389–412

Fitzsimmons NA, Drake W, Hanson TL, Lebedev MA, Nicolelis MA (2007) Primate reaching cued by multichannel spatiotemporal cortical microstimulation. J Neurosci 27:5593–5602

Foerster O (1929) Beitrage zur pathophysiologie der sehbahn und der spehsphare. J Psychol Neurol 39:435–463

Fridman G, Blair H, Blaisdell A, Judy J (2010) Perceived intensity of somatosensory cortical electrical stimulation. Exp Brain Res 203:499–515. doi:10.1007/s00221-010-2254-y

Friedman D, Jones E (1981) Thalamic input to areas 3a and 2 in monkeys. J Neurophysiol 45: 59–85

Gardner EP, Costanzo RM (1981) Properties of kinesthetic neurons in somatosensory cortex of awake monkeys. Brain Res 214:301–319

Gordon J, Ghilardi MF, Ghez C (1995) Impairments of reaching movements in patients without proprioception. I. Spatial errors. J Neurophysiol 73:347–360

Griffin DM, Hudson HM, Belhaj-Saïf A, Cheney PD (2011) Hijacking cortical motor output with repetitive microstimulation. J Neurosci 31:13088–13096. doi:10.1523/jneurosci.6322-10.2011

Histed MH, Bonin V, Reid RC (2009) Direct activation of sparse, distributed populations of cortical neurons by electrical microstimulation. Neuron 63:508–522. doi:10.1016/j.neuron.2009.07.016 S0896-6273(09)00545-5 [pii]

Hochberg LR, Serruya MD, Friehs GM et al (2006) Neuronal ensemble control of prosthetic devices by a human with tetraplegia. Nature 442:164–171. doi:10.1038/nature04970

Hoogerwerf AC, Wise KD (1994) A three-dimensional microelectrode array for chronic neural recording. IEEE Trans Biomed Eng 41:1136–1146

Houweling AR, Brecht M (2008) Behavioural report of single neuron stimulation in somatosensory cortex. Nature 451:65–68. doi:10.1038/nature06447 nature06447 [pii]

Hubel DH, Wiesel TN (1962) Receptive fields, binocular interaction and functional architecture in the cat's visual cortex. J Physiol 160:106–154

Huffman KJ, Krubitzer L (2001) Area 3a: topographic organization and cortical connections in marmoset monkeys. Cereb Cortex 11:849–867

Jennings VA, Lamour Y, Solis H, Fromm C (1983) Somatosensory cortex activity related to position and force. J Neurophysiol 49:1216–1229

Jones KE, Campbell PK, Normann RA (1992) A glass/silicon composite intracortical electrode array. Ann Biomed Eng 20:423–437

Kakei S, Hoffman DS, Strick PL (1999) Muscle and movement representations in the primary motor cortex. Science 285:2136–2139

Kim S, Callier T, Tabot GA, Gaunt RA, Tenore FV, Bensmaia SJ (2015a) Behavioral assessment of sensitivity to intracortical microstimulation of primate somatosensory cortex. Proc Natl Acad Sci 112:15202–15207. doi:10.1073/pnas.1509265112

Kim S, Callier T, Tabot GA, Tenore FV, Bensmaia SJ (2015b) Sensitivity to microstimulation of somatosensory cortex distributed over multiple electrodes. Frontiers Syst Neurosci 9:47. doi:10.3389/fnsys.2015.00047

Koivuniemi AS, Otto KJ (2012) The depth, waveform and pulse rate for electrical microstimulation of the auditory cortex. In: 2012 annual international conference of the IEEE Engineering in Medicine and Biology Society (EMBC), pp 2489–2492

Krause F, Schum H (1932) Die epileptischen Erkrankungen, ihre anatomischen und physiologischen Unterlagen sowie ihre chirurgische Behandlung. Enke, Stuttgart

Krubitzer L, Huffman KJ, Disbrow E, Recanzone G (2004) Organization of area 3a in macaque monkeys: contributions to the cortical phenotype. J Comp Neurol 471:97–111

Lennerstrand G (1968) Position and velocity sensitivity of muscle spindles in the cat. I. Primary and secondary endings deprived of fusimotor activation. Acta Physiol Scand 73:281–299

Lewis PM, Rosenfeld JV (2016) Electrical stimulation of the brain and the development of cortical visual prostheses: an historical perspective. Brain Res 1630:208–224

London BM, Miller LE (2013) Responses of somatosensory area 2 neurons to actively and passively generated limb movements. J Neurophysiol 109:1505–1513. doi:10.1152/jn.00372.2012

London BM, Jordan LR, Jackson CR, Miller LE (2008) Electrical stimulation of the proprioceptive cortex (area 3a) used to instruct a behaving monkey. IEEE Trans Neural Syst Rehabil Eng 16:32–36. doi:10.1109/TNSRE.2007.907544

London BM, Ruiz-Torres R, Slutzky MW, Miller LE (2011) Designing stimulation patterns for an afferent BMI: Representation of kinetics in somatosensory cortex. In: 2011 annual international conference of the IEEE Engineering in Medicine and Biology Society, EMBC, pp 7521–7524

Löwenstein K, Borchardt M (1918) Symptomatologie und elektrische Reizung bei einer Schußverletzung des Hinterhauptlappens. Deutsche Zeitschrift für Nervenheilkunde 58: 264–292. doi:10.1007/BF01629694

Macefield G, Gandevia SC, Burke D (1990) Perceptual responses to microstimulation of single afferents innervating joints, muscles and skin of the human hand. J Physiol 429:113–129

Marasco PD, Kim K, Colgate JE, Peshkin MA, Kuiken TA (2011) Robotic touch shifts perception of embodiment to a prosthesis in targeted reinnervation amputees. Brain 134:747–758. doi:10.1093/brain/awq361

McIntyre CC, Grill WM (2000) Selective microstimulation of central nervous system neurons. Ann Biomed Eng 28:219–233

McIntyre CC, Grill WM (2002) Extracellular stimulation of central neurons: influence of stimulus waveform and frequency on neuronal output. J Neurophysiol 88:1592–1604

Mileusnic MP, Loeb GE (2006) Mathematical models of proprioceptors. II. Structure and function of the Golgi tendon organ. J Neurophysiol 96:1789–1802. doi:10.1152/jn.00869.2005

Mileusnic MP, Brown IE, Lan N, Loeb GE (2006) Mathematical models of proprioceptors. I. Control and transduction in the muscle spindle. J Neurophysiol 96:1772–1788. doi:10.1152/jn.00868.2005
Morrow MM, Jordan LR, Miller LE (2007) Direct comparison of the task-dependent discharge of M1 in hand space and muscle space. J Neurophysiol 97:1786–1798
Murasugi CM, Salzman CD, Newsome WT (1993) Microstimulation in visual area MT: effects of varying pulse amplitude and frequency. J Neurosci 13:1719–1729
Navarro X, Krueger TB, Lago N, Micera S, Stieglitz T, Dario P (2005) A critical review of interfaces with the peripheral nervous system for the control of neuroprostheses and hybrid bionic systems. J Peripheral Nerv Syst 10:229–258
Oby ER, Ethier C, Miller LE (2013) Movement representation in the primary motor cortex and its contribution to generalizable EMG predictions. J Neurophysiol 109:666–678. doi:10.1152/jn.00331.2012
O'Doherty JE, Lebedev MA, Hanson TL, Fitzsimmons NA, Nicolelis MA (2009) A brain-machine interface instructed by direct intracortical microstimulation. Front Integr Neurosci 3:1–10. doi:10.3389/neuro.07.020.2009
O'Doherty JE, Lebedev MA, Ifft PJ, Zhuang KZ, Shokur S, Bleuler H, Nicolelis MAL (2011) Active tactile exploration using a brain-machine-brain interface. Nature 479:228–231
Otto KJ, Rousche PJ, Kipke DR (2005a) Cortical microstimulation in auditory cortex of rat elicits best-frequency dependent behaviors. J Neural Eng 2:42–51
Otto KJ, Rousche PJ, Kipke DR (2005b) Microstimulation in auditory cortex provides a substrate for detailed behaviors. Hear Res 210:112–117
Pons TP, Garraghty PE, Cusick CG, Kaas JH (1985) The somatotopic organization of area 2 in macaque monkeys. J Comp Neurol 241:445–466
Prud'homme MJL, Kalaska JF (1994) Proprioceptive activity in primate primary somatosensory cortex during active arm reaching movements. J Neurophysiol 72:2280–2301
Rajan AT, Boback JL, Dammann JF et al (2015) The effects of chronic intracortical microstimulation on neural tissue and fine motor behavior. J Neural Eng 12:066018
Raspopovic S, Capogrosso M, Petrini FM et al (2014) Restoring natural sensory feedback in real-time bidirectional hand prostheses. Sci Transl Med 6:222ra219. doi:10.1126/scitranslmed.3006820
Romo R, Hernandez A, Zainos A, Salinas E (1998) Somatosensory discrimination based on cortical microstimulation. Nature 392:387–390
Saal HP, Bensmaia SJ (2015) Biomimetic approaches to bionic touch through a peripheral nerve interface. Neuropsychologia 79:344–353
Sainburg RL, Poizner H, Ghez C (1993) Loss of proprioception produces deficits in interjoint coordination. J Neurophysiol 70:2136–2147
Sainburg RL, Ghilardi MF, Poizner H, Ghez C (1995) Control of limb dynamics in normal subjects and patients without proprioception. J Neurophysiol 73:820–835
Salzman CD, Murasugi CM, Britten KH, Newsome WT (1992) Microstimulation in visual area MT: effects on direction discrimination performance. J Neurosci 12:2331–2355
Sanes JN, Mauritz KH, Evarts EV, Dalakas MC, Chu A (1984) Motor deficits in patients with large-fiber sensory neuropathy. Proc Natl Acad Sci USA 81:979–982
Sanes JN, Suner S, Lando JF, Donoghue JP (1988) Rapid reorganization of adult rat motor cortex somatic representation patterns after motor nerve injury. Proc Natl Acad Sci USA 85:2003–2007
Schmidt E, Bak M, Hambrecht F, Kufta C, O'Rourke D, Vallabhanath P (1996) Feasibility of a visual prosthesis for the blind based on intracortical microsimulation of the visual cortex. Brain 119:507–522
Scott SH, Kalaska JF (1995) Changes in motor cortex activity during reaching movements with similar hand paths but different arm postures. J Neurophysiol 73:2563–2567
Serruya MD, Hatsopoulos NG, Paninski L, Fellows MR, Donoghue JP (2002) Instant neural control of a movement signal. Nature 416:141–142
Sherrington C (1906) The integrative action of the nervous system. CUP Archive

Sinclair RJ, Burton H (1991) Neuronal activity in the primary somatosensory cortex in monkeys (Macaca mulatta) during active touch of textured surface gratings: responses to groove width, applied force, and velocity of motion. J Neurophysiol 66:153–169

Soso MJ, Fetz EE (1980) Responses of identified cells in postcentral cortex of awake monkeys during comparable active and passive joint movements. J Neurophysiol 43:1090–1110

Suminski AJ, Tkach DC, Fagg AH, Hatsopoulos NG (2010) Incorporating feedback from multiple sensory modalities enhances brain-machine interface control. J Neurosci 30:16777–16787. doi:10.1523/JNEUROSCI.3967-10.2010 30/50/16777 [pii]

Tabot GA, Dammann JF, Berg JA, Tenore FV, Boback JL, Vogelstein RJ, Bensmaia SJ (2013) Restoring the sense of touch with a prosthetic hand through a brain interface. Proc Natl Acad Sci 110:18279–18284

Talwar SK, Xu S, Hawley ES, Weiss SA, Moxon KA, Chapin JK (2002) Rat navigation guided by remote control. Nature 417:37–38

Tan DW, Schiefer MA, Keith MW, Anderson JR, Tyler J, Tyler DJ (2014) A neural interface provides long-term stable natural touch perception. Sci Transl Med 6:257ra138 doi:10.1126/scitranslmed.3008669

Thomson EE, Carra R, Nicolelis MA (2013) Perceiving invisible light through a somatosensory cortical prosthesis. Nat Commun 4:1482. doi:10.1038/ncomms2497

Tremblay F, Ageranioti-Belanger S, Chapman C (1996) Cortical mechanisms underlying tactile discrimination in the monkey. I. Role of primary somatosensory cortex in passive texture discrimination. J Neurophysiol 76:3382–3403

Vargas-Irwin CE, Shakhnarovich G, Yadollahpour P, Mislow JM, Black MJ, Donoghue JP (2010) Decoding complete reach and grasp actions from local primary motor cortex populations. J Neurosci 30:9659–9669. doi:10.1523/JNEUROSCI.5443-09.2010

Venkatraman S, Carmena JM (2011) Active sensing of target location encoded by cortical microstimulation. IEEE Trans Neural Syst Rehabil Eng 19:317–324. doi:10.1109/tnsre.2011.2117441

Wang J, Liu Y, Qin L, Chimoto S, Nakamoto K, Sato Y (2012) Chronic microstimulation of cat auditory cortex effective to evoke detection behaviors. Neuroscience 206:81–88. doi:10.1016/j.neuroscience.2012.01.012

Weber DJ, London BM, Hokanson JA et al (2011) Limb-state information encoded by peripheral and central somatosensory neurons: Implications for an afferent interface. IEEE Trans Neural Syst Rehabil Eng 19:501–513

Weber DJ, Friesen R, Miller LE (2012) Interfacing the somatosensory system to restore touch and proprioception: essential considerations. J Mot Behav 44:403–418. doi:10.1080/00222895.2012.735283

Weerakkody N, Taylor JL, Gandevia S (2009) The effect of high-frequency cutaneous vibration on different inputs subserving detection of joint movement. Exp Brain Res 197:347–355

Wodlinger B, Downey JE, Tyler-Kabara EC, Schwartz AB, Boninger ML, Collinger JL (2015) Ten-dimensional anthropomorphic arm control in a human brain−machine interface: difficulties, solutions, and limitations. J Neural Eng 12:016011

Yawn R, Hunter JB, Sweeney AD, Bennett ML (2015) Cochlear implantation: a biomechanical prosthesis for hearing loss. F1000Prime Reports 7:45 doi:10.12703/P7-45

Yizhar O, Fenno LE, Davidson TJ, Mogri M, Deisseroth K (2011) Optogenetics in neural systems. Neuron 71:9–34

Zaaimi B, Ruiz-Torres R, Solla SA, Miller LE (2013) Multi-electrode stimulation in somatosensory cortex increases probability of detection. J Neural Eng 10:056013

# Erratum to: Progress in Motor Control

Jozsef Laczko and Mark L. Latash

## Erratum to:
## J. Laczko and M.L. Latash (eds.), *Progress in Motor Control*, Advances in Experimental Medicine and Biology 957, https://doi.org/10.1007/978-3-319-47313-0

The chapter "Encoding Temporal Features of Skilled Movements—What, Whether and How?" is now available open access under a CC BY 4.0 license.

The updated online version of the book can be found at
https://doi.org/10.1007/978-3-319-47313-0



J. Laczko and M.L. Latash (eds.), *Progress in Motor Control*,
Advances in Experimental Medicine and Biology 957,
https://doi.org/10.1007/978-3-319-47313-0_21

MIX
Papier aus verantwortungsvollen Quellen
Paper from responsible sources
FSC® C105338

If you have any concerns about our products,
you can contact us on
**ProductSafety@springernature.com**

In case Publisher is established outside the EU,
the EU authorized representative is:
**Springer Nature Customer Service Center GmbH**
**Europaplatz 3, 69115 Heidelberg, Germany**

Printed by Libri Plureos GmbH
in Hamburg, Germany